Biotechnology of Forest Product Wastes

Biotechnology of Forest Product Wastes

G.Tharun

RANDOM

RANDOM PUBLICATIONS

NEW DELHI (INDIA)

Biotechnology of Forest Product Wastes

ISBN 978-93-5111-826-8

Published in 2016 in India by

RANDOM PUBLICATIONS

4376-A/4B, Gali Murari Lal, Ansari Road
New Delhi-110 002
Phone : +9111-43580356, 011-23289044, 011-43142548
e-mail: sales@randompublications.com,
info@randompublications.com, randomexports@gmail.com

Reprint 2023

Type Setting by : Friends Media, Delhi-110089
Printed at : Replika Press Pvt. Ltd.

Preface

Biotechnology is the application of biological systems in technology that can only be achieved through an integration of the biological, physical and engineering sciences. The current approach of biotechnology, in the forest products industry is to apply biological products such as enzymes rather than live cultures. Enzymatic action is very specific and also easy to control. This paper considers biotechnological applications in different operations including forestry, wood preservation, pulping, bleaching, deinking, papermaking and water treatment. The benefits of biotechnological applications are: improved product quality, production rate or diminished environmental impact. Utilisation of enzymes will become more widespread as biological products become less expensive through large-scale production.

Biotechnology can be applied in forests and especially plantations to improve fibre production and to protect trees. When exotic tree species were initially planted in South Africa, it was observed that stunted growth occurred due to the absence of mycorrhizal symbionts. Soil, containing these fungi, was then introduced into the country to improve growth of trees. Today, it is believed that fibre production can be improved by inoculating seedlings with more efficient mycorrhizal fungi. Mycorrhizal fungi enhance plant growth through improvement of mineral and water absorption, protection against pathogens and even secretion of growth enhancing hormones. A forest product is any material derived from a forest for direct consumption or commercial use, such as lumber, paper, or foragefor livestock. Wood, by far the dominant forest product, is used for many purposes, such as wood fuel (e.g. in form of firewood orcharcoal) or the finished structural materials used for the construction of buildings, or as a raw material, in the form of wood pulp, that is used in the production of paper. All other non-wood products derived from forest resources, comprising a broad variety of other forest products, are collectively described as non-timber forest products.

An attempt is made in this book to explain the principles of economics and management as applied to a development project in the forest sector.

– Author

Contents

1

Applications of Biotechnology in the Forests Products Industry

INTRODUCTION

Biotechnology has been defined as the integrated use of biological, physical and engineering sciences in order to achieve technological application of biological systems. The goal of technological application implies that biotechnology excludes fundamental research and that it must be relevant to industry. Fundamental studies do, however, play a crucial role in the development of the applicable biotechnological processes. Recombinant DNA technology can, for example, be used to exploit or enhance the properties of natural biological systems for industrial purposes.

Most biotechnological processes make use of microorganisms such as bacteria, yeasts and filamentous fungi, but vascular plants, algae and even animal tissue can also be utilised. Classical biotechnological processes have been used for ages in the production of food and beverages such as bread, cheese, beer and wine. These processes are characterised by the direct application of live organisms and the in situ production of enzymes and other products. New biotechnology has a stronger focus on the application of biological products or enzymes and these are often produced ex situ. The advantages of the new biotechnology over classical biotechnology are that mass balances are easier to solve and process control is improved.

The forest products industry is based on the processing of biological raw material and it is, therefore, well suited to the introduction of biotechnology. Biotechnological processes have been developed for most of the unit operations of the industry and a number of them are commercialised. The aim of this paper is to examine the potential of different biotechnological processes and their integration into conventional operations of the forest products industry.

BIOTECHNOLOGY APPLICATIONS FORESTRY

Biotechnology can be applied in forests and especially plantations to improve fibre production and to protect trees. When exotic tree species were initially planted in South Africa, it was observed that stunted growth occurred due to the absence of mycorrhizal symbionts. Soil, containing these fungi, was then introduced into the country to improve growth of trees. Today, it is believed that fibre production can be improved by inoculating seedlings with more efficient mycorrhizal fungi. Mycorrhizal fungi enhance plant growth through improvement of mineral and water absorption, protection against pathogens and even secretion of growth enhancing hormones.

Tree protection can be achieved by using biological control agents such as fungi or bacteria. The biological control agents (BCAs) are most effective against root diseases and especially where they are applied in an enclosed environment such as seedling trays. These BCAs act by competing with pathogens for nutrients and space, by secreting toxic substances or by direct consumption of the pathogens.

TREATMENT OF TIMBER

After felling, logs can be treated with fungi to improve debarking or to protect the wood against blue staining and decay. Inoculation of softwood logs with the white-rot fungus Phlebiopsis gigantea leads to rapid colonisation of the logs during transport or storage. The fungus is able to loosen bark, reduce pitch and protect the wood against sap staining. The loosening of bark is accomplished during the degradation of cambial cells by the fungus. The best-known biological agent to protect timber against blue staining is the fungus Ophiostoma piliferum. A white mutant of this fungus has been selected and the inoculum is produced commercially as Cartapip®. The fungus is a primary coloniser of wood and is able to utilise sugars and extractives in the wood without affecting the strength of the wood. These characteristics enable the fungus to out-compete any sap-stain fungi that subsequently infect the wood.

BIOPULPING

Biopulping is a solid-substrate fermentation (SSF) process where lignocellulosic materials are treated with fungi prior to pulping in order to reduce energy during mechanical pulping or to reduce chemical consumption during chemical pulping processes. Research has focussed on the utilisation of lignin-degrading fungi for biopulping, but the only commercial process currently available, utilises O. piliferum. Wood chips have also been treated successfully with fungal enzymes to improve the penetration of pulping liquor.

The aim of bio-thermomechanical pulping is to reduce the energy consumption during pulping and to improve pulp strength. The beneficial effects of fungal treatment on mechanical pulping can be ascribed to the reduction of the binding capacity of fibres and improved fibrillation. During bio-chemical pulping, the lignin content of the pulp or pulping time can be reduced. These benefits are the result of increased lignin solubility that is caused by fungal degradation. Depitching is one application of biopulping where Cartapip® is applied to stored wood chips and the pitch in the wood is reduced by the fungus metabolising it. It is evident that different fungi are suited to different pulping processes, but these processes must be adapted to achieve the full potential of an environmentally friendly technology.

PULPWOOD PRESERVATION

Cost of wood contributes more than 80 % to the variable cost of unbleached chemical pulp. Contaminating organisms that degrade cellulose or cause staining of wood can, therefore, have a significant impact by reducing yields of chemical pulp or necessitating bleaching of mechanical pulp. Biopulping fungi are inoculated at high dosages and incubated under conditions that favour their ability to colonise and compete with undesirable organisms. These fungi are thus able to protect pulpwood against decay and staining organisms.

BIOBLEACHING

Two different approaches have been used to enhance bleaching through enzymes. The first method utilise hemicellulolitic enzymes such as xylanase. These enzymes are unable to degrade lignin, but cause limited degradation of hemicellulose in kraft pulp to expose lignin to further attack by bleaching chemicals. Xylanase pre-bleaching is used at several kraft mills worldwide. In 1994, eight percent of Canada's bleached kraft pulp production was treated with xylanase. This technology holds environmental and economic advantages, especially since the enzymes have become less expensive as result of increased market size.

The typical chemical saving associated with the use of xylanase is more than 15 % active chlorine. Unfortunately, the limit of approximately 20 % in chemical savings cannot be exceeded. The second approach to enzymatic bleaching is the utilisation of oxidative enzymes, mainly laccase and manganese peroxidase (MnP) to delignify pulp. Of these enzymes, laccase is preferred, because MnP requires hydrogen peroxide, manganese (II) ions and a chelator. Laccase can cause delignification of pulp under slight oxygen pressure, but is considerably more effective when mediators are added. The most important obstacles to commercial application of laccases are the lack of sufficient enzyme stocks and the cost of mediators. Marked progress has been made over the last two years to solve these problems and it is expected that biobleaching with laccase will be able to compete with other TCF bleaching processes.

ENZYMATIC DEINKING

Enzymatic deinking can be applied on old newsprint and office waste, but it is more effective on the latter, because it contains large amounts of laser and xerographic inks that are more difficult to remove by conventional methods.

The commercial cellulases used, are able to hydrolyse fines to release ink particles. The ink can then be removed by chemical or air flotation. A secondary benefit of the enzyme treatment is that the stripping of fines from fibres results in improved drainage without reduction in strength properties.

IMPROVEMENTS IN PAPER MAKING

Most of the applications of enzymes to improve the papermaking characteristics of pulp are aimed at the improvement of secondary fibre. *Application of cellulases and hemicellulases to modify fibres has resulted in the following benefits:*

- Improved beating to save energy;
- Higher freeness;
- Enhanced drainage;
- Improvement of certain strength properties;
- Decreased disintegration time of recycled pulp, and
- Reversal of hornification.

In other applications, enzymes are used to remove the non-fibrous fractions of secondary fibre that hamper drainage. A good example of such material is residual starch that is especially abundant where old corrugated containers are used as raw material. Amylases can be applied to degrade remaining starch that would not contribute to the web characteristics, but would restrict drainage. In one mill, application of amylases resulted in a 6,9 % increase in production of paper.

CONTROL OF MICROBIAL FOULING

Microorganisms in mill water systems cause slime or biofilm build-up, odour and corrosion. The consequences of the contamination are increased web breakages, holes and spots in paper as well as increased cost of maintenance. Biocides are generally used to control microbial build-up, and deposits are removed during boilouts by sodium hydroxide or dispersants. Inoculation of water systems with competing microorganisms or bacteriophages have been proposed, but the most feasible biotechnological approach appears to utilise enzymes to prevent aggregation of biofilm-producing organisms and degrade extra-cellular polysaccharides. These enzymes include proteases and carbohydrate-degrading enzymes and combinations of these enzymes are often applied. The enzymes also expose microorganisms to toxins by reducing protective polysaccharides, and are usually dosed in combination with biocides.

EFFLUENT TREATMENT

The oldest application of biotechnology in the pulp and paper industry is probably that of wastewater treatment. Biological processes are usually employed in secondary or polishing treatments that follow sedimentation or other primary treatment. Biological treatment processes include: Aerated stabilisation basins, activated sludge, oxygen-activated sludge, trickling filters, rotating biological contactors, anaerobic lagoons, upflow anaerobic sludge blankets, anaerobic filters and anaerobic fluidised beds. These systems are all characterised by a complex microbial community that is responsible for the improvement of water quality.

CONCLUSIONS

Biotechnology is applied on a commercial scale in many operations of the forest products industry. Application of enzymes has become very prominent recently, because they are highly selective in their action and have a negligible environmental impact. Wider application of these enzymes are restricted by cost and availability of the enzymes. This problem will soon cease to exist, given that the cost of enzymes has decreased considerably over the last decade as the market increased. A number of companies are now also producing enzymes specifically aimed at the pulp and paper market. One remaining hurdle, is the perceived incompatibility of biotechnology with conventional processes. This problem is more difficult to solve since it is largely due to a lack of understanding of biological systems and a lack of acceptance that biological products are expected to work under extreme conditions in mills. It is, therefore, clear that biotechnology can only be developed through integration and understanding of biological as well as physical sciences.

FOREST PRODUCT

Fig. Wood cut from Victorian Mountain Ash

A forest product is any material derived from a forest for direct consumption or commercial use, such as lumber, paper, or foragefor livestock. Wood, by far the dominant forest product, is used for many purposes, such as wood fuel (e.g. in form of firewood orcharcoal) or the finished structural materials used for the construction of buildings, or as a raw material, in the form of wood pulp, that is used in the production of paper. All other non-wood products derived from forest resources, comprising a broad variety of other forest products, are collectively described as non-timber forest products.

Resource Pressures

Many 0f forest management policies have been implemented that impact forest product economics, including forest access restrictions, harvesting fees, and harvest limits.Deforestation, global warming and other environmental concerns have increasingly affected the availability and sustainability of forest products, as well as the economies of regions dependent upon forestry around the world. In recent years, the idea of sustainable forestry, which aims to preserve crop yields without causing irreversible damage toecosystem health, has changed the relationship between environmentalists and the forest products industry. Stakeholders in the forest products industry include government departments, commercial enterprises, non-governmental organizations (NGOs), policy-makers and analysts, private and international organizations.

Forest Products Details

Since 1947, the Food and Agriculture Organization of the United Nations has published an annual yearbook of forest products. The FAO Yearbook of Forest Products is a compilation of statistical data on basic forest products for all countries and territories of the world. It contains series of annual data on the volume of production and the volume and value of trade in forest products. It includes tables showing direction of trade and average unit values of trade for certain products. Statistical information in the yearbook is based primarily on data provided to the FAO Forestry Department by the countries through questionnaires or official publications. In the absence of official data, FAO makes an estimate based on the best information available.

FAO also publishes an annual survey of pulp and paper production capacities around the world. The survey presents statistics on pulp and paper capacity and production by country and by grade. The statistics are based on information submitted by correspondents worldwide, most of them pulp and paper associations, and represents 85% of the world production of paper and paperboard.

FOREST PRODUCTS AND INDUSTRY

The fourth general category of issues confronting the forestry sector in the Asia Pacific Region include those dealing with the composition, status and strategies for the conversion of forest yields into products. The focus has, of course, been primarily on the conversion of timber into marketable commodities, yielding both consumptive values for local markets as well as the direct and indirect benefits of income generation, employment and, frequently, government revenues. Forest production can be simple and unorganized, as in the case of subsistence collection of firewood, fodder or poletimber for immediate family use, or highly organized and institutionalized through business organizations and formal markets. In either case, the efficiency and competitiveness of this productive process largely determines the long term status of the forest resource base and the degree to which sustainability can be obtained.

FOREST PRODUCTS

As noted in the review of the Asia Pacific Region, the most basic forest product has been roundwood. A fundamental decision is made as to whether roundwood will be 'marketed' in unprocessed form, or rather subjected to further conversion prior to marketing. Non-industrial uses of timber, including fuelwood and much local subsistence use is unorganized and functions without the traditional forms of 'industry' or markets. A major concern throughout Asia Pacific is the extent to which processing of forest products can be competitive in lieu of simple harvesting and export of unprocessed timber. Questions of comparative advantage have historically revolved around issues of natural forest stocks and low-cost labor. These 'advantages' often led to profitable export of roundwood when the establishment of industrial infrastructure was not feasible. "Local" processing has been frequently at small scale, with limited technology, and oriented towards serving limited local needs.

Both producers of roundwood and consumers/importers are vitally interested in the economics of forest products conversion and comparative advantage. Both seek opportunities to 'capture' the benefits of primary and secondary processing for the benefit of local income generation and employment. Traditionally, converted forest products have followed a typical hierarchy from simple forms of solid wood such as sawnwood to partially restructured wood as veneers and panels, and further to various re-constituted products such as particleboard and fiberboard. Alternatively, wood may flow directly into primary fiber utilization through pulping and potentially paper and paperboard. Increasingly, the choice of forest products strategies depends upon both local comparative advantage as well as global concerns related to markets, potential competitors (both traditional forest products and non-wood substitutes), and prices and capital movements. The simple availability of roundwood supply is no longer sufficient to determine forest products strategies. Technology, capital, transportation, and scope and scale of markets continue to erode traditional advantages based on raw material and labor.

NON-WOOD FOREST PRODUCTS

Forests traditionally have yielded many products and values other than timber and commodity products. Uses vary greatly by location and forest type, but all forms of 'minor forest products' have been important to forest dwellers in support of subsistence and 'cottage industry' yielding marketable products and/or goods for community barter. In numerous instances, development of forest products industry has displaced or constrained opportunities for non-wood forest products, with strong implications as to the equity or distributional consequences of forest use. In parts of Asia Pacific where population pressures are extreme and land settlement has resulted in substantial intrusion into forests, social pressures of displacing traditional 'subsistence use' can lead to conflict and continuing problems of forest protection against illegal or unauthorized uses. Since a large proportion of non-wood forest products do not enter formal markets these uses are treated as 'secondary' or 'minor' in spite of the importance to local peoples.

Non-wood forest products are increasingly viewed as options for local development as means for greater self-sustaining economic capacity. Local, small scale 'industry' serving local consumption needs and supporting local employment is increasingly seen as an alternative to large scale industrial development. New procedures for resource assessment, which fully recognize the potential for such uses need to be incorporated into forest sector strategies along with conventional forest products assessment.

FOREST PROCESSING

As noted previously, the capacity of the Asia Pacific Region to produce timber is linked to the size of the forest estate, the local forest conditions, the levels of forest management and protection, and increasingly the sustainable long term productivity based on investments and management as supplements to natural forest yields. Many issues surrounding the determinants of forest capacity, both short and long term, have been noted. Another important issue for the Region relates to where that timber will be processed. Intra-regional as well as external trade beyond the Asia Pacific Region will reflect both national priorities and strategies as well as the comparative economic advantages at each stage of processing prior to end use consumption.

Each country has a nature! incentive to accomplish as much processing as possible within the local economy regardless of the location of final market. Products may be consumed in other Asia Pacific sub-regions and countries in addition to the domestic market.

The export of unprocessed roundwood is increasingly politically unwelcome in producer countries due to considerations of "value added" processing. On the other hand, consumer countries may well prefer access to unprocessed materials for a wide variety of reasons, including the generation of income and employment within their economies. However, only a finite production capacity is required in total to satisfy the needs of conversion of appropriate industrial roundwood into sawnwood products.

Protection measures including both import and export barriers to trade distort markets at each level. When imposed on roundwood trade, such barriers involve subsidies to existing and new capacity which would otherwise yield to greater efficiencies elsewhere. Problems of excess capacity are frequently compounded by lack of competitiveness in technology, economies of scale, marketing and distribution systems that are inefficient, and other chronic inhibitors of processing efficiencies. Where individual country capacity exceeds local market requirements, inefficiencies due to protection of industry processing will ultimately contribute to limited international competitiveness. Alternatively, the opening of markets and reductions in barriers carries the risk of industrial restructuring, changing the comparative advantages for different stages or levels of industrial processing, imposing short term threats to established patterns of timber utilization and trade. Regional cooperation and strategies for mutual benefit needs to supplement national tendencies to seek advantage within the increasingly competitive environment surrounding the forest sector in the Asian Pacific region.

RESEARCH AND INSTITUTIONS

An increasingly competitive environment for the forestry sector in the Asian Pacific region requires new knowledge, new technology, and new forms of organization and administration of forest resources and forest industry strategies. Research and education is critical to maintaining adequate institutional capacity to maintain and increase the performance of the forestry sector.

Much research on technical forestry has been conducted throughout the Asian Pacific Region, often with excellent cooperation and technical exchange. New forms of communication and information technologies facilitates access to knowledge, reducing the need for independent closed systems within each country. Means of cooperation and promotion of mutual interests will require greater focus on development of regional capacity and sharing of knowledge.

Institutional capacity increasingly incorporates new and different disciplines and fields that have not traditionally been seen as 'forestry'. To a great extent, today's issues and problems transcend technical forestry, involving economic, policy and social sciences. Linkages with other disciplines, and sectors, requires new and innovative approaches of research and administration. For example, population growth and economic development have not traditionally been seen as primary forestry matters. Silviculture, forest management, and growth-yield models have taken preference over population models, welfare economics, and public administration.

Critical re-thinking of the institutional framework for forestry sector development will increasingly require expanding the margins of forestry to integrate these 'external' elements. Forestry, and forest products, will increasingly become a business science in the Asian Pacific region. A strong scientific base for forestry and product technology will remain essential but must be supplemented by additional disciplinary dimensions to assure that sound structures are put in place to maximize the opportunities and competitiveness of the region.

Another dimension of national capacity in many countries of Asia Pacific includes significant issues surrounding the economic role of market structures in decision-making and the allocation of resources, including the role of market-like institutions for the production and distribution of forest products. However, the important role this factor plays in shaping the role of the forest products industry needs to be recognized as a major determinant of competitiveness and investments in operating efficiency.

LARGE SCALE FOREST INDUSTRY & INVESTMENT

Issues surround the proper role and scope of large scale industrial development and the sources and requirements for investment. Historically, forest industry was primarily domestically owned/controlled, often with government direct involvement. Technology was frequently simple and inexpensive. Development of modern industry capacity has frequently led to large economies of scale, with larger production capacity, larger raw material input requirements, and newer and more expensive technology. Not infrequently the capacity to develop modern industrial structures has led to foreign investment and the sharing of control and management. Concerns regarding 'foreign control' of forest resources, charges of reckless 'exploitation' without regard to environmental consequences or future conditions are also common.

Where the scale of production exceeds the needs of local markets, the role of marketing and distribution efficiencies also affect the scale of industry. Larger transactions and continuity in supplying international channels requires much more coordination in production and distribution, requiring capacities often beyond the small individual enterprises traditionally found in many parts of the developing countries.

SMALL SCALE FOREST INDUSTRY

Traditional, small scale forest industries are seen as being 'closer to the people' and in many cases more responsive to local consumption needs. Local ownership and management and the participation of local residents in primary employment are seen as primary benefits of forest industry quite apart from the actual production of output and the consumption of products destined for markets elsewhere. The distributional and secondary impacts of industry, rather than the efficiency in operations within an international context, may well dominate industrial policies.

New technology is also often seen as 'inappropriate' for local applications where fears of displacing existing capacity can prevail. Further, the indirect consequences of large scale industry in terms of raw material requirements, extensive control over land and resources, and the potential for unintended environmental risks can all create the desire to retain small scale operations and enterprises in spite of possible adverse economics and inefficiencies.

CATEGORIES OF FOREST BIOTECHNOLOGY

Recent literature varies in how it categorizes forest biotechnology. Owing to the gradual evolution of our understanding of genetics, and the consequent development of new tools and technologies by refining and combining new knowledge with prior research, these categories overlap to some degree.

The author used five major categories, each of which can be applied to a characteristic subset of applications:

- Markers (biochemical and molecular).
- Propagation and multiplication.
- Genomics (functional, structural, comparative, associative, statistical).
- Marker-assisted selection (MAS) and breeding.
- Genetic modification.

These tools and their utility are applicable on different scales, from the individual cell or plant to the landscape level. The only category which results in genetic manipulation of living trees is number 5, although development of some markers and DNA libraries for genomics may involve transformation of bacteria in the laboratory. The impact of these biotechnologies will therefore vary with the specific use, as will the associated benefits and risks. Certification of forest products, forest companies and lands is also potentially affected by the application of biotechnology. There are many certification agencies around the world, and some, such as the Forest Stewardship Council (FSC), have specifically excluded genetically modified (GM) trees from certifiability, or the land on which they grow, or other products obtained from stands containing these trees (FSC 1996). Some clonal forestry operations are certified, but most agencies have no explicit guidelines concerning genetic applications. Industrial processes which utilize genetic transformation of enzymes to chemically break down lignin in the harvested wood, e.g. pulping using enzyme digestion, have been certified on the grounds that this reduces toxic chemical use and discharge. Thus, certification standards vary among agencies, countries, products, processes and applications. Public opinion regarding genetic modification is more amenable to downstream (post-harvest) than upstream (pre-harvest) applications; the main cause for concern appears to be the potential for release of GM trees into the environment, which is an issue only for the latter.

Developed countries and developing countries or countries in transition have different priorities and applications for biotechnology. DNA-based applications (some markers, genomics and genetic modification) require a large initial outlay of resources, may have continuing high costs depending on the project, and entail a highly trained work force. Developed nations have so far been leaders in developing and applying these technologies for a variety of purposes (breeding, commerce and conservation; public and private uses), but it should be noted, based on the agriculture model, that barriers to the flow of capital and expertise associated with the introduction of biotechnology to new territories are virtually non-existent. In other words, the introduction of forest biotechnology tools to developing countries and countries in transition could be rapid as long as economic opportunities are present. Government agencies in developed countries typically provide much of the funding and infrastructure for the development of basic biotechnology research; however, it should be noted that applied research and development associated with any potential commercialization are mainly driven by the private sector.

The shift of funds supporting biotechnology from the public to the private sectors requires a new way of viewing the introduction of these tools into the forest sector. Thus, published information indicates that the majority of known forest biotechnology activities in developing countries and countries in transition have been more restricted in their scope, primarily to the use of markers and propagation, although other tools may have been introduced into the private sectors but, if so, information on the types of biotechnology application, their places of introduction and their frequency are unknown. The FAO Biotechnology in Developing Countries (FAO-BioDeC) database could be a useful storage facility for such forestry biotechnology information.

The flow of some biotechnology through teams including partners both from developed countries and from developing countries and countries in transition, such as CAMCORE (International Cooperative for Tree Conservation and Domestication), have provided increased access to these resources and tools.

Risk assessment is a critical application of many of the new biotechnological tools, and must, in turn, be applied to innovations as they are developed. Targeted gene engineering is of key concern, both with respect to inserting and eliminating gene function. Gene flow assessments are being conducted for poplars, but much more information is required to allay public concerns and deal with ecological hazards surrounding the potential for gene escape. This applies to pollen and seed dispersal, as well as the ability of species to sprout via roots or stumps, become vigorous weeds, or hybridize with wild sympatric congeners. Escapes and introgression of agricultural genes have been documented, as has unauthorized planting of GM crop seeds, leading to assessment, monitoring and regulatory problems. The relative fitness effects conferred by each novel trait must be evaluated in the environmental and genetic context of the field in order to assess risks properly. Promoter genes and other biosensors are being investigated as means of tracking GM material, a necessary precondition prior to approval for field testing or deployment. Experts have widely concurred that a case-by-case examination is necessary for approval of GM trees.

APPLICATIONS

The high cost of biotechnology will probably steer its commercial use towards short-rotation plantations as opposed to less intensively managed forests, but some applications are suitable for all forest types and can guide forest conservation activities.

Markers (Biochemical and Molecular)

Biotechnology has progressed from phenotype to genotype, and has been trying to quantify relationships between the two ever since. Early work focused on biochemistry (e.g. terpene analysis), progressed to biosynthetic constitutive or induced protein expression (isoenzymes) and is now utilizing DNA-based tools (e.g. microsatellites, quantitative trait loci [QTLs]) since the advent of PCR (polymerase chain reaction) technology.

Molecular markers can be either dominant (only the dominant allele is expressed if both parents are genetically different at a given locus) or codominant (the genotypes of both parents are quantifiable at a locus). Dominant markers (amplified fragment length polymorphisms [AFLPs], randomly amplified polymorphic DNAs [RAPDs]) require larger sample sizes for statistical analysis than codominant markers (isoenzymes, microsatellites/simple sequence repeats [SSRs], restricted fragment length polymorphisms [RFLPs], sequence-tagged sites [STSs], expressed sequence tags [ESTs], single nucleotide polymorphisms [SNPs]).

The different markers all have varying benefits, drawbacks, costs and ease of development and all can be applied to a limited range of optimal applications, generally focused on neutral or non-adaptive genetic variation. Molecular tools have been widely used to measure gene flow and genetic diversity of natural and artificial populations of forest trees and associated species, and the impacts of anthropogenic disturbance on their evolutionary potential. The long-term viability of species and endemics subject to influences such as pollution or climate change can be modelled using genetic data. There are still no highly repeatable, easily assayed markers directly linked to quantitative traits. Although progress has been made in identifying QTLs for some adaptive traits in some species, a good genomic linkage map is essential, as are breeding and identification and sequencing of candidate genes for quantitative traits of interest. The application of marker technology in forestry is by far the most extensive use of other tools. It is not restricted to either developed countries or developing countries and countries in transition, but the type of markers, the speed of use and the frequency of utilization differs with the status of marker development and the biometrical methods needed for data analyses.

Propagation and Multiplication

Clonal forestry using both broadleaves and conifers is gaining in popularity due to the resulting uniformity and ease of silviculture, harvesting and processing. More importantly, it allows the capture of levels of genetic gain that cannot be attained through sexual reproduction. This technology is restricted to organogenesis and somatic embryogenesis, which require extremely detailed understanding of cell biology, multiplication, biochemical signalling, differentiation and the production of copies or clones in a laboratory setting. Both of the above methods capitalize on plant species' ability to regenerate an entire genetically identical individual from a single cell or group of differentiated or undifferentiated cells via *in vitro* tissue culture. Individuals with selected traits or the highest performers from testing trials or breeding programmes can be replicated on a large scale, capturing both additive and non-additive genetic variation from traditional breeding. The advantage of these methods is the ability to store regenerated tissues indefinitely in liquid nitrogen or under laboratory conditions (i.e. hold the genotype constant) while field testing is underway to identify elite lines. This allows selected lines with desirable attributes to be further multiplied for operational deployment. Currently some operational scale production is underway both in developed countries and in developing countries and countries in transition, but it should be noted that the scale of production in broadleaves is larger than that in conifers.

Genomics

Our understanding of fundamental biology and evolution is also being enriched through collaborative efforts using model species in genomics research around the world. Evolutionary synteny among taxa, even those as distantly related as algae and angiosperms, can be quantified and candidate genes or groups of proteins (microarrays) involved in biochemical pathways of interest can be identified easily: this process of identifying putative genes controlling certain traits is called gene discovery. Expression of key gene families or microarrays and their relative up- or down-regulation following stress or environmental changes can be gauged, possibly leading to identification of compounds of significance, and their interactions with genotype and the environment. Characterization of genetic components of disease or pest resistance is a rapidly expanding field.

Genomics is a fairly new field with many subdisciplines, which requires substantial investment for start-up and maintenance: high-technology automata, costly supplies and chemicals, PCR and other machines, highly trained laboratory staff, marker (EST, SNP) development, as well as vast bioinformatics and associated statistical capacity are necessary. At present, the majority of activity in this area is at the research level.

Marker-Assisted Selection and Breeding

While the phenotype is the desired end result of breeding, variation associated with differences in expression, genotype-by-environment interaction and non-additive or epistatic genetic variability necessitates large, costly and time-consuming field trials for trees with traits of interest. Mendelian inheritance and our increasing ability to understand and target specific genes have enabled the development of MAS to complement traditional breeding. This technique has been used with tremendous success in agricultural crop breeding for various genes and traits over the past two decades, and is increasing in importance as more comprehensive genetic maps, and the locations of QTLs on those maps, are developed for each species. If MAS can be used to characterize and select tree genotypes, substantial cost savings may be realized by a much shorter breeding cycle (i.e. more rapid turnover between generations). Although based on tools developed using markers, it requires the use of material of known pedigree and integrates the study of specific structural, functional and morphological attributes of species' genomes. Although this work is experimental at this stage, several studies both in developed countries and in developing countries and countries in transition on broadleaves and conifers are underway. Information on the progress of these studies is sketchy owing to their often mixed funding nature (i.e. public and private).

GENETIC MODIFICATION

Research on and the potential for deployment of GM trees have caused widespread public concern. Regulatory agencies have called for more research on gene flow, likelihood of horizontal and vertical gene transfer via escapes, hybridization and introgression, and a range of ecological impacts of GM material in the field. Concerns about GM trees share similarities with those about agricultural crops, but while the latter are grown and used for research and consumer products throughout many developed and developing nations, transforming trees has only recently begun, and is currently mainly at the experimental stage. In the laboratory, transformation has been achieved using biolistics and *Agrobacterium* species for conifer and broad-leaved genera of commercial importance.

Specific genes and regions, primarily conferring insect or disease resistance, sterility, and wood quality attributes, have been the focus of nearly all of the research. Some promoter and marker genes have also been tested. In some regions, pulp and paper processing has achieved dramatic reductions in the use of highly toxic chlorine and other chemicals by employing enzymes genetically modified to digest lignin. The use of genetic modification during industrial processing does not currently appear to cause public concern, especially when the outcome includes substantial environmental benefits. It is difficult to give an accurate account of this type of work since the majority of it is being conducted by the private sector.

There is, to date, no single system systematically collating the different forest biotechnology activities around the world. This would be necessary to gauge current concerns and predict future trends, impacts and needs in this rapidly expanding field. International regulation of biotechnology is inconsistent, and even varies between jurisdictions. Research and development being done in the private sector subject to confidentiality or intellectual property agreements, patent protection and corporate competition preclude or prevent dissemination of some types of information. Other forestry biotechnology applications occur on private lands, or in areas where there is no requirement or framework to report their implementation or extent. Technical development of forestry, in most cases, mirrors that of agriculture. The use of biotechnology tools in forestry has already caused confusion.

2

Ethanol Production from Forest Product Waste

Recently there has been a push by the federal and state governments to increase the production of fuels from biomass as a substitute for fossil fuels such as oil, coal and natural gas. However, a recent headline announced that more fossil fuel energy is used to produce ethanol from corn and wood than the energy provided by the ethanol produced. This is a disturbing thought, given that each year over a billion bushels of corn are converted to ethanol! It would also mean that efforts to make ethanol from wood in Tennessee should be stopped. But is it true?

The controversial headline comes from a recent study by David Pimentel at Cornell University and Tad Patzek at the University of California/Berkley. This study reported that ethanol production from corn required 29% more fossil fuel energy inputs than the energy contained in the ethanol produced. The numbers for other feedstocks were even worse: 50% more for switchgrass and 57% more for wood.

Dr. Pimentel has produced other reports critical of ethanol production from corn in the past. However, there have been a number of other researchers that have pointed out possible errors and shortcomings in his methods, and have contradicted his conclusions. In general, other researchers have concluded that ethanol produced from corn provides a net energy surplus of about 30%.

But what about wood as a source of ethanol? One of the advantages of wood products is that wood requires relatively little energy to grow and process. Thus it is surprising that wood could be a less efficient energy source than other feedstocks. A closer look at Pimentel and Patzek's recent study reveals a number of problems with their analysis of using wood for ethanol production. For example, they assume that large amounts of nitrogen fertilizer (made using natural gas) will be required to replenish the forest soils. However, because wood is very low in nitrogen content, fertilization of forest soils is not necessary. Pimentel and Patzek also assume that the steam used in the ethanol production process would be generated by burning fossil fuels. However, steam for current wood processing steps – e.g. for kiln drying lumber and for pulp and paper making – is usually generated by burning wood waste. Steam for ethanol production from wood is likely to come from wood waste also.

A final, fundamental shortcoming of Pimentel and Patzek's ethanol-from-wood analysis is that they assume that ethanol would be the only product from the trees. This means that all of the energy "costs" are carried by the ethanol produced in their analysis. But this is very unlikely. Currently, harvesting of a forest stand produces a number of different products: veneer, grade lumber, railway ties, pulp chips, hog fuel and mulch. If ethanol production from wood becomes a reality, it will probably be just one more product in the mix – not the only product. This makes the 'energy efficiency' of ethanol production much higher.

The forests of Tennessee cover more than ½ of the state and represent an enormous natural and renewable resource. Efforts are underway to try to convert some of this resource into liquid fuels. Whether the final product is ethanol or bio-oil or something else, wood from Tennessee may become an important substitute for fossil fuels.

CELLULOSIC ETHANOL

Cellulosic ethanol is a biofuel produced from wood, grasses, or the inedible parts of plants.

OVERVIEW

Cellulosic ethanol is a type of biofuel produced from lignocellulose, a structural material that comprises much of the mass of plants. Lignocellulose is composed mainly ofcellulose, hemicellulose and lignin. Corn stover, *Panicum virgatum* (switchgrass), *Miscanthus* grass species, wood chips and the byproducts of lawn and tree maintenance are some of the more popular cellulosic materials for ethanol production. Production of ethanol from lignocellulose has the advantage of abundant and diverse raw material compared to sources such as corn and cane sugars, but requires a greater amount of processing to make the sugar monomers available to the microorganisms typically used to produce ethanol by fermentation.

Switchgrass and *Miscanthus* are the major biomass materials being studied today, due to their high productivity per acre. Cellulose, however, is contained in nearly every natural, free-growing plant, tree, and bush, in meadows, forests, and fields all over the world without agricultural effort or cost needed to make it grow.

According to Michael Wang of Argonne National Laboratory, one of the benefits of cellulosic ethanol is it reduces greenhouse gas emissions (GHG) by 85% over reformulated gasoline. By contrast, starch ethanol (e.g., from corn), which most frequently uses natural gas to provide energy for the process, may not reduce GHG emissions at all depending on how the starch-based feedstock is produced. According to the National Academy of Sciences in 2011, there is no commercially viable bio-refinery in existence to convert lignocellulosic biomass to fuel. Absence of production of cellulosic ethanol in the quantities required by the regulation was the basis of a United States Court of Appeals for the District of Columbia decision announced January 25, 2013, voiding a requirement imposed on car and truck fuel producers in the United States by the Environmental Protection Agency requiring addition of cellulosic biofuels to their products. These issues, along with many other difficult production challenges, led George Washington University policy researchers to state that "in the short term, [cellulosic] ethanol cannot meet the energy security and environmental goals of a gasoline alternative."

HISTORY

The French chemist, Henri Braconnot, was the first to discover that cellulose could be hydrolyzed into sugars by treatment with sulfuric acid in 1819. The hydrolyzed sugar could then be processed to form ethanol through fermentation. The first commercialized ethanol production began in Germany in 1898, where acid was used to hydrolyze cellulose. In the United States, the Standard Alcohol Company opened the first cellulosic ethanol production plant in South Carolina in 1910. Later, a second plant was opened in Louisiana. However, both plants were closed after WWI due to economic reasons.

The first attempt at commercializing a process for ethanol from wood was done in Germany in 1898. It involved the use of dilute acid to hydrolyze the cellulose to glucose, and was able to produce 7.6 liters of ethanol per 100 kg of wood waste (18 US gal (68 L) per ton). The Germans soon developed an industrial process optimized for yields of around 50 US gallons (190 L) per ton of biomass. This process soon found its way to the US, culminating in two commercial plants operating in the southeast during WWI. These plants used what was called "the American Process" — a one-stage dilute sulfuric acid hydrolysis. Though the yields were half that of the original German process (25 US gallons (95 L) of ethanol per ton versus 50), the throughput of the American process was much higher. A drop in lumber production forced the plants to close shortly after the end of WWI. In the meantime, a small but steady amount of research on dilute acid hydrolysis continued at the USFS's Forest Products Laboratory. During World War II, the US again turned to cellulosic ethanol, this time for conversion to butadiene to produce synthetic rubber. The Vulcan Copper and Supply Company was contracted to construct and operate a plant to convert sawdust into ethanol. The plant was based on modifications to the original German Scholler process as developed by the Forest Products Laboratory. This plant achieved an ethanol yield of 50 US gal (190 L) per dry ton, but was still not profitable and was closed after the war.

With the rapid development of enzyme technologies in the last two decades, the acid hydrolysis process has gradually been replaced by enzymatic hydrolysis. Chemical pretreatment of the feedstock is required to prehydrolyze (separate) hemicellulose, so it can be more effectively converted into sugars. The dilute acid pretreatment is developed based on the early work on acid hydrolysis of wood at the USFS's Forest Products Laboratory. Recently, the Forest Products Laboratory together with the University of Wisconsin–Madison developed a sulfite pretreatment to overcome the recalcitrance of lignocellulose for robust enzymatic hydrolysis of wood cellulose.

US President George W. Bush, in his State of the Union address delivered January 31, 2006, proposed to expand the use of cellulosic ethanol. In his State of the Union Addresson January 23, 2007, President Bush announced a proposed mandate for 35 billion US gallons (130,000,000 m^3) of ethanol by 2017. It is widely recognized that the maximum production of ethanol from corn starch is 15 billion US gallons (57,000,000 m^3) per year, implying a proposed mandate for production of some 20 billion US gallons (76,000,000 m^3) more per year of cellulosic ethanol by 2017. Bush's proposed plan includes $2 billion funding (from 2007 to 2017) for cellulosic ethanol plants, with an additional $1.6 billion (from 2007 to 2017) announced by the USDA on January 27, 2007.

In March 2007, the US government awarded $385 million in grants aimed at jump-starting ethanol production from nontraditional sources like wood chips, switchgrass, and citrus peels. Half of the six projects chosen will use thermochemical methods and half will use cellulosic ethanol methods.

The American company Range Fuels announced in July 2007 that it was awarded a construction permit from the state of Georgia to build the first commercial-scale 100-million-US-gallon (380,000 m^3)-per-year cellulosic ethanol plant in the US. Construction began in November, 2007. The Range Fuels plant was built in Soperton, GA, but was shut down in January 2011, without ever having produced any ethanol. It had received a $76 million grant from the US Department of Energy, plus $6 million from the State of Georgia, plus an $80 million loan guaranteed by the U.S. Biorefinery Assistance Program.

PRODUCTION METHODS

The two ways of producing ethanol from cellulose are:

- Cellulolysis processes which consist of hydrolysis on pretreated lignocellulosic materials, using enzymes to break complex cellulose into simple sugars such as glucose, followed by fermentation and distillation.
- Gasification that transforms the lignocellulosic raw material into gaseous carbon monoxide and hydrogen. These gases can be converted to ethanol by fermentation or chemical catalysis.

Fig. Bioreactor for cellulosic ethanolresearch.

One production method that has yet to be utilized is that of the symbiotic microorganisms streptomyces strains B207 and L201 which have lignocellulosic capabilities § and live in the gut of termites to breakdown large quantities of cellulotic material into usable sugars. Current research for this experimental aspect of large scale lignocellulosic breakdown using industrial processes and enzymatic action of the gut of termites within titanium incubation equipment is being performed by scientist, entrepreneur, and politician Troy McComak. This new technique has been called "The McComak Process" although no patents have been currently applied for or awarded. Isolation of the genes which create the lignocelluases and cellulases is currently underway yet a lack of funding has severely hampered the project. According to McComak, "A world where middle east oil is no longer an economic viability is what will stop or start WWIII, therefore this project needs no further attention or funding until such social mechanisms have been put in place as to create a snowballing effect which will cause a world war regardless of middle east oil."

As is normal for pure ethanol production, these methods include distillation.

Cellulolysis (Biological Approach)

The stages to produce ethanol using a biological approach are:

- A "pretreatment" phase, to make the lignocellulosic material such as wood or straw amenable to hydrolysis
- Cellulose hydrolysis (cellulolysis), to break down the molecules into sugars
- Separation of the sugar solution from the residual materials, notably lignin
- Microbial fermentation of the sugar solution

- Distillation to produce roughly 95% pure alcohol
- Dehydration by molecular sieves to bring the ethanol concentration to over 99.5%

In 2010, a genetically engineered yeast strain was developed to produce its own cellulose-digesting enzymes. Assuming this technology can be scaled to industrial levels, it would eliminate one or more steps of cellulolysis, reducing both the time required and costs of production.

Pretreatment

Although lignocellulose is the most abundant plant material resource, its usability is curtailed by its rigid structure. As the result, an effective pretreatment is needed to liberate the cellulose from the lignin seal and its crystalline structure so as to render it accessible for a subsequent hydrolysis step. By far, most pretreatments are done through physical or chemical means. To achieve higher efficiency, both physical and chemical pretreatments are required. Physical pretreatment is often called size reduction to reduce biomass physical size. Chemical pretreatment is to remove chemical barriers so the enzymes can have access to cellulose for microbial destruction.

To date, the available pretreatment techniques include acid hydrolysis, steam explosion, ammonia fiber expansion, organosolv, sulfite pretreatment, AVAP® (SO_2-ethanol-water) fractionation, alkaline wet oxidation and ozone pretreatment. Besides effective cellulose liberation, an ideal pretreatment has to minimize the formation of degradation products because of their inhibitory effects on subsequent hydrolysis and fermentation processes. The presence of inhibitors will not only further complicate the ethanol production but also increase the cost of production due to entailed detoxification steps. Even though pretreatment by acid hydrolysis is probably the oldest and most studied pretreatment technique, it produces several potent inhibitors including furfural and hydroxymethyl furfural (HMF) which are by far regarded as the most toxic inhibitors present in lignocellulosic hydrolysate. Ammonia Fiber Expansion (AFEX) is a promising pretreatment with no inhibitory effect in resulting hydrolysate.

Most pretreatment processes are not effective when applied to feedstocks with high lignin content, such as forest biomass. Organosolv, SPORL ('sulfite pretreatment to overcome recalcitrance of lignocellulose') and SO_2-ethanol-water (AVAP®) processes are the three processes that can achieve over 90% cellulose conversion for forest biomass, especially those of softwood species. SPORL is the most energy efficient (sugar production per unit energy consumption in pretreatment) and robust process for pretreatment of forest biomass with very low production of fermentation inhibitors. Organosolv pulping is particularly effective for hardwoods and offers easy recovery of a hydrophobic lignin product by dilution and precipitation. AVAP® process effectively fractionates all types of lignocellulosics into clean highly digestible cellulose, undegraded hemicellulose sugars, reactive lignin and lignosulfonates, and is characterized by efficient recovery of chemicals.

Cellulolytic Processes

The cellulose molecules are composed of long chains of sugar molecules. In the hydrolysis process, these chains are broken down to free the sugar before it is fermented for alcohol production. There are two major cellulose hydrolysis (cellulolysis) processes: a chemical reaction using acids, or an enzymatic reaction.

Chemical Hydrolysis

In the traditional methods developed in the 19th century and at the beginning of the 20th century, hydrolysis is performed by attacking the cellulose with an acid. Dilute acid may be used under high heat and high pressure, or more concentrated acid can be used at lower temperatures and atmospheric pressure. A decrystalized cellulosic mixture of acid and sugars reacts in the presence of water to complete individual sugar molecules (hydrolysis). The product from this hydrolysis is then neutralized and yeast fermentation is used to produce ethanol. A significant obstacle to the dilute acid process is that the hydrolysis is so harsh that toxic degradation products are produced that can interfere with fermentation. BlueFire Renewables uses concentrated acid because it does not produce nearly as many fermentation inhibitors, but must be separated from the sugar stream for recycle [simulated moving bed (SMB) chromatographic separation, for example] to be commercially attractive.

Agricultural Research Service scientists found they can access and ferment almost all of the remaining sugars in wheat straw. The sugars are located in the plant's cell walls, which are notoriously difficult to break down. To access these sugars, scientists pretreated the wheat straw with alkaline peroxide, and then used specialized enzymes to break down the cell walls. This method produced 93 US gallons (350 L) of ethanol per ton of wheat straw.

Enzymatic Hydrolysis

Cellulose chains can be broken into glucose molecules by cellulase enzymes. This reaction occurs at body temperature in the stomachs of ruminants such as cattle and sheep, where the enzymes are produced by microbes. This process uses several enzymes at various stages of this conversion. Using a similar enzymatic system, lignocellulosic materials can be enzymatically hydrolyzed at a relatively mild condition (50 °C and pH 5), thus enabling effective cellulose breakdown without the formation of byproducts that would otherwise inhibit enzyme activity. All major pretreatment methods, including dilute acid, require an enzymatic hydrolysis step to achieve high sugar yield for ethanol fermentation. Currently, most pretreatment studies have been laboratory-based, but companies are exploring means to transition from the laboratory to pilot, or production scale.

Various enzyme companies have also contributed significant technological breakthroughs in cellulosic ethanol through the mass production of enzymes for hydrolysis at competitive prices. The fungus *Trichoderma reesei* is used by Iogen Corporation to secrete "specially engineered enzymes" for an enzymatic hydrolysis process. Their raw material (wood or straw) has to be pre-treated to make it amenable to hydrolysis.

Another Canadian company, SunOpta, uses steam explosion pretreatment, providing its technology to Verenium (formerly Celunol Corporation)'s facility in Jennings, Louisiana, Abengoa's facility in Salamanca, Spain, and a China Resources Alcohol Corporation in Zhaodong. The CRAC production facility uses corn stover as raw material.

Genencor and Novozymes have received United States Department of Energy funding for research into reducing the cost of cellulases, key enzymes in the production of cellulosic ethanol by enzymatic hydrolysis.

Other enzyme companies, such as Dyadic International, are developing genetically engineered fungi which would produce large volumes of cellulase, xylanase andhemicellulase enzymes, which can be used to convert agricultural residues such as corn stover, distiller grains, wheat straw and sugarcane bagasse and energy crops such asswitchgrass into fermentable sugars which may be used to produce cellulosic ethanol. In 2010, BP Biofuels bought out the cellulosic ethanol venture share of Verenium, which had itself been formed by the merger of Diversa and Celunol, and with which it jointly owned and operated a 1.4-million-US-gallon (5,300 m^3) per year demonstration plant in Jennings, LA, and the laboratory facilities and staff in San Diego, CA.

BP Biofuels continues to operate these facilities, and has begun first phases to construct commercial facilities. Ethanol produced in the Jennings facility was shipped to London and blended with gasoline to provide fuel for the Olympics. KL Energy Corporation, formerly KL Process Design Group, began commercial operation of a 1.5-million-US-gallon (5,700 m^3) per year cellulosic ethanol facility in Upton, WY in the last quarter of 2007. The Western Biomass Energy facility is currently achieving yields of 40–45 US gallons (150–170 L) per dry ton. It is the first operating commercial cellulosic ethanol facility in the nation. The KL Energy process uses a thermomechanical breakdown and enzymatic conversion process. The primary feedstock is soft wood, but lab tests have already proven the KL Energy process on wine pomace, sugarcane bagasse, municipal solid waste, and switchgrass.

Microbial Fermentation

Traditionally, baker's yeast (*Saccharomyces cerevisiae*), has long been used in the brewery industry to produce ethanol from hexoses (six-carbon sugars).

Due to the complex nature of the carbohydrates present in lignocellulosic biomass, a significant amount of xylose and arabinose (five-carbon sugars derived from the hemicellulose portion of the lignocellulose) is also present in the hydrolysate. For example, in the hydrolysate of corn stover, approximately 30% of the total fermentable sugars is xylose. As a result, the ability of the fermenting microorganisms to use the whole range of sugars available from the hydrolysate is vital to increase the economic competitiveness of cellulosic ethanol and potentially biobased proteins.

In recent years, metabolic engineering for microorganisms used in fuel ethanol production has shown significant progress. Besides *Saccharomyces cerevisiae*, microorganisms such as *Zymomonas mobilis* and *Escherichia coli* have been targeted through metabolic engineering for cellulosic ethanol production.

Recently, engineered yeasts have been described efficiently fermenting xylose, and arabinose, and even both together. Yeast cells are especially attractive for cellulosic ethanol processes because they have been used in biotechnology for hundreds of years, are tolerant to high ethanol and inhibitor concentrations and can grow at low pH values to reduce bacterial contamination.

Combined Hydrolysis and Fermentation

Some species of bacteria have been found capable of direct conversion of a cellulose substrate into ethanol. One example is *Clostridium thermocellum*, which uses a complexcellulosome to break down cellulose and synthesize ethanol. However, *C. thermocellum* also produces other products during cellulose metabolism, including acetate and lactate, in addition to ethanol, lowering the efficiency of the process. Some research efforts are directed to optimizing ethanol production by genetically engineering bacteria that focus on the ethanol-producing pathway.

GASIFICATION PROCESS (THERMOCHEMICAL APPROACH)

The gasification process does not rely on chemical decomposition of the cellulose chain (cellulolysis). Instead of breaking the cellulose into sugar molecules, the carbon in the raw material is converted into synthesis gas, using what amounts to partial combustion. The carbon monoxide, carbon dioxide and hydrogen may then be fed into a special kind of fermenter. Instead of sugar fermentation with yeast, this process uses *Clostridium ljungdahlii* bacteria. This microorganism will ingest carbon monoxide, carbon dioxide and hydrogen and produce ethanol and water.

The Process can thus be broken into three steps:

- Gasification — Complex carbon-based molecules are broken apart to access the carbon as carbon monoxide, carbon dioxide and hydrogen
- Fermentation — Convert the carbon monoxide, carbon dioxide and hydrogen into ethanol using the *Clostridium ljungdahlii* organism
- Distillation — Ethanol is separated from water

Fig. Fluidized Bed Gasifier inGüssing Burgenland Austria

A recent study has found another *Clostridium* bacterium that seems to be twice as efficient in making ethanol from carbon monoxide. Alternatively, the synthesis gas from gasification may be fed to a catalytic reactor where it is used to produce ethanol and other higher alcohols through a thermochemical process. This process can also generate other types of liquid fuels, an alternative concept successfully demonstrated by the Montreal-based company Enerkem at their facility in Westbury, Quebec.

HEMICELLULOSE TO ETHANOL

Studies are intensively conducted to develop economic methods to convert both cellulose and hemicellulose to ethanol. Fermentation of glucose, the main product of cellulose hydrolyzate, to ethanol is an already established and efficient technique. However, conversion of xylose, the pentose sugar of hemicellulose hydrolyzate, is a limiting factor, especially in the presence of glucose. Moreover, it cannot be disregarded as hemicellulose will increase the efficiency and cost-effectiveness of cellulosic ethanol production. Sakamoto (2012) et al. show the potential of genetic engineering microbes to express hemicellulase enzymes.

The researchers created a recombinant Saccharomyces cerevisiae strain that was able to:

- Hydrolyze hemicellulase through codisplaying endoxylanase on its cell surface,
- Assimilate xylose by expression of xylose reductase and xylitol dehydrogenase.

The strain was able to convert rice straw hydrolyzate to ethanol, which contains hemicellulosic components. Moreover, it was able to produce 2.5x more ethanol than the control strain, showing the highly effectiveness process of cell surface-engineering to produce ethanol.

Economics

The shift to a renewable fuel resource has been a target for many years now. However, most of its production is with the use of corn ethanol. In the year 2000, there was only 6.2 billion liters produced in the United States and it has expanded over 800% to 50 billion litres in just a decade (2010). Government pressures to shift to renewable fuel resources has been apparent since the U.S Environmental Protection Agency has implemented the 2007 Renewable Fuel Standard (RFS) to use a percentage of renewable fuel in products or face penalties. The shift to cellulosic ethanol production instead of corn has been strongly promoted by the US government. Even with these policies in place and the government attempting to create a market for cellulose ethanol, there was no commercial production of this fuel in 2010 and 2011. The Energy Independence and Security Act originally set goals of 100 million, 250 million and 500 million gallons for the years 2010, 2011 and 2012 respectively. However, as of 2012 it was projected that the production of cellulosic ethanol would be approximately 10.5 million far from its target. In 2007 alone, the US government provided 1 billion US dollars for cellulosic ethanol projects, while China invested 500 million US dollars into cellulosic ethanol research.

Due to the lack of existing commercialized plant data, it is difficult to determine the exact method of production that will be most commonly employed. Model systems try to compare different technologies costs, however these models cannot be applied to commercial-plant costs. Currently, there are many pilot and demonstration facilities open that show cellulosic production on a smaller scale.

Start-up costs for pilot scale lignocellulosic ethanol plants are high. On 28 February 2007, the U.S. Dept. of Energy announced $385 million in grant funding to six cellulosic ethanol plants. This grant funding accounts for 40% of the investment costs. The remaining 60% comes from the promoters of those facilities. Hence, a total of $1 billion will be invested for approximately 140-million-US-gallon (530,000 m^3) capacity. This translates into $7/annual gallon production capacity in capital investment costs for pilot plants; future capital costs are expected to be lower. Corn-to-ethanol plants cost roughly $1–3/annual gallon capacity, though the cost of the corn itself is considerably greater than forswitchgrass or waste biomass.

As of 2007, ethanol is produced mostly from sugars or starches, obtained from fruits and grains. In contrast, cellulosic ethanol is obtained from cellulose, the main component of wood, straw, and much of the structure of plants. Since cellulose cannot be digested by humans, the production of cellulose does not compete with the production of food, other than conversion of land from food production to cellulose production (which has recently started to become an issue, due to rising wheat prices.) The price per ton of the raw material is thus much cheaper than that of grains or fruits. Moreover, since cellulose is the main component of plants, the whole plant can be harvested. This results in much better yields—up to 10 short tons per acre (22 t/ha), instead of 4-5 short tons/acre (9–11 t/ha) for the best crops of grain.

The raw material is plentiful. An estimated 323 million tons of cellulose-containing raw materials which could be used to create ethanol are thrown away each year in US alone. This includes 36.8 million dry tons of urban wood wastes, 90.5 million dry tons of primary mill residues, 45 million dry tons of forest residues, and 150.7 million dry tons of corn stover and wheat straw. Transforming them into ethanol using efficient and cost-effective hemi(cellulase) enzymes or other processes might provide as much as 30% of the current fuel consumption in the United States. Moreover, even land marginal for agriculture could be planted with cellulose-producing crops, such as switchgrass, resulting in enough production to substitute for all the current oil imports into the United States.

Paper, cardboard, and packaging comprise a substantial part of the solid waste sent to landfills in the United States each day, 41.26% of all organic municipal solid waste (MSW) according to California Integrated Waste Management Board's city profiles. These city profiles account for accumulation of 612.3 short tons (555.5 t) daily per landfill where an average population density of 2,413 per square mile persists. All these, except gypsum board, contain cellulose, which is transformable into cellulosic ethanol. This may have additional environmental benefits because decomposition of these products produces methane, a potent greenhouse gas.

Reduction of the disposal of solid waste through cellulosic ethanol conversion would reduce solid waste disposal costs by local and state governments. It is estimated that each person in the US throws away 4.4 lb (2.0 kg) of trash each day, of which 37% contains waste paper, which is largely cellulose. That computes to 244 thousand tons per day of discarded waste paper that contains cellulose. The raw material to produce cellulosic ethanol is not only free, it has a negative cost—i.e., ethanol producers can get paid to take it away.

In June 2006, a U.S. Senate hearing was told the current cost of producing cellulosic ethanol is US $2.25 per US gallon (US $0.59/litre), primarily due to the current poor conversion efficiency. At that price, it would cost about $120 to substitute a barrel of oil (42 US gallons (160 L)), taking into account the lower energy content ofethanol. However, the Department of Energy is optimistic and has requested a doubling of research funding. The same Senate hearing was told the research target was to reduce the cost of production to US $1.07 per US gallon (US $0.28/litre) by 2012. "The production of cellulosic ethanol represents not only a step toward true energy diversity for the country, but a very cost-effective alternative to fossil fuels. It is advanced weaponry in the war on oil," said Vinod Khosla, managing partner of Khosla Ventures, who recently told a Reuters Global Biofuels Summit that he could see cellulosic fuel prices sinking to $1 per gallon within ten years.

In September 2010, a report by Bloomberg analyzed the European biomass infrastructure and future refinery development. Estimated prices for a litre of ethanol in August 2010 are EUR 0.51 for 1g and 0.71 for 2g. The report suggested Europe should copy the current US subsidies of up to $50 per dry tonne.

Recently on October 25, 2012, British Petrolum, one of the leaders in fuel products, announced the cancellation of their proposed $350 million commercial-scale plant. It was estimated that the plant would be producing 36 million gallons a year at its location in Highlands County of Florida. BP has still provided 500 million US dollars for biofuel research at the Energy Biosciences Institute. General Motors (GM) has also invested into cellulosic companies more specifically Mascoma and Coskata. There are many other companies in construction or heading towards it. Abengoa is building a 25 million-gallon per year plant in \ technology platform based on the fungus Myceliophthora thermophila to convert lignocellulose into fermentable sugars. Poet is also in midst of producing a 200 million dollar, 25-million-gallon per year in Emmetsburg, Iowa. Mascoma now partnered with Valero has declared their intention to build a 20 million gallon per year in Kinross, Michigan. China Alcohol Resource Corporation has developed a 6.4 million liter cellulosic ethanol plant under continuous operation.

Also, since 2013, the Brazilian company GranBio is working to become a producer of biofuels and biochemicals. The family-held company is commissioning a 82 million liters per year (22 MMgy) cellulosic ethanol plant (2G ethanol) in the state of Alagoas, Brazil, which will be the first industrial facility of the group. GranBio's second generation ethanol facility is integrated to a first generation ethanol plant operated by Grupo Carlos Lyra, uses process technology from Beta Renewables, enzymes from Novozymes and yeast from DSM. Breaking ground in January 2013, the plant is in final commissioning. According to GranBio Annual Financial Records, the total investment was 208 million US Dollars.

ENZYME-COST BARRIER

Cellulases and hemicellulases used in the production of cellulosic ethanol are more expensive compared to their first generation counterparts. Enzymes required for maize grain ethanol production cost 2.64-5.28 US dollars per cubic meter of ethanol produced. Enzymes for cellulosic ethanol production are projected to cost 79.25 US dollars, meaning they are 20-40 times more expensive. The cost differences are attributed to quantity required. The cellulase family of enzymes have a one to two order smaller magnitude of efficiency. Therefore, it requires 40 to 100 times more of the enzyme to be present in its production. For each ton of biomass it requires 15-25 kilograms of enzyme. There is also relatively high capital costs associated with the long incubation times for the vessel that perform enzymatic hydrolysis. Altogether, enzymes comprise a significant portion of 20-40% for cellulosic ethanol production.

Feedstocks

In general there are two types of feedstocks: forest (woody) Biomass and agricultural biomass. In the US, about 1.4 billion dry tons of biomass can be sustainably produced annually. About 370 million tons or 30% are forest biomass. Forest biomass has higher cellulose and lignin content and lower hemicellulose and ash content than agricultural biomass. Because of the difficulties and low ethanol yield in fermenting pretreatment hydrolysate, especially those with very high 5 carbon hemicellulsoe sugars such as xylose, forest biomass has significant advantages over agricultural biomass. Forest biomass also has high density which significantly reduces transportation cost. It can be harvested year around which eliminates long term storage. The close to zero ash content of forest biomass significantly reduces dead load in transportation and processing. To meet the needs for biodiversity, forest biomass will be an important biomass feedstock supply mix in the future biobased economy. However, forest biomass is much more recalcitrant than agricultural biomass. Recently, the USDA Forest Products Laboratory together with the University of Wisconsin–Madison developed efficient technologies that can overcome the strong recalcitrance of forest (woody) biomass including those of softwood species that have low xylan content. Short-rotation intensive culture or tree farming can offer an almost unlimited opportunity for forest biomass production.

Woodchips from slashes and tree tops and saw dust from saw mills, and waste paper pulp are common forest biomass feedstocks for cellulosic ethanol production.

The following are a few examples of agricultural biomass:

Switchgrass (*Panicum virgatum*) is a native tallgrass prairie grass. Known for its hardiness and rapid growth, this perennial grows during the warm months to heights of 2–6 feet. Switchgrass can be grown in most parts of the United States, including swamplands, plains, streams, and along the shores & *interstate highways*. It is *self-seeding* (no tractor for sowing, only for mowing), resistant to many diseases and pests, & can produce high yields with low applications of fertilizer and other chemicals. It is also tolerant to poor soils, flooding, & drought; improves soil quality and prevents erosion due its type of root system.

Switchgrass is an approved cover crop for land protected under the federal Conservation Reserve Program (CRP). CRP is a government program that pays producers a fee for not growing crops on land on which crops recently grew. This program reduces soil erosion, enhances water quality, and increases wildlife habitat. CRP land serves as a habitat for upland game, such as pheasants and ducks, and a number of insects. Switchgrass for biofuel production has been considered for use on Conservation Reserve Program (CRP) land, which could increase ecological sustainability and lower the cost of the CRP program. However, CRP rules would have to be modified to allow this economic use of the CRP land.

Miscanthus × giganteus is another viable feedstock for cellulosic ethanol production. This species of grass is native to Asia and is the sterile triploid hybrid of *Miscanthus sinensis* and *Miscanthus sacchariflorus*. It can grow up to 12 feet (3.7 m) tall with little water or fertilizer input. Miscanthus is similar to switchgrass with respect to cold and drought tolerance and water use efficiency. Miscanthus is commercially grown in the European Union as a combustible energy source. Corn cobs and corn stover are the most popular agricultural biomass. It has been suggested that Kudzu may become a valuable source of biomass.

ENVIRONMENTAL EFFECTS

The environmental impact from the production of fuels is an important factor in determining its feasibility as an alternative to fossil fuels. Over the long run, small differences in production cost, environmental ramifications, and energy output may have large effects. It has been found that cellulosic ethanol can produce a positive net energy output. The reduction in green house gas (GHG) emissions from corn ethanol and cellulosic ethanol compared with fossil fuels is drastic. Corn ethanol may reduce overall GHG emissions by about 13%, while that figure is around 88% or greater for cellulosic ethanol. As well, cellulosic ethanol can reduce carbon dioxide emissions to nearly zero.

Croplands

A major concern for the viability of current alternative fuels is the cropland needed to produce the required materials. For example, the production of corn for corn ethanol fuel competes with cropland that may be used for food growth and other feedstocks. The difference between this and cellulosic ethanol production is that cellulosic material is widely available and is derived from a large resource of things. Some crops used for cellulosic ethanol production include switchgrass, corn stover, and hybrid poplar. These crops are fast-growing and can be grown on many types of land which makes them more versatile. Cellulosic ethanol can also be made from wood residues (chips and sawdust), municipal solid waste such as trash or garbage, paper and sewage sludge, cereal straws and grasses. It is particularly the non-edible portions of plant material which are used to make cellulosic ethanol, which also minimizes the potential cost of using food products in production.

The effectiveness of growing crops for the purpose of biomass can vary tremendously depending on the geographical location of the plot. For example, factors such as precipitation and sunlight exposure may greatly effect the energy input required to maintain the crops, and therefore effect the overall energy output. A study done over five years showed that growing and managing switchgrass exclusively as a biomass energy crop can produce 500% or more renewable energy than is consumed during production. The levels of GHG emissions and carbon dioxide were also drastically decreased from using cellulosic ethanol compared with traditional gasoline.

Corn-Based vs. Grass-Based

In 2008, there was only a small amount of switchgrass dedicated for ethanol production. In order for it to be grown on a large-scale production it must compete with existing uses of agricultural land, mainly for the production of crop commodities. Of the United States' 2.26 billion acres (9.1 million km^2) of unsubmerged land, 33% are forestland, 26% pastureland and grassland, and 20% crop land. A study done by the U.S. Departments of Energy and Agriculture in 2005 determined whether there were enough available land resources to sustain production of over 1 billion dry tons of biomass annually to replace 30% or more of the nation's current use of liquid transportation fuels. The study found that there could be 1.3 billion dry tons of biomass available for ethanol use, by making little changes in agricultural and forestry practices and meeting the demands for forestry products, food, and fiber. A recent study done by the University of Tennessee reported that as many as 100 million acres (400,000 km^2, or 154,000 sq mi) of cropland and pasture will need to be allocated to switchgrass production in order to offset petroleum use by 25 percent.

Currently, corn is easier and less expensive to process into ethanol in comparison to cellulosic ethanol. The Department of Energy estimates that it costs about $2.20 per gallon to produce cellulosic ethanol, which is twice as much as ethanol from corn. Enzymes that destroy plant cell wall tissue cost 30 to 50 cents per gallon of ethanol compared to 3 cents per gallon for corn. The Department of Energy hopes to reduce production cost to $1.07 per gallon by 2012 to be effective. However, cellulosic biomass is cheaper to produce than corn, because it requires fewer inputs, such as energy, fertilizer, herbicide, and is accompanied by less soil erosion and improved soil fertility.

Additionally, nonfermentable and unconverted solids left after making ethanol can be burned to provide the fuel needed to operate the conversion plant and produce electricity. Energy used to run corn-based ethanol plants is derived from coal and natural gas. The Institute for Local Self-Reliance estimates the cost of cellulosic ethanol from the first generation of commercial plants will be in the $1.90–$2.25 per gallon range, excluding incentives. This compares to the current cost of $1.20–$1.50 per gallon for ethanol from corn and the current retail price of over $4.00 per gallon for regular gasoline (which is subsidized and taxed).

One of the major reasons for increasing the use of biofuels is to reduce greenhouse gas emissions. In comparison to gasoline, ethanol burns cleaner, thus putting less carbon dioxide and overall pollution in the air· Additionally, only low levels of smog are produced from combustion. According to the U.S. Department of Energy, ethanol from cellulose reduces greenhouse gas emission by 86 percent when compared to gasoline and to corn-based ethanol, which decreases emissions by 52 percent. Carbon dioxide gas emissions are shown to be 85% lower than those from gasoline. Cellulosic ethanol contributes little to the greenhouse effect and has a five times better net energy balance than corn-based ethanol. When used as a fuel, cellulosic ethanol releases less sulfur, carbon monoxide, particulates, and greenhouse gases. Cellulosic ethanol should earn producers carbon reduction credits, higher than those given to producers who grow corn for ethanol, which is about 3 to 20 cents per gallon.

It takes 0.76 J of energy from fossil fuels to produce 1 J worth of ethanol from corn. This total includes the use of fossil fuels used for fertilizer, tractor fuel, ethanol plant operation, etc. Research has shown that fossil fuel can produce over five times the volume of ethanol from prairie grasses, according to Terry Riley, President of Policy at the Theodore Roosevelt Conservation Partnership. The United States Department of Energy concludes that corn-based ethanol provides 26 percent more energy than it requires for production, while cellulosic ethanol provides 80 percent more energy. Cellulosic ethanol yields 80 percent more energy than is required to grow and convert it. The process of turning corn into ethanol requires about 1700 times (by volume) as much water as ethanol produced. Additionally, it leaves 12 times its volume in waste. Grain ethanol uses only the edible portion of the plant.

Table. U.S. Environmental Protection AgencyDraft life cycle GHG emissions reduction resultsfor different time horizon and discount rate approaches (includes indirect land use change effects)

Fuel Pathway	100 years +2% Discountrate	30 years +0% Discountrate
Corn ethanol natural gas dry mill)	-16%	+5%
Corn ethanol (Best case NG DM)	-39%	-18%
Corn ethanol (coal dry mill)	+13%	+34%
Corn ethanol (biomass dry mill)	-39%	-18%
Corn ethanol (biomass dry mill with combined heat and power)	-47%	-26%
Brazilian sugarcane ethanol	-44%	-26%
Cellulosic ethanol from switchgrass	-128%	-124%
Cellulosic ethanol from corn stover	-115%	-116%

Notes:

(1) Dry mill (DM) plants grind the entire kernel and generally produceonly one primary co-product: distillers grains with solubles (DGS).

(2) Best case plants produce wet distillers grains co-product.

Cellulose is not used for food and can be grown in all parts of the world. The entire plant can be used when producing cellulosic ethanol. Switchgrass yields twice as much ethanol per acre than corn. Therefore, less land is needed for production and thus less habitat fragmentation. Biomass materials require fewer inputs, such as fertilizer, herbicides, and other chemicals that can pose risks to wildlife. Their extensive roots improve soil quality, reduce erosion, and increase nutrient capture. Herbaceous energy crops reduce soil erosion by greater than 90%, when compared to conventional commodity crop production. This can translate into improved water quality for rural communities. Additionally, herbaceous energy crops add organic material to depleted soils and can increase soil carbon, which can have a direct effect on climate change, as soil carbon can absorb carbon dioxide in the air. As compared to commodity crop production, biomass reduces surface runoff and nitrogen transport. Switchgrass provides an environment for diverse wildlife habitation, mainly insects and ground birds. Conservation Reserve Program (CRP) land is composed of perennial grasses, which are used for cellulosic ethanol, and may be available for use.

For years American farmers have practiced row cropping, with crops such as sorghum and corn. Because of this, much is known about the effect of these practices on wildlife. The most significant effect of increased corn ethanol would be the additional land that would have to be converted to agricultural use and the increased erosion and fertilizer use that goes along with agricultural production. Increasing our ethanol production through the use of corn could produce negative effects on wildlife, the magnitude of which will depend on the scale of production and whether the land used for this increased production was formerly idle, in a natural state, or planted with other row crops.

Another consideration is whether to plant a switchgrass monoculture or use a variety of grasses and other vegetation. While a mixture of vegetation types likely would provide better wildlife habitat, the technology has not yet developed to allow the processing of a mixture of different grass species or vegetation types into bioethanol. Of course, cellulosic ethanol production is still in its infancy, and the possibility of using diverse vegetation stands instead of monocultures deserves further exploration as research continues.

A study by Nobel Prize winner Paul Crutzen found ethanol produced from corn had a "net climate warming" effect when compared to oil when the full life cycle assessmentproperly considers the nitrous oxide (N20) emissions that occur during corn ethanol production. Crutzen found that crops with less nitrogen demand, such as grasses and woody coppice species, have more favourable climate impacts.

CELLULOSIC ETHANOL COMMERCIALIZATION

Cellulosic ethanol commercialization is the process of building an industry out of methods of turning cellulose-containing organic matter into fuel.

Companies such as Iogen,POET, and Abengoa are building refineries that can process biomass and turn it into ethanol, while companies such as DuPont, Diversa, Novozymes, and Dyadic are producing enzymes which could enable a cellulosic ethanol future. The shift from food crop feedstocks to waste residues and native grasses offers significant opportunities for a range of players, from farmers to biotechnology firms, and from project developers to investors.

The cellulosic ethanol industry developed some new commercial-scale plants in 2008. In the United States, plants totaling 12 million liters (3.17 million gal) per year were operational, and an additional 80 million liters (21.1 million gal.) per year of capacity - in 26 new plants - was under construction. In Canada, capacity of 6 million liters per year was operational. In Europe, several plants were operational in Germany, Spain, and Sweden, and capacity of 10 million liters per year was under construction.

Italy-based Mossi & Ghisolfi Group broke ground for its 13 MMgy cellulosic ethanol facility in northwestern Italy on April 12, 2011. The project will be the largest cellulosic ethanol project in the world, 10 times larger than any of the currently operating demonstration-scale facilities.

TREETHANOL

Treethanol is an ethanol fuel (more precisely cellulosic ethanol) made from trees. The biofuel is a contender in the race to find an energy alternative to fossil fuels. Proponents of Treethanol claim that its energy yield is higher compared to the energy required for production when compared with more common sources of ethanol i.e. sugar cane and corn.

PRODUCTION

Cellulosic ethanol is produced using the lignocellulose biomass that comprises much of the mass of plants. Essentially at the core of the plant material is cellulose, which can be broken down into simple carbohydrate sugars. After these sugars have been extracted, they can be then be fermented into an alcohol, which is known as ethanol. The most widely used and promising means of creating cellulosic ethanol is called the cellulolysis process. The process consists of hydrolysis on pretreated lignocellulosic materials. Then enzymes are used to break down cellulose into glucose. This glucose is then fermented and distilled. The pretreatment step mentioned above is necessary when processing cellulosic ethanol because the glucose (sugars) are not readily accessible as they are with other ethanol sources such as corn or sugar cane. Rather, the cellulose in wood must be separated from the encapsulating hemicellulose and lignin.

There are three types of pretreatment: physical, chemical, and biological. Physical treatment involves physically reducing wood particle size. This can be accomplished through chipping, grinding, etc. Biological treatments involve the use of microorganisms to break down the wood. This type is considered favorable to physical pretreatments because it consumes far less energy in comparison, but the biological method has not proven to be scalable to an industrial level. The chemical method utilizes an alkaline or otherwise acidic medium to make the cellulose within wood fibers more accessible. This has shown to be the most efficient and has the lowest energy cost.

Forest trees make up more than 90% of the total terrestrial biomass while performing functions such as carbon sequestration, producing oxygen, and promoting biodiversity. Trees are a promising source of ethanol because they grow all year round, require significantly less fertilizer and water and contain far more carbohydrates (the chemical precursors of ethanol) than food crops (like corn) do.

Sources

Poplar, Willow, and Eucalyptus tree are emerging as favorites in the process of creating Treethanol. This is due to their ability to grow at a fast rate in many parts of the world.

APPLICATIONS

Treethanol is not an energy source that promises to power houses in the future, but rather it can be taken advantage of in applications where combustion engines are used. Approximately 85% of US energy consumption is produced from fossil fuels such as natural gas, coal, and oil. With China, India, and other rapidly developing nations increasing their demand for fossil fuels, the world's total energy use is expected to grow by 57% over the next 20 years. It is estimated that the U.S. alone uses 140 billion gallons of fuel per year for transportation alone. Not only can Treethanol be mixed with ordinary fuels, it can also be burned directly in modified engines to greatly reduce greenhouse gas emissions.

ADVANTAGES

Cellulosic ethanol is an environmentally friendly and renewable transportation fuel that can be produced from forest biomass. Trees are a particularly promising feedstock because they grow all year round, require vastly less fertiliser and water and contain far more carbohydrates (the chemical precursors of ethanol) than food crops do. Also, compared to corn ethanol, cellulosic biofuel does not require the same quantity of fertilizers, pesticides, energy, or water to grow. The most important attribute of this type of ethanol is, like all types of ethanol, it is renewable. If you want or need to make more of it, you just grow more trees.

The development of all types of biofuel, including Treethanol can be of importance for countries looking to decrease their dependence on petroleum, especially those countries that import most of their petroleum and also have plenty of crop/forest land such as New Zealand and Sweden. An important issue is whether Treethanol is a superior alternative to more common forms of ethanol like corn based. The general consensus in an article by Hoover, F., & Abraham, J. (2009), is that most forms of cellulosic ethanol have the potential to yield higher energy outputs and be more sustainable than corn ethanol. They also note that while cellulosic ethanol does not necessarily yield more energy than say, corn based ethanol per unit of measurement, it requires far less energy inputs to produce which could give it a far higher net energy yield at the end of processing. The findings that lignocellulosic biomass has a far greater productivity yield than traditional biofuel sources is supported by Papini, A., & Simeone, M. (2010). Responsible forestry practices do not contribute to greenhouse gases because the forest is allowed to regenerate following fiber harvesting. For this reason wood can be considered to be an essentially carbon-neutral source of energy.

Drawbacks

While it seems reasonable that Treethanol could be an alternative to current ethanol types, it has one flaw, which is the extra processing needed to break down the tough cellulose and hemicellulose within the walls of the cell to isolate the sugars. The creating ethanol from the lignocellulose found in tree biomass requires the extra step of "pre-treatment". It is this pre-treatment that still requires too much energy to make the Treethanol worth the effort.

That being said, many believe that the potential pros far out-weigh the short-term cons. The process of growing the tree biomass is energy efficient compared with growing corn or sugar cane for ethanol. However, it also takes longer to grow trees than to grow corn, and so any accurate research on sustainability and crop rotation (even for fast growing trees) requires a long time commitment, which up to now has been hard to find. It has been estimated that this process, including the building of processing plants and then refining of the growing and processing stage could take at least a decade. Another drawback to the processing of cellulosic ethanol is that there is little known about the waste/by products from the processing. Of particular concern to some is the biological method of pre-treatment. It is estimated that there is the possibility of producing almost as much (if not more) waste than usable ethanol, with waste products including mold, bacteria, yeast, biological toxins and allergens produced by these microorganisms, enzymes, and other chemicals.

AN ECONOMIC AND ECOLOGICAL PERSPECTIVE OF ETHANOL PRODUCTION FROM RENEWABLE AGRO WASTE

Since the late 19th century, the mean temperature on earth has increased with 0.8°C and the major part of this increase is likely due to anthropogenic emissions of greenhouse gases. Carbon dioxide is the type of greenhouse gas with largest emission and this originates from the combustion of fossil fuels as coal, oil and natural gas. In USA, transportation accounts for 30% of the total energy consumption. Burning fossil fuels such as coal and oil releases CO_2, which is a major cause of global warming. With only 4.5% of the world's population, the US is responsible for about 25% of global energy consumption and 25% of global CO_2 emissions. The average price of gasoline in 2005 was $2.56 per gallon, which was $0.67 higher than the average price of gasoline in the previous year. Yet in June 2008, the average price of gasoline in the US reached $4.10 per gallon.

Soaring oil prices associated with concerns of climate change and national energy security are driving us to utilize sustainable alternative energy sources, such as solar energy, nuclear energy, wind energy, hydropower, tidal energy, and so on. With the inevitable depletion of world's energy supply, there has been an increasing interest worldwide in alternative sources of energy. Unlike fossil fuels, ethanol is a renewable energy source produced through fermentation of sugars and used as a partial gasoline replacement in a few countries of the world.

Bioethanol market is expected to reach 100×10^9 liters in 2015. The largest producers in the world are the US, Brazil, and China. In 2009, US produced 39.5×10^9 liters of ethanol using corn as a feedstock while the second largest producer, Brazil, created about 30×10^9 liters of ethanol using sugarcane. China is a country that has invested much in the production of ethanol, and is nowadays one of the largest ethanol producers.

Ethanol contains 35% oxygen, which results in a complete combustion of fuel and thus lowers the emission of harmful gases. Moreover, ethanol production uses energy from renewable sources only; hence, no net carbon dioxide is added to the environment, thus reducing green-house gas emissions. It has also been well established now that ethanol increases the octane number, decreases the Reid vapor pressure and produces fuel with clean burning characteristics. Moreover, neat (unblended) ethanol can be burned with greater efficiency, and is thought to produce smaller amounts of ozone precursors (thus decreasing urban air pollution), and is particularly beneficial with respect to low net CO_2 put into the atmosphere.

The increasing demand for various industrial purposes such as alternative source of energy, industrial solvents, cleansing agents and preservatives, has necessitated increased production of ethanol. Furthermore, ethanol by fermentation offers a more favorable trade balance, enhanced energy security, and a major new crop for a depressed agricultural economy. Ethanol is considerably less toxic to humans than is gasoline (or methanol). Ethanol also reduces smog formation because of low volatility; its photochemical reactivity and low production of combustion products. Furthermore, low levels of smog-producing compounds are formed by its combustion. In addition, the low flame temperature of ethanol results in good engine performance.

Currently, bioethanol is being commercially produced only from edible feedstock such as corn-starch and sugarcane juice. The European Union (EU) had established a goal of 5.75% biomass-derived transportation fuels by December, 2010. The use of fuel ethanol has been quite successful in Brazil, where it is being produced at a very low cost by fermentation of sugarcane. In the US, corn is the dominant biomass feedstock for production of ethanol, and in the EU, straw and other agricultural wastes are the preferred types of biomass for ethanol production. These bioethanol production systems pose a concern about competition with food and feed supplies. To avoid this competition, bioethanol production from non edible lignocellulosic biomass such as wheat straw, rice straw, bagasse, corn stover, wood, peels of fruits and vegetables is attracting keen interest. The current production and use of bioethanol processes are a starting point. It is our belief that the next generational change in the use of bioresources will come from a total integration of innovative plant resources, synthesis of biomaterials, and generation of biofuels and biopower.

The present review is a concise overview of current and latest developments in ethanol production with special emphasis on the choice of lignocellulosic substrates, pretreatment methods and types of microorganisms that have been used for optimal, ecological and economic production of ethanol. Also reviewed are the different fungal and bacterial lignocellulolytic enzymatic systems including the current status of the technology for bioconversion of lignocellulose residues by microorganisms (particularly yeasts and fungi), with focus on the most economical and eco-friendly method for ethanol production.

LIGNOCELLULOSIC BIOMASS

Lignocellulose is a renewable organic material and is the major structural component of all plants.

Lignocellulose consists of three major components:

- Cellulose, the major constituent of all plant material and the most abundant organic molecule on earth, is a linear biopolymer of anhydroglucopyranose-molecules, connected by â-1, 4-glycosidic bonds. Cellulose or â-1-4-glucan is a linear polysaccharide polymer of glucose made of cellobiose units. The cellulose chains are packed by hydrogen bonds in so-called 'elementary microfibrils'. These fibrils are attached to each other by hemicelluloses, amorphous polymers of different sugars as well as other polymers such as pectin, and covered by lignin. The microfibrils are often associated in the form of bundles or macrofibrils. This special and complicated structure makes cellulose resistant to both biological and chemical treatments.
- Hemicellulose, the second most abundant component of lignocellulosic

biomass, is a heterogeneous polymer of pentoses (including xylose and arabinose), hexoses (mainly mannose, less glucose and galactose) and sugar acids. Hemicellulose is less complex, its concentration in lignocellulosic biomass is 25 to 35% and it is easily hydrolysable to fermentable sugars. The dominant sugars in hemicelluloses are mannose in softwoods and xylose in hardwoods and agriculture residues.

- Lignin, the third main heterogeneous polymer in lignocellulosic residues, generally contains three aromatic alcohols including coniferyl alcohol, sinapyl and *p*-coumaryl. Lignin serves as a sort of 'glue' giving the biomass fibers its structural strength. Lignin acts as a barrier for any solutions or enzymes by linking to both hemicelluloses and cellulose and prevents penetration of lignocellulolytic enzymes to the interior lignocellulosic structure. Not surprisingly, lignin is the most recalcitrant component of lignocellulosic material to degrade.

LIGNOCELLULOSE SUBSTRATES USED FOR ETHANOL PRODUCTION

Sweet sorghum bagasse can be converted efficiently into fermentable sugars (and is a new potential raw material for fuel ethanol production) by SO_2 catalyzed steam pretreatment at 190°C for 10 min or 200°C for 5 min followed by enzymatic hydrolysis with a result of 89-92% glucan conversion. Hemp and ensiled hemp can be converted into ethanol with steam pretreatment (2% SO_2 catalyst, 210°C for 5 min) followed by simultaneous saccharification and fermentation at high solid loading (7.5% water insoluble solids[WIS]) with a result of 171–163 g ethanol/kg raw material. In Brazil, ethanol is usually produced from cane juice, whereas in USA, starch-crops such as corn are usually used for ethanol production. Using sugars or corn as the main source for ethanol production caused a great deal of controversy due to its effect on food production and costs, which has made it difficult for ethanol to become cost competitive with fossil fuels. These concerns became a driving force in the generation of new biofuel research using lignocellulosic wastes produced by many different industries. Apart from corn and cane juice; wheat, oat and barley straw has also been routinely used to produce up to 0.52 million gallons of ethanol per year. China is the world's largest sweet potato (*Ipomoea batatas* Lam.) producer (accounting for 85% of global production), with the output exceeded 100 M tons in 2005. The sweet potato as an attractive feedstock for bioethanol production from both economic and environment friendly standpoints.

Lignocellulosic wastes are produced in large amounts by different industries including forestry, pulp and paper, agriculture and food, in addition to different wastes from municipal solid waste (MSW), and animal wastes. Those derived from agricultural activities include materials such as straw, stem, stalk, leaves, husk, shell, peel, lint, seed/stones, pulp or stubble from fruits, legumes or cereals (rice, wheat, corn, sorghum, barley), bagasses generated from sugarcane or sweet sorghum milling, spent coffee grounds, brewer's spent grains, and many others. These potentially valuable materials were treated as waste in many countries in the past, and still are today in some developing countries, which raises many environmental concerns. Significant efforts, many of which have been successful, have been made to convert these lignocellulosic residues to valuable products such as biofuels, chemicals and animal feed. Banana peel, an agro waste can be used as a substrate for ethanol production owing to its rich carbohydrate, crude proteins and reducing sugars. Moreover, banana peels are affordable and renewable low cost raw material which makes it potential feedstock for ethanol production. Similarly pineapple is the second harvest of importance after bananas, contributing to over 20% of the world production of tropical fruits. Thailand, Philippines, Brazil and China are the main pineapple producers in the world supplying nearly 50% of the total output. Other important producers include India, Nigeria, Kenya, Indonesia, México and Costa Rica and these countries provide most of the remaining fruit available (50%). Isitua and Ibeh 2010 assayed the feasibility of obtaining ethanol from pineapple waste with the purpose of obtaining a valuable product from the residues of the juice and canning industries.

Large volume of bagasse is generated during sugarcane processing. Agricultural profitability and environmental protection issues are associated with disposal of bagasse. In recent years, potential efforts have been directed towards the utilization of cheap renewable agricultural resources, such as sugarcane bagasse as alternative substrate for ethanol production. Rice is the major crop grown worldwide with an annual productivity around 800 million metric tonnes that corresponds with the large production of rice straw. In search for viable alternatives of biofuels, paddy straw has been pursued as suitable lignocellulosic waste for ethanol production.

Feasibility of lignocellulosic material for ethanol production has been explored around the world depending upon availability. Production of ethanol from wheat straw, one of the most abundant agricultural wastes, has been extensively studied. The average yield of wheat straw is 1.3–1.4 lb per lb of wheat grain. According to Ballesteros et al. 2006, under the 60% ground cover practice, about 354 millions of tons of wheat straw could be available globally and could produce 104 GL of bioethanol.

Europe production would account for about 38% of this world bioethanol capacity. In Spain, grain industry generates important amounts of wheat straw, a part of which is used as bedding straw and the remainder is burned or left on the land to fertilize the soil. Bioconversion of this residue to fuel ethanol would provide an attractive possibility to boost the development of biofuels in a sustainable way.

OVERVIEW OF LIGNOCELLULOSIC FERMENTATION

Schematic picture for the conversion of lignocellulosic biomass to ethanol, including the major steps can been seen in Pretreatment of the lignocellulosic residues is necessary because hydrolysis of non-pretreated materials is slow, and results in low product yield.

Some pretreatment methods increase the pore size and reduce the crystallinity of cellulose. Pretreatment also makes cellulose more accessible to the cellulolytic enzymes, which in return reduces enzyme requirements and, thus, the cost of ethanol production. The pretreatment not only enhance the bio-digestibility of the wastes for ethanol production, but also results in enrichment of the difficult biodegradable materials, and improves the yield of ethanol from the wastes.

Post pretreatment, the recalcitrant lignocellulosic biomass becomes susceptible to acid and/or enzymatic hydrolysis as the cellulosic microfibrils are exposed and/or accessible to hydrolyzing agents. In the pretreatment process, small amounts of cellulose and most of hemicellulose is hydrolyzed to sugar monomers; mainly D-xylose and D-arabinose. The pretreated biomass is then subjected to filtration to separate liquids (hemicellulose hydrolysate) and solid (lignin and cellulose).

After detoxification, the liquid is sent to a xylose (pentose) fermentation column for ethanol production. Solids are subjected to hydrolysis (also called second stage hydrolysis). This process is mainly accomplished by enzymatic methods using cellulases. Mild acid hydrolysis using sulfuric and hydrochloric acids is an alternative procedure. The hydrolyzed sugars such as D-glucose, D-galactose, and D-mannose, can be readily fermented to ethanol using various strains of *Saccharomyces cerevisae*. The pentoses (D-xylose and D-arabinose) from hemicellulose hydrolysis are not easily utilized by *Saccharomyces*strains; therefore, genetically modified strains of *Pichia stipitis*, *Zymomonas mobilis*, are used for their fermentation. *Candida shehatae* is capable of co-fermenting both pentoses and hexoses to ethanol and other value-added products at high yields.

Numerous pretreatment strategies have been developed to enhance the reactivity of cellulose and to increase the yield of fermentable sugars. *Typical goals of pretreatment include:*

- Production of highly digestible solids that enhances sugar yields during enzyme hydrolysis, avoidance of degradation of sugars (mainly pentoses) including those derived from hemicelluloses.
- Minimization of formation of inhibitors for subsequent fermentation steps.
- Recovery of lignin for conversion into valuable co-products.
- Cost effectiveness by operating in reactors of moderate size and by minimizing heat and power requirements.

PHYSICAL PRETREATMENTS METHODS

Physical pretreatments methods such as ball milling and grinding have been used for degradation of lignocelluloses with limited success. This method of pretreatment being cost effective and ecofriendly, and one on which relatively little work has been done and reported, so far, would form one of the thrust areas of future research.

Waste materials can be comminuted by a combination of chipping, grinding and milling to reduce cellulose crystallinity. The size of the materials is usually 10–30 mm after chipping and 0.2–2 mm after milling or grinding. Vibratory ball milling has been found to be more effective in breaking down the cellulose crystallinity of spruce and aspen chips and improving the digestibility of the biomass than ordinary ball milling. The power requirement of mechanical comminution of agricultural materials depends on the final particle size and the waste biomass characteristics.

Pyrolysis has also been used for pretreatment of lignocellulosic materials. When the materials are treated at temperatures greater than 300°C, cellulose rapidly decomposes to produce gaseous products and residual char. The decomposition is much slower and less volatile products are formed at lower temperatures.

The efficiency of ultrasound in the processing of vegetal materials has been already proved. The known ultrasounds benefits, such as swelling of vegetal cells and fragmentation due to the cavitational effect associated with the ultrasonic treatment, act by increasing the yield and by shortening of the extraction time. The effect of ultrasound on lignocellulosic biomass has been employed in order to improve the extractability of hemicelluloses, cellulose, lignin or to get clean cellulosic fiber from used paper but only few attempts to improve the susceptibility of lignocellulosic materials to biodegradation by using ultrasound have been described. It was found out that ultrasound has a beneficial effect on saccharification processes. Sonication has been reported to decrease cellulase requirements by 1/3 to 1/2 and to increase ethanol production from mixed waste office paper by approximately 20%. It was notice that the effect of ultrasound fragmentation of Avicel (microcrystalline cellulose formed by acid treatment) is similar to that of the enzymes for short incubation intervals. The time needed for ultrasonic treatment could be reduced when increasing the irradiation power.

CHEMICAL PRETREATMENT METHODS

ALKALINE PRETREATMENT

Alkaline pretreatment involves the use of bases, such as sodium, potassium, calcium, and ammonium hydroxide, for the pretreatment of lignocellulosic biomass. The use of an alkali causes the degradation of ester and glycosidic side chains resulting in structural alteration of lignin, cellulose swelling, partial decrystallization of cellulose and partial solvation of hemicelluloses. Sodium hydroxide has been extensively studied for many years, and it has been shown to disrupt the lignin structure of the biomass, increasing the accessibility of enzymes to cellulose and hemicellulose. Another alkali that has been used for the pretreatment of biomass is lime. Lignocellulosic feedstocks that have been shown to benefit from this method of pretreatment are corn stover, switchgrass, bagasse, wheat, and rice straw.

The conditions for alkaline pretreatment are usually less severe than other pretreatments. It can be performed at ambient conditions, but longer pretreatment times are required than at higher temperatures. The advantage of lime pretreatment is that the cost of lime required for pretreatment of a given quantity of biomass is lowest among alkaline treatments. Most commonly used alkali in the alkali pretreatment processes are NaOH and $Ca(OH)_2$. This process results in

- The removal of all lignin and part of hemicellulose, and
- Increased reactivity of cellulose in further hydrolysis steps, especially, enzymatic hydrolysis. Effective removal of lignin minimizes adsorption of enzyme onto lignin and thus allows for effective interactions with cellulose. Between NaOH and $Ca(OH)_2$, pretreatment with $Ca(OH)_2$is preferable because it is less expensive, more safer as compared to NaOH and it can be easily recovered from the hydrolysate by reaction with CO_2.

ACID PRETREATMENT METHODS

Acid pretreatment involves the use of concentrated and diluted acids to break the rigid structure of the lignocellulosic material. The most commonly used acid is dilute sulphuric acid (H_2SO_4), which has been commercially used to pre-treat a wide variety of biomass types-switchgrass, corn stover, spruce (softwood), and poplar. Acid pretreatment (removal of hemicellulose) followed by alkali pretreatment (removal of lignin) has shown to yield relatively pure cellulose. Strong acid allows complete breakdown of the components in the biomass to sugars, but also requires large volumes of concentrated sulfuric acid and can result in the production of furfural, an inhibitory byproduct. Dilute acid allows reduced acid concentrations, but requires higher temperatures, and again gives furfural. A key advantage of acid pretreatment is that a subsequent enzymatic hydrolysis step is sometimes not required, as the acid itself hydrolyses the biomass to yield fermentable sugars. A mixture of H_2SO_4 and acetic acid resulted in 90% saccharification. Hemicellulose and lignin are solubilized with minimal degradation, and the hemicellulose is converted to sugars with acid pretreatment. The major drawback to these acid processes is the cost of acid and the requirement to neutralize the acid after treatment.

WET OXIDATION

Wet oxidation utilizes oxygen as an oxidizer for compounds dissolved in water. Typically, the procedure for wet oxidation consists of drying and milling lignocellulosic biomass to obtain particles that are 2 mm in length, to which water is added at a ratio of 1 L to 6 g biomass. Wet oxidation has been used to fractionate lignocellulosic material by solubilizing hemicellulose and removing lignin. It has been shown to be effective in pretreating a variety of biomass such as wheat straw, corn stover, sugarcane bagasse, cassava, peanuts, rye, canola, faba beans, and reed to obtain glucose and xylose after enzymatic hydrolysis. During wet oxidation, lignin is decomposed to carbon dioxide, water and carboxylic acids. Biomass such as straw, reed and other cereal crop residues have a dense wax coating containing silica and protein which is removed by wet oxidation.

Wet oxidation has been combined with other pretreatment methods to further increase the yield of sugars after enzymatic hydrolysis. Combining wet oxidation with alkaline pretreatment has been shown to reduce the formation of byproducts, thereby decreasing inhibition. Bjerre et al. (1996) used wet oxidation and alkaline hydrolysis of wheat straw (20 g straw/l, 170°C, 5–10 min), and achieved 85% conversion yield of cellulose to glucose. Wet oxidation combined with base addition readily oxidizes lignin from wheat straw, thus making the polysaccharides more susceptible to enzymatic hydrolysis. Furfural and hydroxymethylfurfural, known inhibitors of microbial growth when other pretreatment systems are applied, were not observed following the wet oxidation treatment.

GREEN SOLVENTS

Processing of lignocellulosic biomass with ionic liquids (IL) and other solvents has gained importance in the last decade due to the tunability of the solvent chemistry and hence the ability to dissolve a wide variety of biomass types. Ionic liquid (IL) was found to possess a great potential in dissolving cellulose. Ionic liquids are salts, typically composed of a small anion and a large organic cation, which exist as liquids at room temperature and have very low vapor pressure. The chemistry of the anion and cation has been tuned to generate a wide variety of liquids which can dissolve a number of biomass types-corn stover, cotton, bagasse, switchgrass, wheat straw.

Dadi and coworkers (2007) have studied the enzymatic hydrolysis of Avicel regenerated from two different ILs, 1-*n*-butyl-3-methylimidazolium chloride and 1-allyl-3-methylimidazoliumchloride. Hydrolysis kinetics of the IL-treated cellulose was significantly enhanced. A limitation in using ionic liquids is the fact they tend to inactivate cellulose.

A solvent which has been effective in dissolution of cellulose and has a low vapor pressure similar to that of the ionic liquids is N-methyl morpholine

N-oxide (NMMO). NMMO retains all the advantages of the ionic liquids ability to dissolve a variety of lignocellulosic substrates without the need to chemically modify them and >99% of the solvent can be recovered due to its low vapor pressure. It is also nontoxic and biodegradable as proven by the work of Lenzing and other researchers. Further research is needed to evaluate and improve the economics of usage of ILs and NMMO for pretreatment of biomass. Pretreatment of lignocellulosic materials with acidified organic solvents (mixture of 80% ethylene glycol, 19.5% water and 0.5% HCl at 178°C for 90 min) has also been successfully used. The advantages of these methods include recovery and recycling of organic solvents as they can be easily distilled out. The disadvantages are that the process requires expensive high pressure equipment. Their performances could be improved by heating, microwave, or sonication.

PHYSICOCHEMICAL PRETREATMENT METHODS

STEAM-EXPLOSION

Steam-Explosion pretreatment is one of the most commonly used pretreatment options, as it uses both chemical and physical techniques in order to break the structure of the lignocellulosic material. This hydrothermal pretreatment method subjects the material to high pressures and temperatures for a short duration of time after which it rapidly depressurizes the system, disrupting the structure of cellulose microfibrils. The disruption of the fibrils increases the accessibility of the cellulose to the enzymes during hydrolysis.

Steam explosion is typically initiated at a temperature of 160–260°C (corresponding pressure 0.69–4.83 MPa) for several seconds to a few minutes before the material is exposed to atmospheric pressure. The process causes hemicellulose degradation and lignin transformation due to high temperature, thus increasing the potential of cellulose hydrolysis.

However, some disadvantages have been seen when using this process. Dilute acids are required to be added during softwood pretreatment or even when increased yields are warranted for lower acetylated feedstock. The factors that affect steam explosion pretreatment are residence time, temperature, chip size and moisture content. Recent studies indicate that lower temperature and longer residence time are more favorable.

LIQUID HOT WATER (LHW)

Much like the steam-explosion process, liquid hot water (LHW) pretreatment uses water at elevated temperatures and high pressures to maintain its liquid form in order to promote disintegration and separation of the lignocellulosic matrix. Temperatures can range from 160°C to 240°C over lengths of time ranging from a few minutes up to an hour with temperatures dominating the types of sugar formation and time dominating the amount of sugar formation.

This process has been found to be advantageous from a cost standpoint in that no additives such as acid catalysts are required. Furthermore, expensive reactor systems have not been necessary to use due to the low corrosive nature of this pretreatment technique. Neutralization of degradation products is not needed due to their fractionation and utilization in the liquid fraction. In the same sense, inhibitory products have not been reported to form overwhelmingly in the respective fractions allowing higher yields under specific conditions.

AMMONIA FIBER EXPLOSION (AFEX)

The ammonia fiber/freeze explosion (AFEX) process is another physicochemical process, much like steam explosion pretreatment, in which the biomass material is subjected to liquid anhydrous ammonia under high pressures and moderate temperatures and is then rapidly depressurized. The moderate temperatures (60°C to 100°C) are significantly less than that of the steam explosion process, thus allowing less energy input and overall cost reduction associated with the process.

There have been extensive literature reviews on this type of pretreatment over the last decade, focusing on the advantages and disadvantages of the AFEX process used for different feedstocks. An overview of some of the advantages include lower moisture content, lower formation of sugar degradation products due to moderate conditions, 100% recovery of solid material, and the ability for ammonia to lessen lignin's effect on enzymatic hydrolysis. A smaller number of disadvantages can be seen in the form of higher costs due to recycle and treatment of chemicals that are being used.

AMMONIA RECYCLE PERCOLATION (ARP)

Ammonia recycle percolation (ARP) has been paired with the AFEX pretreatment process by many authors, but it can have some different characteristics that need to be taken into consideration when looking at different pretreatment options. In this process, aqueous ammonia of concentration between 5-15% (wt%) is sent through a packed bed reactor containing the biomass feedstock at a rate of about 5 ml/min. The advantage with this process over AFEX is its ability to remove a majority of the lignin (75–85%) and solubilize more than half of the hemicellulose (50–60%) while maintaining high cellulose content. Primarily, herbaceous biomass has been most treated with this process: 60-80% delignification has been achieved for corn stover and 65–85% delignification for switchgrass.

SUPERCRITICAL FLUID (SCF) PRETREATMENT

A supercritical fluid is a material which can be either liquid or gas, used in a state above the critical temperature and critical pressure where gases and liquids can coexist. It shows unique properties that are different from those of either gases or liquids under standard conditions-it possesses a liquid like density and exhibits gas-like transport properties of diffusivity and viscosity. Thus, SCF has the ability to penetrate the crystalline structure of lignocellulosic biomass overcoming the mass transfer limitations encountered in other pretreatments.

The lower temperatures used in the process aids in the stability of the sugars and prevents degradation observed in other pretreatments. Kim and Hong 2001 investigated supercritical CO_2 pretreatment of hardwood (Aspen) and southern yellow pine with varying moisture contents followed by enzymatic hydrolysis. SCF pretreatment showed significant enhancements in sugar yields when compared to thermal pretreatments without supercritical CO_2. Alinia and coworkers investigated the effect of pretreatment of dry and wet wheat straw by supercritical CO_2alone and by a combination of CO_2 and steam under different operating conditions (temperature and residence time in the reactors). It was found that a combination of supercritical CO_2 and steam gave the best overall yield of sugars.

BIOLOGICAL PRETREATMENT METHODS

Biological pretreatment uses microorganisms and their enzymes selectively for delignification of lignocellulosic residues and has the advantages of a low-energy demand, minimal waste production and a lack of environmental effects. In biological pretreatment processes, microorganisms such as brown-, white- and soft-rot fungi are used to degrade lignin and hemicellulose in waste materials Schurz (1978).

White-rot basidiomycetes possess the capabilities to attack lignin. *Penicillium chrysosporium*, for example, has been shown to non-selectively attack lignin and carbohydrate. *P.chrysosporium* has been successfully used for biological pretreatment of cotton stalks by solid state cultivation (SSC) and results have shown that the fungus facilitates the conversion into ethanol. Brown rots mainly attack cellulose, while white and soft rots attack both cellulose and lignin.

White-rot fungi are the most effective basidiomycetes for biological pretreatment of lignocellulosic materials. Other basidiomycetes such as *Phlebia radiata*, *P. floridensis* and *Daedalea flavida*, selectively degrade lignin in wheat straw and are good choices for delignification of lignocellulosic residues. *Ceriporiopsis subvermispora*, however, lacks cellulases (cellobiohydrolase activity) but produces manganese peroxide and laccase, and selectively delignifies several different wood species.

The advantages of biological pretreatment include low energy requirement and mild environmental conditions. However, the rate of hydrolysis in most biological pretreatment processes is very low.

HYDROLYSIS OF PRETREATED BIOMASS

After pretreatment, the released cellulose and hemicelluloses are hydrolyzed to soluble monomeric sugars (hexoses and pentoses) using cellulases and hemicellulases, respectively. The initial conversion of biomass into sugars is a key bottleneck in the process of biofuel production and new biotechnological solutions are needed to improve their efficiency, which would lower the overall cost of bioethanol production. Enzymatic hydrolysis has been considered key to cost-effective bioethanol in the long run, and the reaction is carried out with mainly cellulase and hemicellulase for cellulose and hemicellulose, respectively. The advantages of using enzyme (cellulase) over acid is to eliminate corrosion problems and lower maintenance costs with mild processing conditions to give high yields.

Despite the fact that some fungal strains have the advantages of being thermostable and producing cellulases, most of these fungal strains do not produce sufficient amounts of one or more lignocellulolytic enzymes required for efficient bioconversion of lignocellulosic residues to fermentable sugars. In addition, plant cell walls are naturally resistant to microbial and enzymatic (fungal and bacterial) deconstruction, collectively known as 'biomass recalcitrance'. These rate-limiting steps in the bioconversion of lignocellulosic residues to ethanol remain one of the most significant hurdles to producing economically feasible cellulosic ethanol. Improving fungal hydrolytic activity and finding stable enzymes capable of tolerating extreme conditions has become a priority in many recent studies.

FUNGAL EXTRACELLULAR CELLULASES

Enzymatic saccharification of lignocellulosic materials such as sugarcane bagasse, corncob, rice straw, *Prosopis juliflora*, *Lantana camara*, switch grass, saw dust, and forest residues by cellulases for biofuel production is perhaps the most popular application currently being investigated. Both bacteria and fungi can produce glucanases (cellulases) that hydrolyze of lignocellulosic materials. These microorganisms can be aerobic or anaerobic and mesophilic or thermophilic. Bacteria belonging to genera of *Clostridium*, *Cellulomonas*, *Bacillus*, *Thermomonospora*,*Ruminococcus*, *Bacteriodes*, *Erwinia*, *Acetovibrio*, *Microbispora*, and *Streptomyces* are known to produce cellulase. Anaerobic bacterial species such as *Clostridium phytofermentans*, *Clostridium thermocellum*, *Clostridium hungatei, and Clostridium papyrosolvens* produces cellulases with high specific activity. Most commercial glucanases (cellulases) are produced by*Trichoderma ressei* and â-D-glucosidase is produced from *Aspergillus niger*. Fungi known to produce cellulases include *Sclerotium rolfsii*, *Phanerochaete chrysosporium* and various species of*Trichoderma*, *Aspergillus*, *Schizophyllum and Penicillium*. Among the fungi, *Trichoderma* species have been extensively studied for cellulase production.

High temperature and low pH tolerant enzymes are preferred for the hydrolysis due to the fact that most current pretreatment strategies rely on acid and heat. In addition, thermostable enzymes have several advantages including higher specific activity and higher stability which improve the overall hydrolytic performance. Ultimately, improvement in catalytic efficiencies of enzymes will reduce the cost of hydrolysis by enabling lower enzyme dosages. Some fungal strains such as *T.emersonii*, *Chaetomium thermophilum* and *Corynascus thermophilus* can produce thermostable enzymes which are stable and active at elevated temperatures (60°C) well above their optimum growth temperature (30-55°C). Due to the promising thermostability and acidic tolerance of thermophilic fungal enzymes, they have good potential to be used for hydrolysis of lignocellulosic residues at industrial scales.

The anaerobic bacteria *Clostridium thermocellum* and *Clostridium cellulovorans* and the filamentous fungus*Trichoderma reesei* are well known as strongly cellulolytic and xylanolytic microorganisms. *C.thermocellum* and *C. cellulovorans* produce a cellulosome complex consisting of cellulase and hemicellulase organized on the cell surface; *T. reesei*, meanwhile, extracellularly secretes three types of cellulolytic enzyme, including five endoglucanases (EG [EC 3.2.1.4]), two cellobiohydrolases (CBH [EC 3.2.1.91]), and two β-glucosidases (BGL [EC 3.2.1.21]) (Sang-Mok and Koo 2001). Endoglucanases act randomly against the amorphous region of the cellulose chain to produce reducing and nonreducing ends for cellobiohydrolases, which produce cellobiose from reducing or nonreducing ends of crystalline cellulose. Cellulose chains are thus efficiently degraded to soluble cellobiose and cellooligosaccharides by the endo-exo synergism of EG and CBH (Hebeish and Ibrahim 2007). In the last step of enzymatic cellulose degradation, cellooligosaccharides are hydrolyzed to glucose by β-glucosidase. In addition to endo-exo synergism, exo-exo synergism between two cellobiohydrolases has also been reported.

FUNGAL HEMICELLULASES

Several different enzymes are needed to hydrolyze hemicelluloses, due to their heterogeneity (Saha 2003). Xylan is the most abundant component of hemicellulose contributing over 70% of its structure. Xylanases are able to hydrolyze β-1,4 linkages in xylan and produce oligomers which can be further hydrolyzed into xylose by β-xylosidase. Not surprisingly, additional enzymes such as β-mannanases, arabinofuranosidases or β-L-arabinases are needed depending on the hemicellulose composition which can be mannan-based or arabinofuranosyl-containing. Also similarly to cellulases, most of the hemicellulases are glycoside hydrolases (GHs), although some hemicellulases belong to carbohydrate esterases (CEs) which hydrolyze ester linkages of acetate or ferulic acid side groups. A mixture of hemicellulases or pectinases with cellulases exhibited a significant increase in the extent of cellulose conversion. Many fungal species such as *Trichoderma*,*Penicillium*, *Aspergillus* and *T. emersonii* have been reported to produce large amounts of extracellular cellulases and hemicellulases.

FUNGAL LIGNINASES

Fungi degrade lignin by secreting enzymes collectively termed "ligninases".

These include two ligninolytic families;

- Phenol oxidase (laccase) and
- Peroxidases [lignin peroxidase (LiP) and manganese peroxidase (MnP)]. White-rot basidiomycetes such as *Coriolus versicolor*, *P. chrysosporium* and *T. versicolor* have been found to be the most efficient lignin-degrading microorganisms studied. Interestingly, LiP is able to oxidize the non-phenolic part of lignin, but it was not detected in many lignin degrading fungi. In addition, it has been widely accepted that the oxidative ligninolytic enzymes are not able to penetrate the cell walls due to their size. Thus, it has been suggested that prior to the enzymatic attack, low-molecular weight diffusible reactive oxidative compounds have to initiate changes to the lignin structure and hemicellulose, fungal cellulosomes are much less well characterized compared to bacterial cellulosomes.

FERMENTATION

In the fermentation process, the hydrolytic products including monomeric hexoses (glucose, mannose and galactose) and pentoses (xylose and arabinose) will be fermented to valuable products such as ethanol. Among these hydrolytic products, glucose is normally the most abundant, followed by xylose or mannose and other lower concentration sugars.

The last two steps of bioconversion of pretreated lignocellulolytic residues to ethanol (hydrolysis and fermentation) can be performed separately (SHF) or simultaneously (SSF). In the separate hydrolysis and fermentation (SHF), the hydrolysate products will be fermented to ethanol in a separate process. The advantage of this method is that both processes can be optimized individually (e.g. optimal temperature is 45-50°C for hydrolysis, whereas it is 30°C for fermentation). However, its main drawback is the accumulation of enzyme-inhibiting end-products (cellobiose and glucose) during the hydrolysis. This makes the process inefficient, and the costly addition of â-glucosidase is needed to overcome end-product inhibition.

Further process integration can be achieved by a process known as consolidated bioprocessing (CBP) which aims to minimize all bioconversion steps into one step in a single reactor using one or more microorganisms. CBP operation featuring cellulase production, cellulose/hemicellulose hydrolysis and fermentation of 5- and 6- carbon sugars in one step have shown the potential to provide the lowest cost for biological conversion of cellulosic biomass to fuels, when processes relying on hydrolysis by enzymes and/or microorganisms are used.

The simultaneous saccharification and fermentation (SSF) process was first studied by Takagi et al. for cellulose conversion to ethanol. The SSF process was originally developed for lignocellulosic biomass by researchers at Gulf Oil Company in 1974. The SSF process eliminates expensive equipment and reduces the probability of contamination by unwanted organisms that are less ethanol tolerant than the microbes selected for fermentation.

SSF combines the enzymatic saccharification of polymeric cellulose to simple monomeric forms such as glucose and its eventual fermentation by yeast to ethanol in the same vessel. In simultaneous saccharification and fermentation (SSF), however, the end-products will be directly converted to ethanol by the microorganism. Therefore, addition of high amounts of β-glucosidase is not necessary and this reduces the ethanol production costs. Rapid conversion of the glucose into ethanol by yeast results in faster rates, higher yields, and greater ethanol concentrations than possible for SHF. The presence of ethanol in the fermentation broth also makes the mixture less vulnerable to invasion by unwanted microorganisms. However, the main drawback of SSF is the need to compromise processing conditions such that temperature and pH are suboptimal for each individual step. The development of recombinant yeast strains with improved thermotolerance can enhance the performance of SSF. It is reported that the major inefficiencies of biochemical process for lignocellulosic bioethanol production were identified as the simultaneous saccharification and fermentation (SSF) process accounting for 27% of the lost energy by thermodynamic analysis. Alkasrawi et al. (2003) reported that addition of surfactants as an additive in SSF can significantly lower the operational cost of the process because it increases the conversion rate of cellulose to glucose. Addition of Tween-20, 2.5 g/l not only reduces the time required to attain maximum ethanol concentration, but also enhances enzyme activity in the liquid fraction at the end of SSF, probably by preventing unproductive binding of the cellulases to lignin, which could facilitate enzyme recovery.

Over the years, various groups have worked on the SSF process to improve the choice of enzymes, fermentative microbes, biomass pretreatment, and process conditions. Extensive studies on SSF have since been conducted focusing on the production of ethanol from cellulosic substrates. Phillipidis et al. have studied the enzymic hydrolysis of cellulose in an attempt to optimize SSF performance. Ghose et al. have increased ethanol productivity by employing a vacuum cycling in an SSF process using lignocellulosic substances. Zhu et al. evaluated the suitability of production of ethanol from the microwave-assisted alkali pretreated wheat straw, the simultaneous saccharification and fermentation (SSF) of the microwave-assisted and conventional alkali pretreated wheat straw to ethanol.

Candida brassicae is accepted as the yeast of choice as far as SSF is considered, although both*Saccharomyces cerevisiae* and *S. carlsbergensis* have been found to offer similar rates. Several other yeasts as well as the bacteria *Zymomonas mobilis* have been studied with cellulose from *T. ressei* mutants for SSF processes. Researchers have also examined several combinations of enzymes with *Z. mobilis*, *S. cerevisiae*, and other ethanol producer, but they have only considered substrate levels lower than necessary to prove economic viability. Wyman et al. evaluated the cellobiose-fermenting yeast *Brettanomyces clausenii*for the SSF of cellulose to ethanol.

There are number of different methods to quantitate ethanol in samples. HPLC has been utilized to monitor the fermentation process This method has the advantage of being able to monitor not only the production of ethanol, but also the reaction substrates and byproducts. Fourier transform infrared spectroscopy, gas chromatography, and Infrared technologies have also been used to detect and quantitate ethanol in samples. While FTIR requires a large investment in instrumentation, the use or less expensive IR technology has been demonstrated to be just as accurate. Gerchman et al. developed a cheap and rapid approach for ethanol quantification in aqueous media during fermentation steps as part of the conversion of biomass to ethanol. The suggested method requires a sample of a small volume and consists of organic extraction, followed by direct use of gas chromatography with a flame ionization detector (GC-FID). The feasibility of such approach is obvious since there is no need for the head-space system, distillation, expensive reagents and sophisticated equipment. The proposed method was also tested for its 'real-life' applicability for ethanol quantification from fermentation process.

According to Keller and Bryan, distillation is still a "formidable competitor" as a major separation method even though much research has been thrust on its alternatives. Hence, distillation, especially simple distillation, tends to be the first choice in industry for separating a liquid mixture; other methods, including complex distillation, e.g., azeotropic distillation, come into play only when simple distillation is deemed to be technically infeasible or economical inviable because of typically three large stainless steel distillation towers, stainless steel heat exchangers and price of stainless up 400% in last six years, high operating costs because 280 MMBTU/hr energy is consumed (100 MGPY ethanol). Mole sieve drying adds to energy costs and that's why energy costs up significantly with price of crude oil.

Under certain circumstances, retrofitting of an existing process can be economically far more viable than constructing a new process, especially when the financial resources are limited and/or when short term needs are to be met under a tight time constraint. Developing economically viable fermentation processes requires efficient downstream processing: selective product removal and avoiding byproduct streams. "ESepis a modular, low-energy process for the recovery of ethanol from fermentation broth with an estimated reduction of up to 60% in both capital and operating costs versus conventional distillation. Use of non-stainless steel components also results in a substantial reduction in construction time". It is applicable to new ethanol plants (corn, sugar and cellulosic). It replaces whole distillation train and mole sieve dryer. With new plants it reduces overall energy consumption by >60%.

The utilization of pervaporation for the production of absolute (anhydrous) ethanol through its coupling with the previous distillation step has been reported. The modeling and optimization of the process using MINLP tools showed 12% savings in the production costs considering a 32% increase in membrane area and the reduction in both reflux ratio and ethanol concentration in the distillate of the column. Through pilot-plant studies, the integration of distillation process with the pervaporation has been achieved resulting in good indexes in terms of energy savings. These savings are due to the low operation costs of pervaporation and to the high yield of dehydrated ethanol, typical of pervaporation processes. The comparison between azeotropic distillation using benzene and pervaporation system using multiple membrane modules showed that, at the same ethanol production rate and quality (99.8 wt.%), operation costs, including the membrane replacement every 2–4 years, are approximately 1/3–1/4 of those of azeotropic distillation.

METHODS USED TO IMPROVE FUNGAL ENZYME PRODUCTION, ACTIVITY AND/OR STABILITY

In order to produce ethanol industrially, the fermentative microorganism needs to be robust. The utilization of all the sugars generated from lignocellulosic hydrolysate is essential for the economical production of ethanol. The conventional ethanol fermenting yeast (*Saccharomyces cerevisiae*) or bacterium (*Zymomonas mobilis*) cannot ferment multiple sugar substrates to ethanol. A major technical hurdle to converting lignocellulose to ethanol is developing an appropriate microorganism for the fermentation of a mixture of sugars such as glucose, xylose, arabinose, and galactose. A number of recombinant microorganisms such as *Escherichia coli*, *Klebsiella oxytoca*, *Z. mobilis*, and *S. cerevisiae* have been developed over the last 25 years with a goal of fermenting mixed sugars to ethanol. Saha and Cotta's research unit has developed a recombinant *E. coli* (strain FBR5) that can ferment mixed multiple sugars to ethanol. The strain carries the plasmid pLOI297, which contains the genes for pyruvate decarboxylase (pdc) and alcohol dehydrogenase (adh) from *Z. mobilis* necessary for efficiently converting pyruvate into ethanol. Technologies required for bioconversion of lignocelluloses to ethanol and other valuable products are currently available but need to be developed further in order to make biofuels cost competitive compared to other available energy resources such as fossil fuels. The most recent and important improvements in production/activity of fungal enzymes using different techniques such as mutagenesis, co-culturing and heterologous gene expression of cellulases.

MUTAGENESIS

Many fungal strains have been subjected to extensive mutagenesis studies due to their ability to secrete large amounts of cellulose-degrading enzymes. Cellulolytic activity of *T. reesei* QM6a has been improved by using different mutagenesis techniques including UV-light and chemicals, resulting in the mutant QM 9414 with higher filter paper activity (FPA). *T. reesei* RUT-C30 is one of the best known mutants, producing 4–5 times more cellulase than the wild-type strain (QM 6a). A recent study by Kovacs et al. 2008 has shown that wild-type *Trichoderma atroviride* (F-1505) produces the most cellulase among 150 wild-type *Trichoderma*. Moreover, *T. atroviride* mutants were created by mutagenesis using *N*-methyl-*N*'-nitro-*N*-nitrosoguanidine (NTG) as well as UV-light. These *T. atroviride* mutants (e.g. *T.atroviride* TUB F-1724) produce high levels of extracellular cellulases as well as β-glucosidase when they are grown on pretreated willow. Cellulase and xylanase activities in *Penicillium verruculosum* 28 K mutants were improved about 3-fold using four cycles of UV mutagenesis. The enzyme production was further improved by 2- to 3-fold in a two-stage fermentation process using wheat bran, yeast extract medium and microcrystalline cellulose as the inducer.

Site-directed mutagenesis (SDM) plays a central role in the characterization and improvement of cellulases including their putative catalytic and binding residues. The application of SDM it was found that Glu 116 and 200 are the catalytic nucleophile and acid–base residues in *Hypocrea jecorina* (anamorph *T. reesei*) Cel12A, respectively. In the study, mutant enzymes were produced where Glu was replaced by Asp or Gln at each position (E116D/Q and E200D/Q). The specific activity of these mutants was reduced by more than 98%, suggesting the critical role of these two residues in the catalytic function of the enzyme. In another study, the thermostable endo-1,4-β-xylanase (XynII) mutants from *T. reesei* were further mutated to resist inactivation at high pH by using SDM. All mutants were resistant to thermal inactivation at alkaline pH. For example, thermotolerance for one mutant (P9) at pH 9 was increased approximately 4–5°C, resulting in better activity in sulphate pulp bleaching compared to the reference. Also, the catalytic efficiency and optimum pH of *T. reesei* endo-β-1,4-glucanase II were improved by saturation mutagenesis followed by random mutagenesis and two rounds of DNA shuffling. The pH optimum of the variant (Q139R/L218H/W276R/N342T) was shifted from 4.8 to 6.2, while the enzyme activity was improved more than 4.5-fold. Moreover, the stability of *T. reesei* endo-1,4-β-xylanases II (XynII) was increased by engineering a disulfide bridge at its N-terminal region. In fact, two amino acids (Thr-2 and Thr-28) in the enzyme were substituted by cysteine (T2C:T28C mutant) resulting in a 15°C increase in thermostability.

CO-CULTURING

Fungal co-culturing offers a means to improve hydrolysis of lignocellulosic residues, and also enhances product utilization which minimizes the need for additional enzymes in the bioconversion process. In the case of cellulose degradation, for example, all three enzymatic components (EG, CBH and β-glucosidase) have to be present in large amounts. However, none of the fungal strains, including the best mutants, are able to produce high levels of the enzymes at the same time. *T. reesei* for example produces CBH and EG in high quantities whereas its β-glucosidase activity is low. *A. niger*however, produces large amounts of β-glucosidase, but has limited EG components. In addition, hemicellulose hydrolysis must also be considered when lignocellulosic residues are subjected to biomass conversion.

However, this will be determined by the pretreatment methods. Specifically in an alkali pretreatment method, a part of lignin will be removed and thus hemicellulose has to be degraded by the use of hemicellulases, whereas in acid-catalyzed pretreatment, the hemicellulose layer will be hydrolyzed. Again, some fungal strains have been shown to work more efficiently on cellulosic residues whereas others produce more hemicellulolytic enzymes and efficiently hydrolyze hemicellulosic portions. Conversion of both cellulosic and hemicellulosic hydrolytic products in a single process can be achieved by co-culturing two or more compatible microorganisms with the ability to utilize the materials. In fact, in nature, lignocellulosic residues are degraded by multiple co-existing lignocellulolytic microorganisms.

Mixed fungal cultures have many advantages compared to their monocultures, including improving productivity, adaptability and substrate utilization. Improving fungal cellulolytic activity of *T. reesei* and *A.niger* by co-culturing was the subject of extensive research including studies done by Maheshwari et al., Ahamed et al. and Juhasz et al. 2003. Moreover, other fungal strains have been co-cultured to obtain better cellulolytic activity such as co-culturing of *T. reesei* RUT-C30 and *A. phoenicis* or *A. ellipticus* and *A. fumigatus*. There are a few examples of co-culturing fungal strains for the purpose of combining cellulose and hemicellulose hydrolysis such as co-culturing *T. reesei*D1-6 and *A. wentii* Pt 2804 in a mixed submerged culture or co-culturing *T. reesei* LM-UC4 and *A. phoenicis* QM329 using ammonia-treated bagasse. In the both cases, enzyme activity for cellulases and hemicellulases was significantly increased. The main drawback of co-culturing however is the complexity of growing multiple microorganisms in the same culture.

METABOLIC ENGINEERING

Metabolic engineering is a powerful method to improve, redirect, or generate new metabolic reactions or whole pathways in microorganisms. This enables one microorganism to complete an entire task from beginning to end. This can be done by altering metabolic flux by blocking undesirable pathway(s) and/or enhancement of desirable pathway(s). For example by application of homologous recombination, the production of *T. reesei* β-glucosidase I was enhanced using xylanase (*xyn3*) and cellulase (*egl3*) promoters which improved β-glucosidase activity to 4.0 and 7.5 fold compared to the parent, respectively. This will permit one fungal strain such as *T. reesei* to be more efficient on hydrolysis of cellulose to glucose which improve the yield and therefore lower the cost. Becker and Boles described the engineering of a *Saccharomyces cerevisiae* strain able to utilize the pentose sugarL-arabinose for growth and to ferment it to ethanol. Expanding the substrate fermentation range of *S.cerevisiae* to include pentoses is important for the utilization of this yeast in economically feasible biomass-to-ethanol fermentation processes. After overexpression of a bacterial L-arabinose utilization pathway consisting of *Bacillus subtilis* AraA and *Escherichia coli* AraB and AraD and simultaneous overexpression of the L-arabinose-transporting yeast galactose permease, we were able to select an L-arabinose-utilizing yeast strain by sequential transfer in L-arabinose media. High L-arabinose uptake rates and enhanced transaldolase activities favor utilization of L-arabinose.

Shaw et al. 2008 engineered *Thermoanaerobacterium saccharolyticum*, a thermophilic anaerobic bacterium that ferments xylan and biomass-derived sugars, to produce ethanol at high yield. Knockout of genes involved in organic acid formation (acetate kinase, phosphate acetyltransferase, and L-lactate dehydrogenase) resulted in a strain able to produce ethanol as the only detectable organic product and substantial changes in electron flow relative to the wild type. Glucose and xylose are co-utilized and utilization of mannose and arabinose commences before glucose and xylose are exhausted.

HETEROLOGOUS EXPRESSION

Heterologous expression is a powerful technique to improve production yield of enzymes, as well as activity. In order to make a robust lignocellulolytic fungal strain, many different fungal cellulases with higher and/or specific activity based on the need for a functional cellulase system in the organism have been cloned and expressed. For example, thermostable â-glucosidase (*cel3a*) from thermophilic fungus *T. emersonii* was expressed in *T. reesei* RUT-C30 using a strong *T. reesei cbh1*promoter. The expressed enzyme has been shown to be highly thermostable (optimum temperature at 71.5°C) with high specific activity. In the study for the improvement of biofinishing of cotton, *T. reesei* cellobiohydrolase (I & II) were overexpressed using additional copy(s) of the genes cloned under *T. reesei cbh1* promoter. The results have shown that the expression of CBHI was increased to 1.3- and 1.5-fold with one or two additional copies of the gene, respectively.

IMMOBILIZATION

Immobilization of microbial cells and enzymes have showed certain technical and economical advantages over free cell system. Using immobilized enzymes not only leads to greater product purity, cleaner processes, and economic operational costs but also makes the use enzyme cost effective and recoverable. The immobilized biocatalysts have been extensively investigated during last few decades. An immobilized cellobiase enzyme system has been used in the enzymatic hydrolysis of biomass for the generation of cellulosic ethanol. Production of alcohol and biodiesel fuel from triglycerides using immobilized lipase has been carried out using porous kaolinite particle as a carrier.

The use of an immobilized yeast cell system for alcoholic fermentation is an attractive and rapidly expanding research area because of its additional technical and economical advantages compared with the free cell system. A reduction in the ethanol concentration in the immediate microenvironment of the organism due to the formation of a protective layer or specific adsorption of ethanol by the support may act to minimize end product inhibition. The most significant advantages of immobilized yeast cell systems are the ability to operate with high productivity at dilution rates exceeding the maximum specific growth rate, the increase of ethanol yield and cellular stability and the decrease of process expenses due to the cell recovery and reutilization. Other advantages of immobilized cell system over presently accepted batch or continuous fermentations with free-cells are: greater volumetric productivity as a result of higher cell density; tolerance to higher concentrations of substrate and products; lacking of inhibition; relative easiness of downstream processing etc. in different types of bioreactors, such as packed bed reactor, fluidized bed reactor, gaslift reactor and reactor with magnetic field.

Perspective techniques for yeasts immobilization can be divided into four categories: attachment or adsorption to solid surfaces (wood chips, delignified brewer's spent grains, DEAE cellulose, and porous glass), entrapment within a porous matrix (calcium alginate, k-carrageenan, polyvinyl alcohol, agar, gelatine, chitosan, and polyacrilamide), mechanical retention behind a barrier (microporous membrane filters, and microcapsules) and self-aggregation of the cells by flocculation.

PROCESS INTEGRATION

One of the most important approaches for the design of more intensive and cost-effective process configurations is process integration. Process integration looks for the integration of all operations involved in the production of fuel ethanol. This can be achieved through the development of integrated bioprocesses that combine different steps into one single unit. Thus, reaction–separation integration by removing ethanol from the zone where the biotransformation takes place, offers several opportunities for increasing product yield and consequently reducing product costs. Other forms of integration may significantly decrease energetic costs of specific flowsheet configurations for ethanol production. Process integration is gaining more and more interest due to the advantages related to its application in the case of ethanol production: reduction of energy costs, decrease in the size and number of process units, intensification of the biological and downstream processes. Integration of fermentation and separation processes for reduction of product inhibition, development of efficient cogeneration technologies using cane bagasse, development of CBP, application of membrane technology (e.g. for ethanol removal or dehydration) are examples of process integration.

INTEGRATION OF FOREST INDUSTRIES AND REDUCTION OF WASTE

Wood waste is a term loosely applied to material from logging or manufacturing operations which is put to no commercial use, and also to material burned for fuel which might be used in the manufacture of products of higher value. In some countries there is practically no waste; in others where timber is more abundant, waste may be as high as 75 to 80 percent of the volume of a forest stand.Much waste could be eliminated by integration of forest management and utilization, and by better integration of wood-using industries.

STATISTICS

If FAO is properly to fulfill its functions in the fields of forestry and forest products, it must have at its disposal up-to-date and accurate information respecting the extent and capacities of the forest resources of the world and the supply of and demand for forest products. International statistical series, instituted by several organizations before the war, should be resumed and consolidated by FAO at the earliest possible moment to serve the organization itself, forest authorities, and the forest-products industries of the world. These series covered production, distribution, and consumption in the major wood-using countries, but there were many gaps as to production, and information on stocks was not entirely adequate. During the war greatly improved information has been collected and it is highly desirable that such statistical series should not be discontinued. It will be necessary to make arrangements whereby corresponding information will be made available for other regions where it is lacking. Attention is called to the fact that in some countries very valuable statistics, notably those concerning industrial output, stocks, and prices, are compiled by trade associations rather than by governments, and it is hoped that suitable arrangements can be made whereby these will become available to FAO through the proper governmental channels.

Through international cooperation centered in FAO, it is expected that general agreement can be reached on the most suitable framework for a body of world-wide forestry statistics. While it is recognized that different terms and units of measurement established by preference and long custom in different countries will continue to be used, each country should be asked to submit definitions of terms and to recommend suitable factors for the conversion of its units to whatever units may be adopted by FAO for regional and worldwide compilations.

MARKETING

Marketing has both national and international aspects. There is known to be great diversity in methods of grading closely similar species used for the same purposes and in acceptable sizes of products of forest industries as between different countries. These are of long standing, and uniformity will not easily be reached; but confusing trade names and lack of well-conceived standard series of sizes and grades result in high costs of production, inefficiency in the use of wood as a material, and unnecessarily high costs to consumers.

The first step for FAO will be to act as a clearinghouse for the collection and distribution of all available information on grading practices and units of measurement. Subsequently, progress toward standardization for improved utilization may be possible through international conferences to be organized by FAO.

Much useful work is being conducted by research organizations and other bodies in various parts of the world on the development of new uses for wood and in the utilization of unfamiliar species of timber. This provides a wide field in which FAO could act as a center for the collection and distribution of information and thus provide a basis for a scientific study of consumers' needs. It could also act as a source from which trade organizations interested in extending the use of forest products could obtain authoritative and reliable information. The Committee has reviewed paragraphs 104 to 107 of the Report of the Technical Committee on Forestry and Forest Products 1 dealing with international trade in forest products and in general endorses the views expressed. In this connection it is stressed that the compilation and publication of adequate estimates of international supplies and requirements of lumber and other forest products would of itself exercise a stabilizing influence upon the market, as well as facilitate adjustments in production and demand.

IMPLEMENTATION OF RECOMMENDATIONS

These recommendations have been kept to a minimum, but it may not be possible to carry out all of them during the formative period of FAO. The

Committee on Forestry and Forest Products urges the appointment of a strong advisory committee on forestry and forest products at the earliest possible moment to help the Director-General in applying the recommendations of this Conference and putting into effect the Committee's suggestions for a world forest policy, which is outlined in the pages that follow.

A WORLD FOREST POLICY

The principles of a world policy together with a specific program for FAO have already been considered by the Interim Commission and laid down in a Report of the Technical Committee on Forestry and Forest Products accompanying the Interim Commission's Third Report to Governments as well as in the report published as one of the five Technical Committee reports of the Interim Commission. The Forestry Committee approves and endorses the former as well as the general principles and recommendations contained in Parts I and II of the latter. The recommendations in Part III of the Technical Committee's report are replaced by the recommendations for early action contained in the report to Commission A by the Committee on Forestry and Forest Products.

Delegates from many nations have presented before the Forestry Committee reports dealing with conditions and problems in the forests and forest industries of their countries as well as in a number of dependent areas.

These reports have suggested the following broad groupings:

- Countries which forests have not the capacity to supply domestic wood needs and which are forced to manage their forests more intensively and expand their forest area in order to increase their wood production and improve the quality of forest products.
- Countries still endowed with vast forests which should control cutting in order to ensure the continued use of their forest resources at a high level, since the experience of the last twenty years has amply demonstrated that forests once considered inexhaustible can rapidly be exhausted.
- Regions now suffering acutely from the results of deforestation with its attendant disastrous influences on climate, soil, and rural economy and which therefore need afforestation policies.
- Countries, especially in the tropics, having vast and dense forests which constitute a very important reserve of forest products and which must be improved by means of proper silviculture and managed efficiently in order to check serious deterioration in their composition, avoid excessive waste in their utilization, and ensure their future productivity.

It is clear that each nation must set its own forest policy, but FAO should define for forestry and forest products certain basic principles worldwide in application. In order to achieve such a world policy, it will be necessary to determine periodically both the annual productive capacity of the world's forests and the annual wood requirements of the world's peoples. The war has created a special problem of rehabilitation of forests which have been damaged in the course of the last few years. This is particularly urgent and calls for immediate recommendations leading to remedial action.

It is equally' desirable that the social problems involved be given full consideration. Measures to improve working conditions in the forests and in forest industries should be studied with the purpose of establishing a basis for general policy recommendations by FAO. With the information now available it is believed possible to enunciate fundamental principles for a broad policy in the field of forestry and forest products; it should deal with physical, demographic, and economic factors.

FOREST CONSERVATION

Forest management presents essentially different aspects in different parts of the world. In temperate regions - Europe, for example - forest management no longer presents any great technical difficulties; but there remain problems of economics, especially with regard to private forests. In many countries public action has been necessary in order to protect forests adequately against the disruptive consequences of fluctuations in supply and demand and of changing prices for forest products.

In the more newly settled countries, forest management still involves a number of technical problems which forestry research must solve. Here too the economic factors are of great importance because the opening up of undeveloped forests usually involves a preliminary phase of extensive utilization and crude methods of exploitation that are wholly incompatible with good forest management. In this situation both educational and administrative action is indispensable, as demonstrated by the examples of Canada, the United States, and South Africa. Even in forest stands that have not been disturbed by felling operations, the forest is often exposed to destructive agencies which can imperil its existence. These agencies are:

Forest fires. Forest fires cause grave losses, especially in the vast areas of resinous forests of Canada, the United States, and the U.S.S.R. Fire fighting is essentially a technical problem calling for proper equipment and well-trained specialized personnel. In addition it involves an administrative problem since the government is usually forced to assume responsibility for organizing adequate fire protection and for bearing all or part of the costs of protection.

Insects and disease. Here, additional scientific research and the development of new techniques are the prime concern. Grazing. In some types of forest, grazing of livestock gives a supplementary income and is not incompatible with good forest management. In other situations livestock grazing is a serious threat to forests, especially in arid countries as in some Mediterranean and subtropical regions where during parts of the year forests provide the only reserve of moisture and green vegetation. Forestry Committee members from Greece, India, and China emphasized the seriousness of the grazing problem and stressed the necessity of educational and administrative measures in dealing with it.

Nomadic agriculture (shifting cultivation). In many subtropical and tropical regions, the people who practice shifting cultivation continue to destroy the forest in order to clear land for their primitive agriculture and then move on and destroy new areas of forest. The solution of this problem requires a broad policy involving agriculture, forestry, and land utilization based on a clear classification of land best suited for agriculture and land which should in the general interest be maintained or brought under forest crops.

FOREST IMPROVEMENT

In temperate countries, improvement of the forest growing-stock is an inseparable part of forest management, since the large-scale introduction of new species with attendant higher yields is not entirely indispensable. However, highly intensified silviculture involving large-scale plantations and even the use of fertilizer, as well as the application of modern techniques in genetics and plant breeding has been used successfully - as demonstrated by reports from Belgium and Denmark. In equatorial regions, forest improvement is particularly important because the more valuable species are usually widely dispersed. Here silviculture should be directed toward building up forest stands composed of economically useful trees. Here again, the technical aspect is the prime consideration, but administrative action - e.g., tax reduction - has also been recommended since it may encourage private forest owners to adopt measures for improving their forests.

AFFORESTATION

Afforestation of land suitable for agriculture and for the protection of agricultural crops can be greatly encouraged by public action. In this connection examples have been quoted from France (subsidies), Norway and certain newer countries (afforestation loans), and China, where small agricultural communities are under legal obligation to create forest nurseries and establish plantations.

SOIL CONSERVATION

This is one of the most serious problems that foresters face because it involves not only economic consequences but the actual maintenance of populations and has a direct influence on living conditions and rural life. In this work foresters join hands with soil experts and develop programs to their mutual advantage.

WAR DAMAGE

War damage to the forests has been particularly serious in Greece, where one fourth of the forest area has been destroyed by war operations or enemy action. To a lesser degree similar damage has been suffered by many countries in Europe and in the Far East.

This situation will call for the three following measures:

- Reduced cuttings in certain countries coupled with increased imports of forest products.
- Better utilization of forest products, especially the use of small and low-grade woods previously neglected by industries and the processing of waste heretofore discarded. Such utilization has become possible through technical improvements which should now be generally introduced and applied.
- Afforestation on forest soils damaged by war operations. This involves treatment similar to that suggested for the denuded areas.

FOREST UTILIZATION

Proper forest management is based on the removal of growth at periodic intervals. The harvesting of forest products should aim at securing their maximum utilization. In certain temperate countries, it is estimated that only 25 percent of the standing tree volume is industrially used. This enormous waste is in great part due to the fact that logging operators and forest industries sacrifice maximum volume utilization in order to operate at a profit.

To achieve efficient forest utilization, the following should be suggested by FAO to Member nations:

- Adoption of measures encouraging industries to make maximum use of the wood they cut.
- Technical education in order to make industries and other wood users realize that a higher degree of wood utilization is to their interest.
- Technical forest products research in specialized laboratories.
- Integration of forest industries to reduce waste in logging and processing.

It is stressed that research should concern itself not only with the properties of wood but also with new wood-using techniques. Closer wood utilization is even more important in tropical and equatorial forests. Here, utilization has been mainly confined to trees of the more valuable species widely dispersed throughout the forest. This highly selective form of felling makes the whole operation very expensive and definitely limits its possibilities and leads to progressive degradation of the forest stand itself as well as reducing its economic value. Utilization of a greater number of tree species would greatly reduce production costs and make it possible for these regions to contribute more importantly to world requirements. This has not been done so far because there is a market for only a few tropical species. They are little known by the consumers and are handicapped by prohibitive transportation costs.

Where these tropical areas suffer from manpower shortage, intensive mechanization of logging operations is necessary in opening up forests. Industries must be developed within the forest itself in order to process on the spot all species and those parts of the tree which otherwise could not carry transportation charges. All these measures of mechanization and industrialization should be accompanied by the application of silvicultural methods designed to transform low-grade forests into high-quality stands. This might be done with a view to such definite uses as the manufacture of veneers and plywood, or the production of pulp, paper, and textiles.

DISTRIBUTION OF PRODUCTS

In order to avoid obstructions to the orderly distribution of forest products, as well as extreme price variations, FAO should make it a major task to keep governments constantly informed concerning import requirements for and export supplies of forest products. This should be done on a world scale, and to that effect it might be advisable to develop further the recommendations contained in paragraph 107 of the Report to the Interim Commission by the Technical Committee on Forestry and Forest Products.

The foregoing problems, or groups of problems, are basic considerations in the formulation of a world forest policy. In certain respects work towards these objectives was being conducted by a number of international organizations prior to the war. In accordance with decisions taken at this conference, FAO will carry on the activities of these organizations in so far as their purposes fall within its scope.

FISHERIES

Fisheries have been one of the first food-producing industries to be affected by the sudden ending of the war. The war thrust upon one half of the world's fish-producing nations the burden of attempting to maintain production levels formerly achieved by the whole world. Now the sudden reversion - the liberation of the seas and the liberation of effort - has sharply emphasized the problems of distribution and focused attention on faults in the distribution mechanism.

Yet, despite the potentiality of so-called surpluses owing their existence to maldistribution, it is a fact that in some parts of the world deficiencies in protein could be eased by a well-planned development of fisheries resources within roach of the people. Often the reason that this is not already being done is due to lack of knowledge and ignorance of technique. The Food and Agriculture Organization is young - its beginnings foreshadow its ultimate achievements. Therefore suggestions for the initial activities of the Organization in relation to fisheries have been framed with caution. They have been limited to those activities that are within the competence of FAO during its early stages and are at the same time desirable and practical from the point of view of Member nations.

In general the proposals in this report embody broad principles which will allow sufficient flexibility of action by governments that wish to apply them under local conditions. There are, however, certain fields in which uniformity of action is desirable. Among them are the adoption of uniform methods of collecting and reporting basic fisheries data and the standardization of quality, packaging, weight, and designation of fish commodities. Lack of uniformity in these respects constitutes a form of restriction upon the free flow of commodities between nations which injures the interests of both producer and consumer.

On the other hand, FAO has direct responsibilities. Among them are studies contributing to fisheries knowledge in its international aspects, as well as the distribution of such information in usable forms. FAO should also be directly responsible for bringing about such degrees of cooperation among international organizations as may be necessary to ensure the optimum yield from fisheries in the high seas. It is estimated that the sea fisheries yield over 30 thousand million pounds of food fish each year in areas where, in accordance with international law, every nation has the right to fish. It is also true that fisheries can be exhausted to below economic levels and that, unless agreement can be reached among nations upon a mode of behavior that will assure adequate protection, stocks of fish which are now so plentiful will once again rapidly become depleted. For this reason international action on this problem, such as could be obtained by FAO, is regarded as of great importance.

There are also parts of the world where fisheries could be developed to a much greater extent by the use of the proper techniques to catch fish in greater quantities or in new places, or by improving the conditions for fish reproduction and growth. In this way long-term production is enlarged by exploring new avenues, including the full use of lakes and ponds. Progress in this direction would be largely through national rather than international action but could be greatly facilitated by the exchange of information and expert advice and the encouragement of research by FAO. These and other considerations, which are more fully dealt with in the Report to the Interim Commission by the Technical Committee on Fisheries, have led to the following proposals, in which the term "fish and fisheries products" means any products of the aquatic flora or fauna as the context may require.

COLLECTION, ANALYSIS, INTERPRETATION, AND DISSEMINATION OF INFORMATION

The collection of information on fisheries of various nations is of great importance, and the establishment of systems for the collection and publication of fishery data should be encouraged in all countries. Furthermore? Member nations and institutions concerned with fisheries should have knowledge of all published reports the summaries of which would be valuable to research workers

In general? reports of the physical and nutritional sciences are available in standard publications. Economic and sociological publications and those dealing with fish-handling techniques? however, al e not so widely distributed. There is no doubt that both administrative officials of various governments and policy makers would be greatly assisted if they had available, in some form digests of significant contributions to the knowledge of world fisheries. Fishery statistics are so vital that a special effort should be made to encourage the collection and publication by Member nations of basic fishery data. Such data are more easily understood if collected and reported in a uniform manner.

SCIENTIFIC, TECHNOLOGICAL, SOCIOLOGICAL, AND ECONOMIC RESEARCH

Fundamental to the intelligent consideration of fishery resources are investigations to determine

- The natural history, distribution, migrations, and environmental relationships of fishery species;
- The size, extent, and annual and seasonal variations in abundance of fish populations;
- The effect on abundance of continuing fishing operations;
- The most efficient methods of obtaining maximum production without endangering future supply; and
- Effective methods of artificial propagation, stocking! and disease and pollution control.

The methods and results of these coordinated phases of biological and hydrographical research had begun to attain exactness and fruitfulness immediately prior to the war. The scope and magnitude of such research varied considerably among the various primary fishing nations of the world. and some conducted none at all. In no country was the extent of such research commensurate with the magnitude of the fishery resources.

Research done heretofore, designed to identify and appraise the nutritional components of fishery products, appears to be fairly adequate with respect to protein, fat, mineral content, certain of the essential vitamins, and digestibility. large volume of this information is available, and scientists continue to investigate all new phases in this field as advances in international knowledge and techniques are made. Fish are an excellent source of proteins, minerals, and certain of the essential vitamins which would contribute to the well-being of the people in many countries. Better knowledge of the preparation of fish for culinary purposes would increase the consumption of fish.

Research done on the development of pharmacological fishery products is less adequate, but such development contributes to diversification of the uses for fishery products.

In recent years a great mass of information has been assembled on the technological phases of fish production and processing covering the handling of fish aboard the boat or vessel; the preparation of fish for market by icing, freezing, salting, drying, canning, etc.; and the warehousing, storage, and transportation of fishery products. Much work has also been done in the field of fishery by-products, such as fish meal and oil, and in the development of mechanical devices for their preparation. While much still remains to be accomplished in this field, it is believed that existing knowledge is far in advance of application.

Existing facilities are inadequate for the needs of fishery research if it is to be prosecuted on a scale which will develop fishery resources to the full. Further development of centers for all phases of fishery research is required. Since, in many instances, fishermen and shore workers are in the low-income group of labor, more attention should be given to helping them improve their general well-being. The problem of full employment is also vital to the postwar world. Very few studies have been made in the field of fishery economics, but the solution of many fishery problems must depend upon such knowledge.

IMPROVEMENT OF EDUCATION RELATING TO FISHERIES AND FISHERY INDUSTRIES

Available facilities for the training of fishery personnel in all phases of production, processing, and distribution are very limited, and improvement of education relating to fisheries and fishery industries is important to the full development of fishery resources.

CONSERVATION AND DEVELOPMENT OF FISHERY RESOURCES

Fishery conservation problems on the high seas are international in character, but because the problems of conservation are different in the many areas involved, it is considered preferable for any international action for conservation and management to be established on a regional basis. There should, however, be a free interchange of ideas and information between such regional authorities in order to assist in bringing about a wider degree of coordination and interest.

The full use of fishery resources depends to a large degree on the development of fishery techniques best adapted to the many different conditions prevailing. Progress in such development might be accelerated by a better exchange of information. The full use of fishery resources depends not only on the management of fisheries to obtain the maximum yield in perpetuity and improvement of fishing techniques, but also on the improvement of conditions for fish reproduction and growth.

IMPROVEMENT OF THE PROCESSING, MARKETING, AND DISTRIBUTION OF FISHERY PRODUCTS

The fundamental problem of irregularity of supply should be the concern of all nations. More efficient methods of catch must be employed and, above all, work must continue on the application of newly developed methods of preservation which can act as a buffer against fluctuations in the supply of raw material. These, coupled with improvements in transportation and distribution systems, would mean a more regular Row of fishery products to the consumer one of the factors essential to any considerable expansion in consumption.

Processing covers the entire field of fish preservation, including freezing, canning, drying, salting, smoking, and the manufacture of fish byproducts. As a wealth of information is available on newer and more efficient methods of processing fishery products.

There is a wide spree-d between the landed value of fish and its retail price. Fish, one of the least expensive food products at the point of production, becomes one of the more expensive foods in the retail store. Many reasons have been advanced for this situation, but the fact remains that it retards consumption. Some studies of causative factors in the chain of marketing and distribution have been made but they have not led to a solution of the problem. Thus further studies might be undertaken to ensure the production of wholesome products, standardized, where possible, with respect to quality, packaging, weight, and designation.

ADOPTION OF POLICIES FOR THE PROVISIONOF ADEQUATE FISHERY CREDITS, NATIONAL, AND INTERNATIONAL

Fishery industries in general are undercapitalized; however, technical advances should go a long way toward removing some of the great risks that have militated against the investment of capital. The pursuit of technical progress will be national in scope. On the other hand, countries where lack of protein is an outstanding national deficiency may stand in need of international credits in order to develop their fisheries.

3

Biological Control of Forest Pests

FOREST PEST CONTROL

INSECTS

Insects are the most destructive agents affecting forest and shade trees in the South. Tree roots, stems, limbs, needles, leaves of healthy or weakened trees, or logs waiting to be sawed into lumber are all subject to attack. Insects (Class: Insecta) are by far the most numerous animal life inhabiting the forest. They have become well adapted to their surroundings and occupy a wide variety of ecological niches. Although the majority of insect species are either beneficial or innocuous, some are exceedingly harmful. Insect outbreaks that cause economic damage to forests vary greatly in frequency, size and duration. Fortunately, most outbreaks are small and short-lived, and usually consist of one or a few spots in a stand or region. Others, however, may expand and encompass hundreds or thousands of acres and can last for several years.

Managers can reduce the risks and incidence of insect attack by maintaining healthy and vigourously growing stands and trees. Research is being conducted to determine what conditions are conducive to forest insect outbreaks. This research may lead to improved control measures.

BARK BEETLES

Bark beetle populations vary tremendously between years and between locations. The important bark beetles in the South attack pines and belong to the Family Scolytidae. However, some species such as the native elm bark beetle Hylurgopinus rufipes, the small European elm bark beetle Scolytus multistriatus (Marsham), and the hickory bark beetle S. quadrispinosus Say are hardwood pests. The remainder of this bark beetle discussion will be about southern pine beetles, black turpentine beetles and Ips engraver beetles that attack southern pines. Large numbers of attacking bark beetle adults can often overwhelm a tree's natural defences, lay eggs and successfully initiate an infestation. After the eggs hatch, the grub-shaped larvae can girdle the tree by feeding under the bark. Blue-stain fungi are also carried on the bodies of most species and introduced into the tree during adult attack. Proliferation of these introduced fungi in the water-conducting tissues hastens the death of the infested trees.

Pines under stress are particularly susceptible to bark beetle attacks, especially if beetle outbreak conditions exist. Bark beetle attacks can be recognised by boring dust and pitch tubes on the outside of the bark, characteristic galleries under the bark, and adult beetles and larvae in the inner bark.

After a tree is successfully attacked, the foliage fades from bright-green to yellowish-green to red. These "faders" are generally the first apparent sign of a bark beetle attack. Unfortunately, foliage colour change often does not occur until well after the tree is dead and the beetles have completed their development and have left the tree.

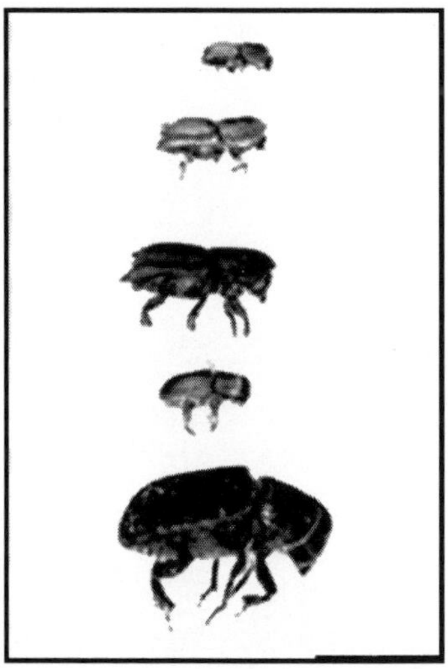

Fig. The Southern Pine Bark Beetles. Top to Bottom: *Ips Avulsus, Ips Grandicollis, Ips Calligraphus, Dendroctonus Frontalis* (SPB), *Dendroctonus Terebrans* (BTB), Gerald J. Lenhard, LSU.

Bark beetle brood development time ranges from 25 to 120 days, depending upon species and temperature. It is common for more than one species of bark beetles to infest individual trees. Since many other insects are associated with dead and/or dying trees, make positive identification of the insects before you take any remedial actions. Reduce the potential for bark beetle attack by ensuring that trees are rapidly growing and healthy. Removing or treating lightning and storm damaged trees promptly and maintaining proper stand densities can reduce the likelihood of bark beetle attack. For specific management and control practices, contact your county Extension Service agent or State Forestry office.

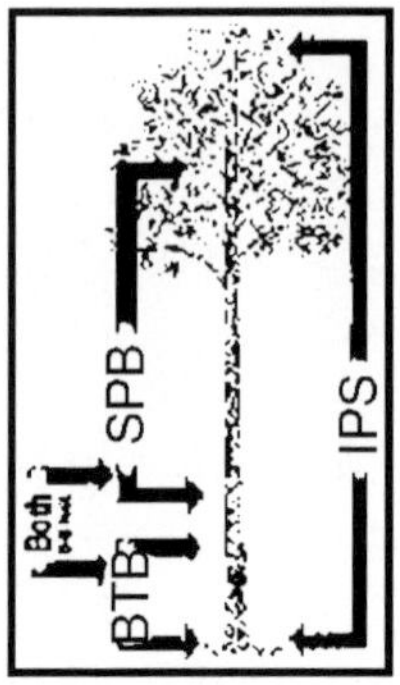

Fig. Sections of the Trunk which Three Types of Bark Teetles Attack.

SOUTHERN PINE BEETLE

The southern pine beetle (SPB), Dendroctonus frontalis Zimmermann, is the most destructive of the eastern species of pine bark beetles. It is a small reddish-brown to black beetle, about 1/8 of an inch long. The rear end of the body is rounded. SPB normally infest boles of trees from the base to the crown, with initial attacks at mid-bole or higher. The life cycle generally requires 35-60 days to complete. There may be up to six generations per year. Small pitch tubes, usually less than 2 inch in diameter, are often present at the site of adult attacks.

Fig. *Southern Pine Beetle*: Pine Killed by Beetle with Blues Stain Fungus, A Cross Section of Log, Ron Billings, Texas Forest Service.

Adult SPB bore directly through the bark and mate. The females excavate the characteristic S-shaped, crisscrossing egg galleries in the inner bark. Eggs, which are deposited in niches on either side of these galleries, hatch into small, legless grubs within 4-9 days. The grubs mine for a short distance before boring into the outer bark where they pupate. Galleries are usually filled with larval fecal material and boring dust. Soon after a tree is attacked, all the needles turn yellow and then brown. Drought seems to trigger major outbreaks of this insect. In addition to direct damage caused by the larval feeding, SPB introduce stain-causing fungi, which invade the tree and hasten death. Once southern pine beetles have successfully infested a tree, no remedial treatments are available to prevent the death of the tree.

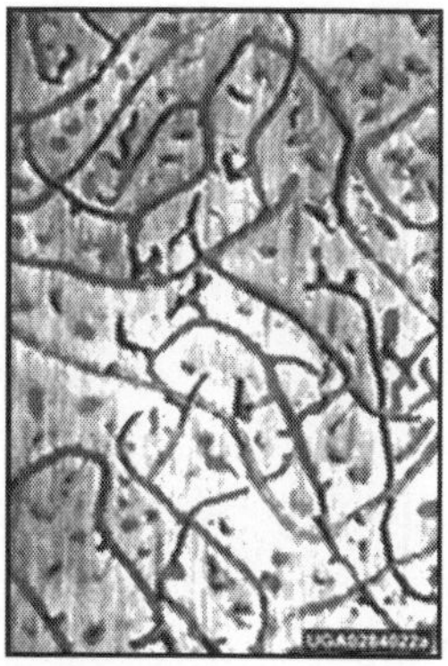

Fig. *Southern Pine Beetle*: 'S' Shaped Galleries Under Bark, Ron Billings, Texas Forest Service.

BLACK TURPENTINE BEETLE

The black turpentine beetle (BTB), Dendroctonus terebrans (Oliver) is the largest of the major southern bark beetles. It is ¼ of an inch or more in length, with a rounded rear end, and is reddish-brown to black in colour. Large purplish coloured pitch tubes (50 cent piece in size) are often present at the site of adult attacks.

Fig. Southern Pine Beetle Pitch Tubes on Loblolly Pine, Jim Meeker, Florida Dept. of Agriculture and Consumer Services.

BTBs attack fresh stumps and living trees by boring through the bark and constructing galleries on the face of the sapwood where 50-200 eggs are laid in a group. BTBs attacks usually only occur on the trunks of trees up to a height of eight feet.

After hatching, the white larvae feed on the inner bark. Unlike southern pine beetles and Ips engraver beetles, black turpentine beetles do not introduce blue stain fungi into the tree and developing larvae feed in "patches" rather than completely encircling the tree.

However, when several broods occur at about the same height, the feeding larvae may completely girdle and kill the tree. Without prompt treatment, from 70 to 90 per cent of the trees attacked by the BTB die. The BTB life cycle takes from 2 ½ to 4 months, depending on the temperature. In the South, there are usually two generations and part of a third each year.

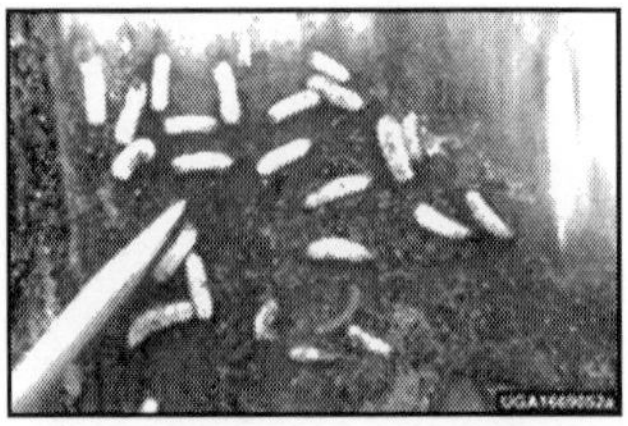

Fig. Black Turpentine Beetle Larval Feeding Patch, NC State University Archives.

Fig. Black Turpentine Beetle Pitch Tube, NC State University Archives.

IPS BEETLES

The four species of Ips beetles commonly found in the South (Ips grandicollis, I. calligraphus, I. avulus, and I. pini) vary from 1/10 to ¼ of an inch long, and are yellowish, dark reddish-brown to black. They are easily recognised by their scooped out posteriors which are surrounded by varying numbers of tooth-like projections. In hot weather, it may take as few as 25 days to complete one generation. Populations of these beetles increase rapidly during favourable conditions. Ips spots usually contain only one to a few trees killed, but under favourable conditions Ips can become epidemic and kill many trees. The fully grown, grublike larvae are yellowish-white and vary from ¼ to 1/3 inch long. Gallery patterns are more or less Y- or H-shaped except for I. avulus, which deviates from these patterns. Small, white eggs are laid singly in small egg niches cut along the main tunnels. Larval feeding tunnels are usually filled with boring dust and frass (excrement).

Fig. *Ips Avulsus* Adult and Larval Galleries, Ron Billings, Texas Forest Service.

Adults girdle trees quickly as they construct their egg galleries in the inner bark. The tree's death is usually hastened by the introduction of blue-stain fungi which blocks the flow of sap. Small reddish pitch tubes are frequently the first sign of an attack. These tubes are usually absent in trees suffering from drought. As with SPB, once Ips beetles have successfully infested a tree, the tree cannot be saved!

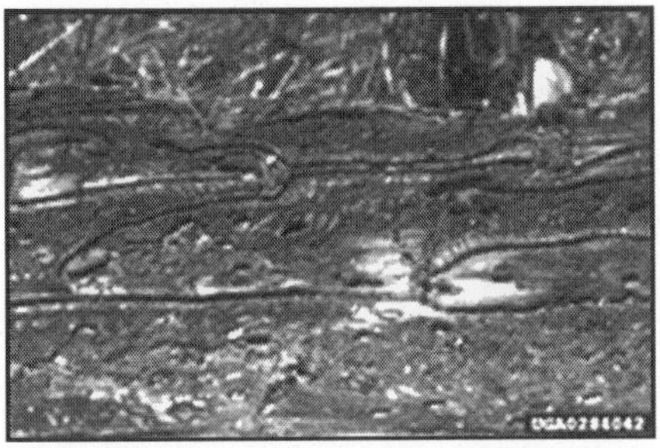

Fig. Eastern Fivespined *Ips* Adult and Larval Galleries, Gerald J. Lenhard, LSU.

LOSSES CAUSED BY FOREST INSECTS

These data are summarised from Special Bulletins published annually by The University of Georgia, College of Agricultural and Environmental Sciences, Department of Entomology, Insect Survey and Losses Committee. These figures have not been adjusted for inflation.

Table. Estimated Average Yearly Losses and Control Costs of Forest Insects in Georgia

Rank	Insect	Cost of Control	Damage Loss	Total Cost
1.	Southern pine beetle	$317,600	$4,680,400	$4,998,000
2.	Pine tip moths[1]	$916,000	$2,310,000	$3,226,000
3.	Defect & degrade[2] causing insects	$90,000	$2,874,000	$2,964,000
4.	Seed & cone[3] insects	$80,400	$2,514,200	$2,594,600
5.	Ips spp. beetles[4] and black turpentine beetle	$485,000	$2,016,400	$2,501,400
6.	Reproduction weevils[5]	$1,153,000	$928,000	$2,081,000
7.	Other insects[6]	$83,800	$1,215,000	$1,298,800
8.	Gypsy moth	$94,000	$--	$75,200
	Totals	$3,201,000	$16,538,000	$19,739,000

Notes:

1. Includes Nantucket pine tip moth and pitch pine tip moth.
2. Includes carpenter ants, ambrosia beetles, lepidopterous oak borers, shothole borers and various other cerambycid, buprestid and scolytid beetles.
3. Includes coneworms, seedworms, seed bugs and cone beetles.
4. Ips avulsus, I. grandicollis, I. calligraphus and I. pini.

5. Pales weevil and pitch-eating weevil.
6. Primarily aphids, scale insects, sawflies and lepidopterous hardwood defoliators including eastern tent caterpillar, forest tent caterpillar, fall webworm, oak skeletoniser and various Anisota spp.

PINE BARK BEETLE CONTROL

Several methods reduce the likelihood of expanded attacks by bark beetles. Most of these are based on good silvicultural practices to keep trees rapidly growing or by reducing stress on trees.

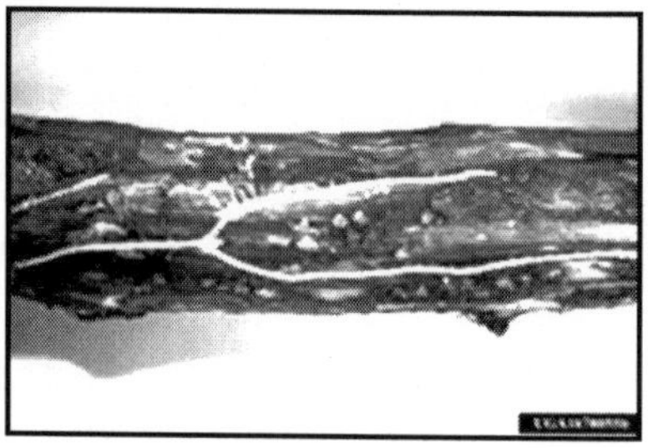

Fig. *Ips Pini* gallery on Small Pine, David McComb, USDA Forest Service.

Techniques currently used are:

- Prompt removal (and/or treatment) of trees damaged by lightning, storms or by construction.
- Salvage removal.
- Cut and leave.
- Pile and burn.
- Insecticide treatments.
- Cut and spray.
- Behavioural modifying chemicals.

Prompt Removal of any damaged trees, whenever possible, significantly reduces the likelihood of successful bark beetle attacks. Since bark beetles are attracted to odours exuded from damaged trees, initial attacks in an area often occur on damaged trees. As the beetle brood matures and exits the infested tree, the infestation frequently expands to other trees in the area. *Salvage Removal* is only feasible when a relatively large volume of wood is available and makes the operation cost effective for the logger. Prompt removal of dead and dying trees is essential to prevent significant degradation of the wood.

Steps involved in a successful salvage removal operation include:

- The removal of a 50-100 feet wide buffer strip of green uninfested trees around the most recently attacked trees;
- Removal of recently attacked trees containing developing beetle brood; and
- Removal of older standing trees from which the brood has already emerged.

Cut and Leave is best for controlling small spots of trees (10-50) when salvage is not practical or cost effective. Attacked trees and a border of healthy trees are felled towards the centre of the spot. It is helpful to cut the limbs on the underside of the felled trees so that the trunks are lying on the ground. The increased sunlight and subsequent higher temperatures, along with increased humidity resulting from the trunks lying on the ground, are thought to cause high mortality of the developing brood.

In *Pile and Burn*, trees with live brood are felled, piled and burned. Although effective, this technique requires heavy equipment to pile the trees so that they can be burned. Additionally, managers must be aware of appropriate weather conditions and prescribed fire and smoke management issues.

Insecticides in bark beetle control are more preventative than corrective. Consequently, with the possible exception of the BTB which does not introduce blue-stain fungi into the tree, trees that have been successfully attacked by bark beetles cannot be saved by insecticide applications. However, bark beetles still in or under the bark can be killed by spraying the tree with appropriate insecticides to prevent spread of the attack. Additionally, non-infested high value trees, judged to be at high risk, can be sprayed with an insecticide as a preventative measure against attack. The area of the tree requiring insecticide treatment depends upon the insect species for which the application is being made. The appropriate area of the tree should be thoroughly wetted with the insecticide spray mixture.

Applications made for BTB only require that the lower 8-10 feet of the main bole be treated and can be accomplished with a small hand or backpack sprayer. If applications are being made for SPB or any of the Ips beetles, spray to wet the entire bole of the tree from ground level up to the upper crown, including the base of large scaffold limbs. On larger trees, this requires high-pressure sprayers for thorough coverage.

A *Cut and Spray* method may stop further spread of bark beetle attack. If beetle infested trees can be felled but cannot be hauled from the site or burned, they can be limbed and bucked into workable lengths. Once cut into workable lengths, the tree sections can be turned as they are thoroughly sprayed with an appropriate insecticide.

Behavioural Chemicals are being tested by researchers to develop SPB control tactics by manipulating the various chemicals that the beetles (and their natural enemies) use to orient, attack or disperse their populations. To date, these promising, non-pesticide-based tactics are not yet ready for large scale field implementation.

BORING INSECTS

Included in this group are the insects that infest terminals, shoots, twigs and roots of living trees as well as those that obtain food and shelter from wood.

Terminal and shoot insects are of particular importance in the initial stages of forest regeneration and early stand growth. These insects are also of great importance in forest nurseries and ornamental trees. We will discuss in detail the Nantucket pine tip moth, and the white pine and deodar weevils which are frequently encountered in and cause significant damage to pine stands in the South.

Other insects in this category damage or destroy trees that would otherwise produce quality lumber or other wood products. Most insects that cause this damage are borers, either adult or larval stages or both. Most borers are secondary invaders, attacking bark and wood of trees that are seriously weakened, dying, or recently cut. Carpenterworms, ambrosia beetles, oak clearwing borers, metallic wood borers, Columbia timber beetles and southern pine sawyers are some of the pests that cause damage in this category.

Trees attacked by these pests are usually scattered so that most control measures are difficult and not economically feasible. Additionally, there are several species of insects that attack, infest and damage wood and wood products that are discussed in the Wood Treatment manual and in other manuals devoted to Pest Control Operator training and are not covered in this manual. Included among the wood product pest group are certain beetles in the family Cerambycidae (notably, the old house borer *Hyloupes bajulus*), ambrosia beetles (Scolytidae and Platypodidae), and the powderpost beetle complex (includes members of the Families Lyctidae, Anobiidae, and Bostrichidae).

APPLICATION METHODS

Foliar and soil-active materials are often broadcast over the entire area to be treated. They can be applied to the foliage or soil by either aerial or ground mechanical equipment. Broadcast applications are common for site preparation. In some areas this method is used for herbaceous weed control and woody release.

FOLIAR

Many forestry herbicides enter the plant through the green foliage and young stems. Plants that are shielded from foliar sprays by taller or adjacent plants will not be controlled as well as those fully exposed. Adjuvants are added to the spray mixture to aid in the coverage effectiveness or safety of these herbicides. However, some formulations already include adjuvants. Always follow label directions. Adjuvants may be particularly useful for late-season use as foliage becomes waxy and difficult to penetrate.

SOIL

Soil-active herbicides may be applied to the soil as liquid or granular formulations. Control will not occur until there is adequate rainfall or sufficient soil moisture. After rainfall dissolves and moves the herbicide into the soil, it is taken up by the roots of established plants. Pre-emergence herbicides applied to the soil kill vegetation as seeds germinate or new plants grow through the treated ground. Season of the year, soil moisture, texture and pH, as well as organic matter and rainfall, greatly affect soil-active materials.

AERIAL APPLICATION

Aerial application is commonly used to apply pesticides in forestry. This is because tract size is often large, access is difficult; and the vegetation is often tall and dense. Large acreage can be treated more economically and in less time by air.

Untreated buffers are established around the perimeter of the treatment area. Firebreaks, flagging, or Global Positioning Systems (GPS) are used to mark treatment boundaries and flight lines. Both fixed- wing aircraft and helicopters are used for forestry applications on insecticides and biological control agents.

Forest herbicides are labelled for helicopter application and not by fixed-wing aircraft. Since most states require a separate training and testing for aerial certification, only a brief discussion will follow here.

The use of control droplet aerial (CDA) spray equipment and orienting the nozzles with the air flow causes large droplets. Boom length should be 75 per cent of the total wing or blade span. Nozzles located on booms longer than this can cause excessive drift to occur. The larger the droplet, the less chance of drift to non-target sites. Drift-reducing agents and invert emulsions that change the physical composition of spray mixtures can be used to reduce chemical drift. However, when they are large, droplets may reduce the effectiveness of foliar-absorbed herbicides. Larger droplet sizes display a tendency to bounce off of leaf surfaces. Large droplets tie-up much more spray volume per drop than do smaller droplets.

This may cause inadequate coverage, unless greater volumes per acre are applied. Spray carrier volume should be adjusted to insure effective coverage of target vegetation. Water-based formulations require 5 to 20 gallons per acre (GPA) with the higher carrier volumes necessary when treating multi-story canopies and dense vegetation. Oil emulsions use carrier volumes of 5 to 10 GPA due to costs and deposition efficiency. Aerial applications for midstory and understory hardwood control requires 15 to 20 GPA to insure good coverage beneath closed pine canopies. Solid formulations of soil active materials are also applied aerially. Uniform distribution of solid materials is more difficult than with liquid formulations. Fine particles and dust from the granules can increase the risk of off-site drift. To minimize streaks or skips in the treatment area and off-site movement, apply solid formulations only when wind speeds are less than 5 miles per hour.

FOREST PATHOGENS AND INSECTS RISKS OF INTERCONTINENTAL SPREAD

Previous records of intercontinental spread of forest pathogens and insects discussed at this Symposium verified the extreme damage and devastation that can result from an introduction of a pathogen or insect to another continent. Chestnut blight (*Endothia parasitica*) was cited as an example of an introduced disease that eliminated a commercially important tree species in the United States and the Symposium was reminded frequently that white pine blister rust (*Cronartium ribicola*) and Dutch elm disease (*Ceratocystis ulmi*) continue to cause severe annual losses in plantations, natural forests, woodlots, and urban areas in many parts of the world. Examples of the many destructive introduced insects included the larch sawfly (*Pristiphora ericsonii*), balsam woolly aphid (*Adelges piceae*), European pine shoot moth (*Rhyacionia buoliana*), and the fall webworm (*Hyphantria cunea*).

It was concluded that there are many recognised pests at present of restricted geographic distribution that constitute a great potential danger to the forests of other continents. Further, pathogens and insects of minor significance in their native habitats may become serious pests in a new environment, for example, European *Adelges piceae* and Asiatic *Endothia parasitica* in North America. In addition, "new" pests may be revealed in certain countries following adequate scientific exploration.

INTERCONTINENTAL DISPERSAL

Both natural and artificial means of intercontinental dispersal of pests were discussed. Natural dispersal methods, although obviously largely beyond human control, were assessed because of their possible influence in weakening or nullifying exclusionary efforts by man. The natural spread of pathogens and insects can occur in many ways: through the air, water, soil, birds, and mammals. These may play vital roles in continental or more local disease and insect epidemiology, but it was concluded that they have little significance in intercontinental pest dispersal except in contiguous areas of Europe and Asia where the host tree may be uninterrupted by natural barriers such as deserts, oceans or mountains.

Human activities provide the greatest hazard to the spread of pathogens and insects through the intercontinental exchange of living plants and products, utilising rapid methods of transportation. The most dangerous carriers for diseases and insects include living plants, soil, seeds, cuttings, and wood products such as logs, lumber, and manufactured products.

In the commerce of living plants, exotics present special hazards. A pest of an exotic may be destructive to a native tree or a pest of minor significance in its native environment may become extremely damaging to its host in a new habitat. Because of these dangers it was strongly recommended that a tree should be introduced to a new habitat only if it is superior to native species and cannot be reproduced by seed.

The exchange of seeds between countries was regarded as the safest way to provide propagative material if seeds are kept clean of foreign debris and effectively treated by fumigants and fungicides to eliminate injurious seed inhabiting insects and pathogens.

This examination of intercontinental dispersal methods led to considerations of procedures to exclude, delay, or limit the seriousness of exotic pests. While these procedures were considered in several presentations, the following general approaches were recommended as the most important activities requiring co-ordination at both national and international levels. An effective and continuing forest insect and disease survey should be developed by countries for detection, appraisal, and as a basis for research; programmes of research need to be intensified, and in many cases, initiated, to identify the forest pests of economic consequence as a prerequisite for effective quarantines, control, or eradication; more effective quarantines are required, developed and maintained on the most recent available knowledge; and further, countries should maintain facilities for prompt control action so that introductions can be eliminated or contained. To assist in the promotion and development of these activities the Symposium recommended that FAO appoint to its forestry staff, specialists in forest entomology and pathology.

APPRAISAL AND PREDICTION

There is no comprehensive method of developing hazard ratings for potential forest pests although helpful data are provided by literature contributions from various countries that list and rate both native and introduced pests and report regularly on research activities. However, these are only starting points in the accumulation of data necessary for effective appraisal and prediction. One approach to appraising international pest hazards is the strategic establishment of test plantations of the most important trees to expose them to the world's pests. This approach has considerable value and much can be learned by thorough inspections of existing exotic plantations. Such plantings are also useful for determining how and where a tree species grows best, and for providing sources of breeding stock for genetic studies.

However, while much useful information on the appraisal of pest hazards can be derived from plantations, there are too many weaknesses to make it the sole method of hazard appraisal. For example, disease and insect incidence fluctuates widely from place to place, and the replication required to incorporate the spectrum of variability for so many pests and host trees to obtain comprehensive and valid appraisals is so formidable as to be impractical and prohibitive. Therefore, the need for improved means of evaluating pest hazards on an international basis was recognised.

In the field of forest diseases an interesting proposal for appraising and predicting disease hazards (abbreviated APPRE) was presented to assist any country to define which of the world's tree pathogens are of potential danger.

The two main objectives of this method are to permit pathologists of any country to determine:

1. Which native tree diseases pose a threat to other countries with reference to import decisions, quarantines, control measures, and needed research.
2. Which diseases in other lands pose a threat to their own forests.

Since this system would depend upon ready access to the world's accumulated knowledge on a pathogen, an information retrieval procedure was described as a basic requirement and entitled, International Tree Disease Information Register (INTREDIS). This register would accumulate published information on tree diseases on punched cards suited to electronic sorting and would rapidly provide a printed record of available literature on a pest according to host tree and other particulars.

The APPRE procedure, therefore, has its foundation in a centralised system of information retrieval, a system not yet established. An important soilborne pathogen, *Phytophthora cinnamomi*, was used as an example in predicting the hazard to forests in various parts of the world from a specific pathogen. A very useful detailed description was provided comparing the present distribution of this pathogen with its expected distribution as determined by the published knowledge of the organism and based on climate, soil, hosts, and other factors known to limit the pathogen or make aggressive attacks unlikely.

Another illustration showed how the threat of disease to a given species in one part of the world could be estimated. *Pinus radiata* in New Zealand was used as the example, along with the knowledge available of its diseases elsewhere and of differences and similarities in climate, hosts and alternate hosts, soil, and other factors between New Zealand and the areas where these diseases are active. The system would provide, in simple tabular form, the pertinent hazard data, including a list of pathogens and information for each on its presence or absence in the country, where to expect damage, and the reason why damage is to be anticipated.

BIOLOGICAL CONTROL OF FOREST INSECT PESTS

Forest ecosystems provide production systems that are much longer lived, more diverse, and generally more stable, than most agroecosytems. Thus, many of the most long lasting and effective biological control programmes have been established against forest pests. Combined with the facts that applying insecticides over such large areas and over such a tall canopy is rarely economical, and that society demands other uses of forests, such as recreation, watershed, wildlife, and aesthetics, in addition to fibre production, the reason for a strong historical emphasis on biological control becomes even more clear. Some recent developments in the biological control of forest pests may be of interest to readers of Midwest Biological Control News.

Westward Migration of the Gypsy Moth. The gypsy moth is the most damaging pest of deciduous trees in North America. It affects production forestry, but also interferes with recreation, reduces home values in urban and suburban settings, and alters environmental quality. The gypsy moth was deliberately brought into New England by a French entrepreneur, escaped, and has been progressing westward and southward ever since. It is now established west into most of Wisconsin, and south into North Carolina. Female European strain gypsy moths are flightless, so the major means of dispersal is by human transport. A number of biological control agents were introduced against the gypsy moth, including predaceous beetles and parasites. These helped somewhat, by reducing the frequency, duration, and impact of gypsy moth outbreaks, but the outbreaks persist. The most spectacular recent development has been the emergence of the fungal pathogen, *Entomophaga maimaiga*. This pathogen was introduced during the early part of this century, presumably became established, but subsequently had little impact. During 1989, the gypsy moth erupted into one of its most extensive and potentially damaging outbreaks ever. But the populations collapsed, due to what appeared to be an epizootic caused by nuclear polyhedrosis virus. Closer examination, however, revealed it was due to *E. maimaiga*, which had gone largely unreported for 80 years. The reasons for this sudden success remain a mystery, and we may never know if it was due to a mutation, subsequent accidental introduction, or some other factor. *E. maimaiga* has been recently introduced into a number of states along the gypsy moth's leading edge, with dramatic results. Some of the most successful releases have been in Michigan. This pathogen is almost entirely specific to the gypsy moth, yet it can be effective at lower gypsy moth densities than are typically needed for virus to exert effects.

At the University of Wisconsin, we have just completed a two year survey of the natural enemies of the gypsy moth. The good news is that *E. maimaiga*, the larval parasitoid *Cotesia melanoscela*, the egg parasitoid *Ooencyrtus kuvanae*, some parasitic flies, and NPV were found to be present. Unfortunately, they were here in extremely low numbers-only 1 larva out of over 1000 larval and soil samples had *E. maimaiga*, and only a few larvae were parasitised. Based on these results, the Department of Natural Resources conducted releases of *E. maimaiga* and *C. melanoscela* in the autumn of 1997. They also disseminate *O. kuvanae* whenever possible. We will conduct follow-up surveys of these releases over the next two summers.

Some additional projects are actively underway on biological control of the gypsy moth throughout the Midwest. Many of these are aimed at determining how the tree species on which this polyphagous insect feeds will affect various natural enemies. We now have a fairly good appreciation of how tree species affects the performance of *Bacillus thuringiensis kurstaki*, two strains of *C. melanoscela*, *O. kuvanae*, and others. There has also been substantial work, particularly at Michigan State University, on how various biological control agents, including Btk, might affect endangered species and non-target insects. Work at the USDA APHIS laboratory in Niles, Michigan, is evaluating some new parasitic flies that seem to be effective at relatively low gypsy moth densities, yet are host specific. Work in Illinois and Indiana continues to develop new insect pathogens such as microsporidia, and Minnesota is experimenting with various types of anticipatory biological control. Efforts among the various states are coordinated through a couple of groups, such as the North Central Regional Committee on Gypsy Moth, NC-212, the USDA Forest Service "Slow the Spread" programme, and the various state agencies.

Chemical Ecology of Bark Beetles. Bark beetles are the most damaging group of insects affecting North American forests. All of the major pests in North America are indigenous, with the exception of the smaller European elm bark beetle, the vector of Dutch elm disease pathogen. Most bark beetles colonise previously killed trees, but a few attack live trees, especially conifers. They kill trees by collectively exhausting the trees' defence systems, through mass attacks. These mass attacks are coordinated by aggregation pheromones.

Entomologists have known for some time that predators of bark beetles, such as clerid beetles and hister beetles, locate their prey by cueing in on the bark beetles' pheromones. This improves the efficiency of these natural enemies, but also poses some management challenges. For example, one tactic that has been tried repeatedly throughout the world is using aggregation pheromones to "trap out" the bark beetle population. This method results in lots of trapped beetles, but insignificant tree protection, largely because it also reduces predator populations. In some cases it has even been reported to extend outbreaks.

Bark beetle pheromones come in a variety of configurations, with different isomers and additives. Work at the University of Wisconsin has demonstrated that alterations in these secondary features can allow bark beetles partial escape from predators while maintaining intraspecific functionality. That is, by producing different versions of the same pheromone, beetles can maintain the communication needed to attack trees and attract mates, yet reduce their attractiveness to natural enemies. Presumably this "shell game" leads to counter-adaptations by the predators, followed by new innovations by the bark beetles, etc.

This coevolutionary race has some important implications to biological control. First, forest managers make control decisions based on the relative abundance of pests and predators captured in pheromone traps. But such lures were developed to maximize attraction by the target bark beetle. We now know that such estimates greatly underestimate the numbers of natural enemies, which prefer different forms of these chemicals. Based on these findings, it will soon be possible to make appropriate corrections in our estimates. Behavioural disparities between predators and prey also offer some opportunities to reduce the negative impacts of pheromonal trap out programmes on beneficial insects. Depending on the combination of chemicals in the pheromone lure, the ratio of pest to beneficial insects in these traps can vary enormously.

So we may soon be able to simultaneously maximize pest catch while excluding most predators. This approach might be applicable to a broad range of forest and agricultural pests. Augmentation of natural enemies to foci of bark beetle attack could also be improved by deploying the appropriate lures. Additional work is evaluating the roles of host tree chemistry, and volatiles emitted by bark beetle fungi, in the chemical signals attracting predators, and parasites. These projects are being conducted in collaboration with the University of California - Berkeley. Biological Control of Chestnut Blight. Not all biological control in forestry is directed at insect pests. One of the worst disturbances to North American forests was the introduction of the chestnut blight fungus. Chestnut blight decimated what was a predominant species in the Northeast and portions of the Midwest. Native trees had no appreciable resistance against infection.

Recent work has demonstrated that a "hypovirulent" form of the fungus has a very different type of relationship with chestnut trees. Unlike the aggressive strains of the pathogen which can quickly colonise and kill trees, hypovirulent strains may form only limited, superficial lesions. However, infection with the hypovirulent strain does not provide systemic protection; these trees can still be infected and killed by the typical virulent fungi. The hypovirulent fungi are not a true strain, however, but rather this condition is induced by the presence of virus-like agents with the hypovirulent fungus. Thus, this biocontrol agent may play a key role in tipping the host-pathogen relationship in favour of the host tree. Workers at Michigan State University are currently among the most active groups studying the molecular biology of this system.

A major challenge lies in determining how hypovirulent fungi can be distributed and maintained in the wild. Current emphasis is largely on high value trees such as in parks, where inoculation is feasible. Some of the more futuristic ideas include the possibility of developing certain insects as vectors. New Arrivals. Some recent introductions pose serious challenges to the forest products industry and forest health.

The pine shoot beetle, *Tomicus piniperda*, a native of Europe, entered the US through ports in Ohio. It attacks the tips of trees, causing disfigurement. It could especially impact Christmas tree production, a major industry in the

Midwest. This insect is widely distributed in Michigan and northern Illinois, and was first detected in Wisconsin last summer.

The USDA organised a three-pronged management approach: determine the host range among North American trees, evaluate cultural controls, and consider releasing a European predator, a clerid beetle. The major cooperators were the APHIS laboratory in Niles, Michigan, the USDA-FS, and Michigan State University. The first line of research showed that this insect could build up populations in downed native tree species, but could only have appreciable effects on live Scots pine. The second line of research developed practical cultural methods that fit well into Christmas tree production, such as removing and destroying tree bases in which beetles overwinter. The third avenue considered evidence that this exotic clerid could conceivably interfere with the native clerids that are so important to the natural control of native bark beetles (described above). Thus, it was decided to not introduce this predator, at least for as long as conditions remain stable. Such decisions should be viewed as successes, not setbacks, for biological control. They demonstrate that critical evaluations precede releases, and that these evaluations take a broad view of pest management, environmental safety, and integrated approaches.

Similar analyses are underway for another introduced pest, the hemlock woolly adelgid. This represents an extremely serious threat to northeastern and midwestern hemlock forests, just when a return to old growth hemlock ecosystems are being promoted as a cornerstone of forest health. Native eastern hemlocks appear totally susceptible to this Chinese insect, which threatens to have an impact equivalent to the balsam woolly adelgid in the southeast. A number of predators from China are being evaluated by USDA Forest Service personnel.

Ongoing Successes. It's easy to become preocuppied with current and new threats to forest ecosystems. But it is also important to recognise that these cases represent unusual exceptions to the general pattern of natural control of most tree-feeding species. It is worth pointing out that a number of successful biological control projects are inconspicuous precisely because of their spectacular success. Some of the best examples include the conifer sawflies, several species of which were introduced, caused severe outbreaks, and were subsequently controlled by introduced parasitoids. Most of these species remain at innocuous levels during most years, particularly on the native tree species that comprise our commercial and unmanaged forests.

PEST CONTROL POTENTIAL OF NEEM

The pest control potential of neem in developing countries, however, remained largely untapped due to the advent of DDT and other and other broad-spectrum synthetic insecticides. Also, wide publicity given to slogans such as "the only good bug is a dead bug" and identifying traditional uses of neem as backward, gradually influenced people away from using neem. It is only in the past decade, that the pest control potential of neem, which does not kill pests but affects their behaviour and physiology, has been recognised. Though subtle, neem's effects such as repellence, feeding and oviposition deterrence, growth inhibition, mating disruption, chemo-sterilisation, etc., are now considered far more desirable than a quick knock-down in integrated pest management programmes as they reduce the risk of exposing pests natural enemies to poisoned food or starvation.

In spite of high selectivity, neem derivatives affect ca. 400 to 500 species of insects belonging to Blattodea, Caelifera, Coleoptera, Dermaptera, Diptera, Ensifera, Hetroptera, Homoptera, Hymenoptera, Isoptera, Lepidoptera, Phasmida, Phthiraptera, Siphonoptera, and Thysanoptera, on species of ostracod, several species of mites and nematodes, and even noxious snails and fungi, including aflatoxin-producing Aspergillus flavus. Results of field trials in some major food crops in tropical countries will illustrate the value of neem-based pest management for enhancing agricultural productivity in Asia and Africa.

PEST OF STORED PRODUCTS

Postharvest losses are notoriously high in developing countries. Worldwide annual losses in store reach up to 10 per cent of all stored grain, *i.e.*, 13 million tons of grain lost due to insects or 100 million tons to failure to store properly. Dr. R.C. Saxena has recently reviewed the potential of neem against pest of stored products grain legumes, maize, sorghum, wheat rice and paddy, potato tubers. At farm level storage and warehouses, the application of neem derivatives to bags and stored grains has provided protection against insect pests. Powdered neem seed kernel mixed with paddy (1 to 2 per cent) significantly reduced infestation and damage to damage to grain during a 3 month storage period; the effectiveness capacity jute bag (100 x 60 cm) controlled 80 per cent of the population of major insects and checked the damage to wheat up to 6 months. The treatment with untreated control. The neem seed extract treatment was as effective as that of 0.0005 per cent primiphos methyl mixed with the grain. Using this technology in Sind, Pakistan, high benefit-cost ratios were obtained by small, medium, and large-scale farmers.

EFFECTIVENESS OF NEEM OIL

The effectiveness of neem oil alone or in combination with fumigation was evaluated against five major species of stored grain pests infesting rice and paddy grains in a warehouse trials conducted in the Philippines. Rice grain treated with 0.05 to 0.1 per cent neem oil or treated with neem oil after fumigation with 'Phostoxin', and stored for 8 months had significantly less Tribolium castaneum adults than in untreated control. Both kinds of neem treatments were as effective as the bag treatment with 'Actellic' at 25ug/cm2 or grain treatment with Actellic at 0.0005 per cent, and suppressed the pest population by 60 per cent. The population build-up also was reduced when either fumigated or non-fumigated rice was stored in bags treated with neem oil at > 1 mg/cm2.

Rhizopertha dominica, Sitophilus oryzae, Oryzaephilus surinmensis, and Corcyra cephalonica were similarly affected by neem treatments alone or in combination with prior grain fumigation. Fumigation and Phostoxin were effective only for about 2 months against R. dominica, and for up to 6 months against other pest species, while neem oil treatments were effective up to 8 months. Compared with the pest damage to untreated or fumigated rice, neem oil treatment significantly reduced the damage to rice grain. At 8 months after storage, weevil attacked grains in neem treatments were 50 per cent of those in the fumigated rice and 25 per cent of those in the untreated rice. Neem treatments also reduced the pest populations and damage in paddy. In studies conducted in Kenya, the growth and development of 1st instars of the maize weevil, Sitophilus zeamais, was completely arrested in maize grain treated with neem oil at 0.02 per cent, while the weight loss of treated cobs was less than 1 per cent as compared with a 50 per cent reduction in weight of untreated cobs stored for 6 months.

While neem treatments cannot replace completely chemical pesticides used in stored products preservation, the amounts of pesticides needed could be reduced, thereby decreasing the pesticide load in food grains. With proper timing and innovative methods of application, their use could be integrated in stored products management.

BLOOD-SUCKING PESTS

Ascher and Meisner have reviewed the effects of neem on hematophagous insects affecting humans and livestock. Application of a paste made from neem leaves and turmeric in 4:1 proportion to the skin cured 97 per cent of the patients suffering from scabies caused by the mite Sarcoptes scabei in 3-15 d.

Monthly sprays of ethanoilic extracts of neem or weekly bathing in azadirachtin-rich aqueous 1:20 'Green Gold' controlled the bush tick, Ixodes holocylus, and the cattle tick, Boophilus microplus in Australia, but were less effective against the brown dog tick, Rhipicephalus sanguineus. In Jamaica, neem kernel extract controlled ticks on cattles and dogs. Neem products repel and affect the development of mosquitoes. Two per cent neem oil mixed in coconut oil, when applied to exposed body parts of human volunteers, provided complete protection for 12 h from bites of all anophelines. Kerosene lamps containing 0.01-1 per cent neem oil, lighted in rooms containing human volunteers, reduced mosquito biting activity as well as catches of mosquitoes resting on walls in the rooms; protection was greater against Anopheles than against Culex.

Effectiveness of mats with neem oil against mosquitoes has also been demonstrated; the vaporising repelled mosquitoes for 5-7 h at almost negligible cost. The sandfly, Phleobotumus argentipes, also was totally repelled by neem oil, mixed with coconut or mustard oil, throughout the night under field conditions in India. Application of neem cake @ 500 kg/ha, either alone or mixed with urea, in paddy fields in southern India reduced the number of pupae of Culex tritaeniorhynchus, the vector of Japanese encephalitis, and also resulted in higher grain yield.

PEST RESISTANCE TO NEEM MATERIALS

A few herbivorous insects, including Homoptera, Coleoptera, and Lepidoptera do survive on neem but, largely, it is free from serious pest problems. Although Taylor indicated that insects may possibly adapt to limonoid rather quickly, but Vollinger demonstrated that two genetically different starins of P. xylostella treated with a neem seed extract showed no sign of resistance in feeding and fecundity tests up to 35 generations. In contrast, deltamethrin-treated lines developed resistance factor of 20 in one line and 35 in the other.

There was no cross resistance between deltamethrin and neem seed extract in the deltamethrin-resistant lines. Also, the esterase and multi-function oxidase enzyme activity did not change during the 35 generations. The diversity of neem allelochemicals and their combined behavioural and physiological effects on insect pests seem to confer a built-in resistance prevention mechanism in neem. However, wisdom demands that users should refrain from exclusive and extended application of single bioactive materials, such as azadirachtin.

CONTROL MANAGEMENT OF PESTS AND DISEASES

HEALTHY FORESTS: THE OBJECTIVE

The objective of managing pests and diseases in forests and forest plantations should be to keep them in a healthy, productive condition. What is a healthy forest? The concept of a healthy forest became a popular concept during the mid- to late 1980s. In economic terms, a "healthy" forest can be defined as "a forest in which pests and diseases remain at low levels and do not interfere with management objectives." In ecological terms, a healthy forest is "a fully functional ecosystem; one in which all of its parts can interact in a mutually beneficial way."

The healthy forest concept directs forest managers to focus on the forest rather than its pests and diseases and takes into account the natural role of insects, fungi, fire and other so called "damaging agents" and their interactions in forest dynamics. Under an overall policy of forest health protection, pests and diseases are looked upon as a symptom of an unhealthy forest rather than as the problem. This directs forest managers and forest protection specialists to address the underlying causes of the pest or disease - factors such as overstocking, over-maturity, poor site/species matching, excessive fuels and single species forests with little diversity. Striving for healthy forests also involves anticipating pests and diseases based on historical records of their occurrence and the knowledge of forest and climatic conditions that favour their abundance. This allows time to implement management practices that will make these forests inhospitable for build-up of damaging pests and disease (Ciesla 1998).

The healthy forest concept may sound like an idyllic approach to managing forest pests and diseases, especially in cases where an introduced pest is causing widespread damage in a simple plantation ecosystem in the absence of its normal complex of natural enemies. Yet even the most damaging of exotic pests and diseases respond to stand and site conditions and there is evidence that they focus their activities on "unhealthy" forests. A classic example is that of *Sirex noctilio* in exotic pine plantations in the Southern Hemisphere. In New Zealand, it was noted that stressed trees in unthinned plantations were the first to be attacked by this insect. Therefore, by its action, the plantations were effectively thinned (McLean 1998). Thinning is now an integral part of managing plantations threatened by *S. noctilio* in places where this insect has been introduced. In another example, when a cypress aphid, originally identified as *Cinara cupressi*, appeared in eastern and southern Africa, most severe damage occurred in plantations of *Cupressus lustianica* that were either overmature or had been established on low nutrient soils (Claude and Fanstin 1991, Ciesla *et al.* 1995, Obiri 1994).

INTEGRATED PEST MANAGEMENT: THE TOOLS

The healthy forest concept provides the umbrella or objective for addressing damage caused by pests and diseases. Integrated pest management (IPM), on the other hand, provides the tools to accomplish this task.

The concept or philosophy of IPM as a "rational" approach to pest control was formalised during the 1960s. Crop protection specialists had become aware of the adverse side effects of dependence on chemical pesticides, including pesticide resistance, occurrence of secondary pests, environmental damage and human health hazards. This led to the realisation that alternative approaches, including cultural, biological and genetic tactics, used either alone or in combination, were also needed to provide long-term, effective protection against damaging pests.

Numerous definitions of IPM appear in the literature. Smith *et al.* (1976) refer to IPM as "*a process based on ecological principles and integrates multi-disciplinary methodologies in developing agro-ecosystem management strategies that are practical and effective and protect both public health and the environment.*" Pimentel (1986) describes IPM as a "*pest control method that includes judicious use of pesticide and non-chemical technologies - all of which are based on sound ecological principles.*"

IPM can be looked upon as consisting of two basic elements; a decision process and an action process. The decision process establishes the basis for any subsequent actions to be undertaken, including no action. The action programme may consist of one or more ecologically, economically and socially acceptable tactics designed to reduce pest populations to non-damaging levels (Ciesla 1982).

The Decision Process

The decision process is often the most time consuming and complex aspect of IPM. It requires careful consideration of the pest, its host, resource management objectives and the ecological, economic and social consequences of the various available tactics. Population levels of pests are estimated and anticipated resource losses are projected, as are the costs of treatment and its anticipated benefits. If treatment costs exceed losses, a rational decision may be to not treat and accept the losses. Other questions to address include: will natural controls take over within a short enough time so that artificial controls will be unnecessary; or will the effects of proposed treatments be so adverse that they would outweigh the benefits of treatment?

Monitoring of forest pests and diseases and their resultant damage is a critical input to the IPM decision process. Pest monitoring is becoming a sophisticated process that makes use of many technologies. Pheromones and other chemical regulators are often used to monitor insect population levels. Remote sensing technologies such as aerial sketch-mapping, aerial photography and airborne video are used to map and assess forest damage. Geographic information systems (GIS) can be used to relate the location of affected areas to key resource values, terrain features, land ownerships and environmentally sensitive areas. Mathematical models can predict resultant damage caused by certain levels of pest numbers and their consequences. In some cases, pest, growth and yield and economic models are linked to make projections of pest and disease impacts. Data visualisation techniques can display the expected results of alternative action tactics.

THE ACTION PROCESS

IPM action programmes consist of two overall strategies: prevention or direct control (suppression). Specific pest management tactics exist under each of these strategies.

Prevention

Prevention consists of tactics designed to either reduce the probability of the occurrence of a pest or disease or to create environmental conditions inhospitable for its buildup into damaging numbers. Regulatory, cultural or genetic tactics are examples of prevention strategies.

Regulatory tactics are designed to prevent introductions of exotic pests and diseases and to prevent their spread once established. Examples include inspection of wood products and wooden containers at ports of entry to intercept pest species, conduct of pest risk analyses when new trade agreements are made and establishment of quarantine zones when a pest species is first discovered in a new location.

Cultural tactics are designed to create conditions inhospitable for the development of damaging numbers of pests and diseases. These include matching tree species selected for planting to suitable growing sites, controlling stocking through intermediate harvests to maintain tree vigour and timely harvesting of plantations when they reach maturity. A drastic but sometimes necessary cultural approach is to simply eliminate a tree species from a plantation programme because of its high susceptibility to certain pests and diseases. An example is Kenya's policy to discontinue planting of *Cupressus macrocarpa* due to the stem canker fungus, *Monochaetia unicornis*, and *P. radiata* because of its susceptibility to *Sphaeropsis sapinea* (Odera and Arap Sang 1975, 1980).

Genetic tactics make use of varieties of host plants that are either more tolerant to damage or less palatable to the pest. Identification and testing of varieties of *Leucaena leucocephala* and hybrids with other species of *Leucaena* for resistance or tolerance to the psyllid, *Heteropsylla cubana*, was a major line of investigation following this insect's introduction into the Asia-Pacific region (Banpot Napompeth 1994). In the southeastern United States, screening and development of pine seedlings resistant to fusiform rust is an integral part of the management of this disease (Manion 1991). Tree breeding programmes in Brazil address resistance to the canker *Cryphonectria cubensis* on eucalypts and breeding of *P. radiata* in New Zealand includes a rating for resistance to the needle fungus *Dothistroma pini* (McLean 1998).

Suppression

Tactics directed against the pest or disease are referred to as direct control or suppression tactics. Examples include various types of biological, mechanical or chemical methods.

Biological control involves the use of natural enemies of a pest or disease to help keep its numbers in check. Classic biological control, a technique widely used in agriculture, involves the importation of natural enemies to help control an exotic pest. In Colombia, the geometrid *Oxydia trychiata*, an indigenous species attacking pines and cypress, was successfully controlled by the introduction and release of an egg parasitoid *Telenomus alsophilae* (Bustillo and Drooz 1977). In another successful example of classic biological control, releases of a parasitoid, *Pauesia bicolour*, resulted in the collapse of an aphid, *Cinara cronartii*, a North American species accidentally introduced into pine plantations in South Africa (Kfir *et al.* 1985, Mills 1990). In Chile, state-of-the-art mass rearing facilities have been established by forest industry for production of the European pine shoot moth parasitoid, *Orgilis obscurator*. One such laboratory, *Controladora de Plagas Forestales*, was established in 1992. In 1996, this facility produced 1.2 million parasitised larvae and 22,600 adult female parasitoids for field release (Controladora de Plagas Forestales 1997).

A key concern about biological control is the possibility that the introduced natural enemy might also attack innocuous or beneficial insects in the ecosystem. Therefore it is necessary to thoroughly evaluate candidate species prior to release to ensure their relative host specificity. Another concern is the hazard of accidentally introducing hyperparasites, natural enemies of the biological control agents, which might eventually affect the agent's efficacy. For example, colonies of the leaucana psyllid predator *Olla v-nigrum*, released on the Indian Ocean island of Reunion, were subsequently discovered infested by three species of hyperparsites (Quilici *et al.* 1995). These could severely affect the ability of this predator to function as an effective biocontrol agent.

Augmentative biological control is a process designed to increase the efficiency of natural enemies already in place. Mass release of *Trichogramma dendrolimi*, an indigenous egg parasitoid of the defoliator *Dendrolimus punctatus* in pine plantations in China is an example of this tactic (Yan and Liu 1992). Another biological control tactic is the use of biological insecticides such as the bacterial agent, *Bacillus thuringiensis* (Bt), nuclear polyhedrosis viruses or fungal preparations to control insects. Bt is widely used for control of lepidopterous defoliators in both natural and plantation forests. In China and Vietnam, a fungus, *Beauvaria bassiana*, is used for control of the pine caterpillar, *Dendolimus punctatus* (Anon 1993).

Mechanical tactics include removal and destruction or rapid removal of infested or infected trees with the objective of destroying the pest. Examples include cutting and burning of trees infested by bark beetles or rapid salvage of infested trees and destruction of infested bark at the sawmill. When the pine woolly aphid, *Pineus borneri*, was first discovered in Kenya, the initial response was to destroy the infested pines. This insect is easily carried on air currents, however, and it was soon realised that the infestation had spread far beyond the designated area where trees were being destroyed (Owour 1991). Use of chemical pesticides applied either from the ground or by low flying aircraft is still considered to be an integral part of IPM. However these materials are used more sparingly, are applied at reduced intervals and with greater precision. In IPM, chemical pesticides are often considered to be the tactic of last resort.

INTEGRATION OF NEW TECHNOLOGIES

An important part of IPM is that no matter how advanced, sophisticated or effective an IPM system for a specific pest or pest complex may be, there is always room for the introduction of new technologies. These may include more accurate pest monitoring, prediction of new pest management tactics, and more efficacious treatments with fewer undesirable side effects.

INTEGRATED PEST MANAGEMENT SYSTEMS

IPM systems consist of a combination of decision-making and pest management tools directed against a pest or pest complex and are in various stages of development. An example of an evolving IPM system is the approach being taken to manage the European wood wasp, *Sirex noctilio*, in pine plantations in southern Brazil. Early detection of this insect is accomplished by baiting suppressed trees in plantations with an herbicide, a procedure that attracts attacking wasps. Infestations are treated either through inoculation of a parasitic nematode, *Daladenus siricidola*, which renders the female wood wasps incapable of producing eggs, or thinning plantations to reduce stocking and maintain tree vigour (Iede and Ciesla 1993). Damage assessment technologies such as aerial sketch-mapping and aerial photos (Disperati *et al.* 1998, Ciesla *et al.* 1999) are currently being evaluated. Landsat satellite imagery is being tested to determine its capacity to map the location of small, non-industrial private pine plantations that might serve as reservoirs for populations of this damaging insect. Introduction of additional parasitic insects is another potential tactic for managing this insect (Iede *et al.* 1998).

INSECT PESTS OF FORAGE TREE LEGUMES IN THE TROPICS

Table summarises the data available on known pests of several important forage tree legumes. Several general papers contain additional information (*e.g.* Sen-Sarma 1987, Kumar 1990). Undoubtedly the list will grow with time. However, it is significant that none of the insects mentioned has been comprehensively studied to unravel their host plant relationships and other aspects of their ecology. For some, even their pest status is unclear. This lack of understanding impedes development of acceptable control measures. Our knowledge of the biology and ecology of the most studied pest of forage tree legumes, the leucaena psyllid (*Heteropsylla cubana*), is still insufficient to achieve acceptable control (Napompeth and MacDicken 1990).

The leucaena psyllid does, however, demonstrate the devastation that insect herbivores can inflict. Even in more natural situations than Agroforestry, insects may consume more than other, larger herbivores. For example, over 60 per cent of total leaf material consumed by herbivores in a South African savannah was eaten by insects, mostly lepidopteran larvae (Gander 1982). Such data must be considered as being only indicative as, in nature, damage levels are likely to vary among systems and among situations. In any case, the extent to which herbivorous insects have reduced the impact of weed species is a good measure of their capabilities against plant species, legume trees included (Hoffmann 1988, Moran and Hoffmann 1989, Hoffmann *et al.* 1990, Dennill and Donnelly 1991).

In dealing with specific examples of insect pests, their adaptations and ecology must be considered. Each species presents us with its own challenge, which may vary with changing environmental conditions from area to area and from time to time (Walter 1994). For example, trees may vary in their physiological status and this may in turn influence insect behaviour and abundance. Some herbivorous insects 'perform' better on stressed trees and the environmental cause of the stress thus induces herbivore outbreaks (White 1969, 1970a,b). Other insects (*e.g. Acizzia russellae* on *Acacia karroo*) may be primarily adapted to flushing growth, the production of which may enhance 'performance' (Webb and Moran 1978). Little is known about such influences on tropical forage tree legumes, but Singh (1986) found that the amount of damage to *Acacia nilotica, Dalbergia sissoo* and *Prosopis cineraria* varied with soil type. This presumably influenced the physiological status of the tree in relation to the pests, which were bagworms (*Cryptothelea cramerii*) and several species of chafer beetles (*Holotricha consanguinea, H. serrata, Adorata* spp. and *Anomala* spp.).

Although virtually all insect species exhibit some degree of host specificity and many are indeed species-specific in their host requirements, some herbivores can feed and reproduce on several host species, sometimes even on ones that are not closely related. Such polyphagous species present problems when populations transfer from one host species to a cultivated species. For instance, many borers have a wide host range. The teak sapling borer, *Sahyadrassus malabaricus,* feeds on many trees besides teak, including *Eucalyptus* spp. and the legumes *Paraserianthes falcataria* and *Calliandra calothyrsus.* Forage tree legumes planted in close proximity to teak plantations, for example, may be subjected to rates of colonisation by this pest that are higher than on trees further from teak plantations. Care should therefore be taken when planting tree legumes to ensure that such situations are avoided. However, in researching and interpreting the population biology of polyphagous species it is imperative to eliminate the possibility that one is combining species from different host plants, simply because they look alike.

Mahon *et al.* (1982) provide an example involving such sibling species of leaf miners on jarrah (*Eucalyptus marginata*), flooded gum (*E. rudis*) and prickly bark (*E. todtiana*). The principles and appropriate techniques applied in the unravelling of sibling species complexes have been outlined by Paterson (1991). Also of possible relevance to the cultivation of tree legumes is the phenomenon of 'induced resistance'. This arises when herbivory induces chemical changes in the plant, which may cause negative feedback on insects of the same species and sometimes even on other species (Haukioja 1991). The effect is so strong that it may influence population numbers of insects, even to the point of playing a role in population cycles (Haukioja 1991). The extent to which insect herbivory affects the nutritive value of tree legumes grown for forage is not known and requires evaluation (Bryant *et al.* 1991).

INSECT AND MITE PESTS OF STONE FRUITS IDENTIFYING AND UNDERSTANDING MAJOR STONE FRUIT PESTS

It is important for growers to recognise all stages of the insects and mites that attack stone fruits. Proper identification is critical to making the correct management decisions. In addition, growers should develop a basic understanding of the pests' habits. The more you know about the pest, the better equipped you will be to make sound and effective management decisions. The following books contain colour photographs of pests of stone fruits, as well as information on their biology.

ORIENTAL FRUIT MOTH

Damage: The oriental fruit moth is one of the most serious pests of peaches in the Midwest. Early in the season, this insect damages succulent terminal growth. It attacks fruit in midsummer. Although there may be four to five generations each year, it is the second and third generations that cause most of the damage. In addition to peaches, this insect attacks apples, plums, cherries, pears, and nectarines.

- *Appearance:* The oriental fruit moth is a small (1/4-inch), charcoal-coloured moth. Fine alternating bands of light and dark lines on the wings give it a mottled appearance. The small, flat, oval eggs are laid individually or in small clusters on the foliage primarily, usually on the upper leaf surfaces of the terminal growth. The larva is pinkish white with a brown head and is 1/2-inch long when fully grown. The larva pupates in a silk cocoon in crevices in the bark or in litter at the base of the tree. The larva is similar to the codling moth larva but can be distinguished by the presence of an anal comb.
- *Life Cycle and Habits:* The oriental fruit moth overwinters as full-grown larvae in cocoons in protected places on the trunk or around the base of the tree. They pupate beginning in late March, and moth emergence usually coincides with peach bloom. First-generation larvae tunnel into the young, tender terminal growth near the base of a leaf early in the season. The larvae may tunnel down the center of the twig for 2 to 6 inches before completing development or exiting and moving to another shoot. This injury causes dieback or flagging of these twigs. Larvae often damage two or three twigs before reaching maturity. These larvae can complete their development in less than a month.

  Larvae of subsequent generations feed on fruit, when available, and twigs. As the twigs harden, the partially grown larvae leave them and enter the fruits to feed. Larvae commonly bore to the center of the fruit and feed around the pit. Larvae feeding on the fruit often causes it to drop. Young fruit often exude gum from the entrance and exit holes left by the larvae. Damaged fruit that remain on the tree are distorted. Occasionally, the larvae may tunnel into the fruit through the stem. In such instances, there may not be any apparent evidence of how the larvae entered the fruit after it was harvested.

- *Monitoring and Thresholds:* Pheromone traps are available for this insect to monitor moth activity and effectively time sprays. Traps are placed in the interior of the tree canopy at eye level or higher just before bloom. One trap per 10 acres is recommended for commercial orchards, with a minimum of two traps. Sprays for the first generation should be applied 6 days after peak flight, which coincides with peak egg laying. Examine trees regularly in the early spring for signs of wilted shoots. Examine wilted shoots carefully to determine if oriental fruit moth was the cause. Young trees with vigourous new growth are often very susceptible to this injury. Detection of early season shoot damage indicates the potential for fruit damage by later generations.
- *Mating Disruption:* Some producers in Michigan and on the West Coast have successfully adopted mating disruption strategies for oriental fruit moth management in peaches. With mating disruption, pheromone dispensers are placed throughout the orchard. As the pheromone is released from the dispensers, male moths that normally use the pheromone to locate females become confused. This interferes with the mating process. Mating disruption is recommended only for orchards of 5 acres or larger. Mating disruption is expensive and does not eliminate the need for insecticide sprays to control other pests.
- *Chemical Control:* Sprays for the first generation should be applied 6 days after peak flight. This often coincides with the time for plum curculio control. Sprays for the second and third generations need to be applied 3 days after peak flight. Depending on the anticipated harvest date for the fruit, sprays for the third generation may need to be adjusted or omitted in order to meet the necessary pre-harvest intervals (PHI) requirements for certain insecticides.

PLUM CURCULIO LARVA, ADULT

- *Damage:* Plum curculio attacks stone fruits as well as apples and pears. Surface feeding and egg laying by the overwintering adults can scar or misshape the fruit by harvest, while feeding by the larvae causes premature drop of the fruit.

- *Appearance:* The adult is a typical snout beetle, 1/4-inch long and dark brown with patches of white or gray. There are four prominent humps on the wing covers. The snout is one-quarter the length of the body, with mouth parts located at the end. The larva is a legless, grayish white grub with a brown head. Its length is about 1/3-inch when full grown.
- *Life Cycle and Habits:* Plum curculio overwinters in the adult stage in ground litter or soil usually outside the orchard. Adults migrate into the orchards each spring, with border rows near woods often showing the initial injury. The first signs of damage typically coincide with the onset of 60°F nighttime temperatures. Eggs are laid on crescent-shaped flaps cut in the skin of young fruit. Often border rows near woods are the first to show injury. Fruits attacked by plum curculio will drop from the tree early in the season, along with unpollinated fruit. When the larvae are fully developed, they exit the fruit through clean-out holes, drop to the ground, and pupate 1 to 2 inches below ground. There is one generation per year.
- *Monitoring and Thresholds:* Currently there are no methods to accurately predict when plum curculio damage will occur. However, plum curculio pyramid traps are currently being tested in several midwestern states.
- *Chemical Control:* Plum curculio is usually controlled with petal-fall and shuck-split sprays directed at the adult prior to egg laying. Considerable egg-laying damage can occur over a short period of time. Where plum curculio has been a problem in the past, use preventive sprays at petal fall and shuck-split to reduce damage. Cool weather during petal fall may delay the immigration of the adults into the orchard. Under these conditions, a first cover and possibly a second cover spray may be needed.

TARNISHED PLANT BUG ADULT, NYMPH

- *Damage:* Tarnished plant bug damages young peach buds, causing blossom drop and early fruit drop. After bloom, feeding by this insect causes cat-facing and increased fruit drop. Tarnished plant bug cat-facing is characterised by sunken corky areas on the surface of the fruit.
- *Appearance:* Adults are 1/4-inch long, mottled brown insects with wings folded over the abdomen. A yellow-tipped triangle is present in the middle of their backs. Nymphs are small and greenish and resemble the adult without wings. Both nymphs and adults have a beak used for sucking plant juices.
- *Life Cycle and Habits:* Adult tarnished plant bugs overwinter under

bark, in leaf litter, and other such protected places. The adult feeds on opening buds or flowers in early spring, and later on developing fruit. They lay eggs in the plant tissue of their many hosts. Nymphs emerge about 1 week later and feed for about 3 weeks before reaching adulthood. Several generations of this insect occur each year.

- *Monitoring and Thresholds:* Pay particular attention to plant bugs when buds are in the pre-bloom stage. Adults are difficult to find and will fly when disturbed. They can be monitored with 6-by-8-inch, non-UV-reflecting, white sticky traps. Use one to two traps for every 5 acres. Hang traps in border rows near wooded areas. Treatment is recommended if an average of one tarnished plant bug per trap per week is captured.
- *Cultural Control*: Tarnished plant bugs overwinter in, feed on, and may build up in number on groundcover plants. Cover crop management is important in preventing tarnished plant bugs from moving into fruit trees. Because tarnished plant bug is attracted to flowering broadleaf weeds, management of annual weeds through regular mowing is an important practice.
- *Chemical Control:* Spray peaches at pink stage to reduce blossom drop. Sprays targeting tarnished plant bug at petal fall and shuck-split will reduce cat-facing injury to the fruit.

PEACHTREE BORERS MALE, FEMALE

- *Damage:* These borers attack the trunk and limbs of stone fruit trees, especially peaches. The peachtree borer is a pest of young trees, and a single borer can kill a tree. The lesser peachtree borer prefers to attack mature trees. Peachtree borer injury occurs a few inches above or below ground, and lesser peachtree borer injury may occur anywhere on the trunk or limbs where larvae can get under dead bark.
- *Appearance:* Adult peachtree borers are moths that look more like wasps. The adult female has a dark blue-black body with an orange band on the abdomen, dark blue front wings, and clear hind wings. The male is blue-black, marked with narrow yellow bands on the abdomen, thorax, head, and legs; front wings and hind wings are clear, but the edges and veins are outlined with blue-black scales. The male is 3/4- to 1 1/4-inches, the transparent portions of his wings are tinted with yellow, and at least three to four narrow bands of yellow are usually visible on the abdomen. Eggs are small, oval, reddish brown, and hard. The larva is dull white with a brown head and three pairs of short jointed legs. Larvae are 1 1/4-inch when fully grown.

  Adult male and female lesser peachtree borers are similar to each other in appearance and also look like wasps. Lesser peachtree borer moths are slender and dark blue with some pale yellow markings; both pairs of wings are clear, except for the edges and veins that have blue-black scales. Lesser peachtree borers resemble the male peachtree borer. Larvae of the lesser peachtree borer are similar to other clearwing borer larvae. They are about 1 inch when mature. The head is light brown and the body is creamy white, but some individuals may be pinkish.

- *Life Cycle and Habits:* The peachtree borer overwinters as larvae under the bark and resumes feeding and completes its larval stages in spring and early summer. When fully grown, the larva pupates under bark or in the soil near the tree base, then emerges as an adult. Adults start to emerge in mid-June; emergence peaks in midsummer and extends into September. Emergence is greater on days after a rain. Soon after adults emerge, the female moths lay eggs under bark scales or on rough bark. Each female lays about 400 eggs. Eggs hatch in 8 to 10 days into larvae that bore into the tree. It has only one generation per year.

  Lesser peachtree borer overwinters as larvae underneath the bark. Larvae of all stages except the first may be found during the winter. The larvae feed for a period in the spring before burrowing just below the surface of the bark to pupate. Borers remain in the pupal stage from 18 to 30 days before emerging as adults. Female moths lay eggs in small clusters in cracks and crevices near wounds between ground level and 8 feet high. Females lay an average of 400 small oval, reddish brown eggs. Larvae begin to hatch in 8 to 10 days and burrow into the bark, often entering through cracks caused by other factors such as winter injury, pruning scars, or machinery wounds. Moths emerge from early May until late September. There are two generations a year, with adult emergence in May and June, then again in August and September.
- *Monitoring and Thresholds:* While pruning in early spring, look for symptoms of borer. If symptoms are found, then an intensive control strategy is needed. To determine the most appropriate time to apply insecticide, a sticky trap or bucket trap baited with a pheromone lure can be used to monitor activity of peachtree borer or lesser peachtree borer. Traps for peachtree borer should be hung 3 feet from the ground and set up in late May. Traps for lesser peachtree borer should be hung 4 to 5 feet above ground and set up in late April. It is important to notice when emergence begins (when the first moth is trapped) and when emergence reaches a peak (when the number of moths trapped per week is highest).

The pheromone lure for each of these pests also can attract other similar clearwing borers such as the lilac-ash borer. Trapped moths thus should be examined carefully to be sure the correct species is being counted. If a trap for lesser peachtree borer is set up in the same planting as a trap for peachtree borer, the two traps should be separated by at least 60 feet to minimize trapping of non-target species.

- *Chemical Control:* Control needs to target small borer larvae before they bore into trees. Control may also be achieved by fumigant action of the insecticide, which can kill larvae already in trees at the time of application. An insecticide with long residual action gives the best control of peachtree borer.

Protection from peachtree borer is most critical during the first 3 to 5 years after planting. When new trees are planted, the roots and crowns should be dipped in insecticide before planting; this will protect them from borers during their first year. In established plantings, insecticide should be applied as a bark drench at a rate of 1/2 to 1 gallon of spray mix per tree. Thorough coverage is necessary. The insecticide should run down the trunk and soak the ground at the base of the tree. Any prunings, debris, or weeds at the base of trees should be removed so that they do not block the treatment. With lesser peachtree borer, an insecticide should be applied as a bark drench to the trunk and scaffold branches at a rate of at least 1/2 to 1 gallon of spray mix per tree.

The best time to treat and the number of insecticide applications needed for borer control depend on whether trees are known to be infested with these pests. One insecticide treatment is adequate in orchards where trees show some peachtree borer infestation; the best time to treat is at the time of peak adult flight, which is usually in early August. In orchards where borer injury is found on most trees, two treatments should be made: the first about 10 days after adults begin to emerge (late June) and the second at peak emergence about 6 weeks later (early August). In orchards where trees show signs of infestation by lesser peachtree borer, then an early application is needed to target the first generation in June, as well as an early-September application that targets the second generation. The first treatment should be applied 10 days after adults begin to emerge, which will probably be in mid-May.

CHERRY FRUIT FLIES

- *Damage:* Cherry fruit fly maggots attack the fruit of sweet and sour cherries. Infested fruit often appear normal until the maggots are nearly fully grown. Damage appears as sunken, shriveled areas on the surface of the fruit. Fruit may be blemished by the egg-laying punctures made by the female near the bottom of the fruit.
- *Appearance:* The adults are black flies with yellow heads and are

somewhat smaller than a house fly. Between the base of the wings is a white- or cream-coloured dot. The dark markings on the wings are used to distinguish the species. The abdomen of the black cherry fruit fly is entirely black, while the cherry fruit fly is marked with a series of white crossbands.

- *Life History and Habits:* Cherry fruit flies spend 10 months of the year in the soil beneath the trees. Adults emerge from late May to early July and lay their eggs on the fruit. The black cherry fruit fly generally begins to emerge about 10 to 14 days earlier than the cherry fruit fly. There are usually 10 days between the fly emergence and egg laying. The eggs hatch in about a week, and the maggots feed for about 2 weeks before exiting the fruit and dropping to the ground. They pupate 1 to 2 inches beneath the soil surface. There is only one generation of each fly.
- *Monitoring and Thresholds*: Monitor cherry fruit flies with yellow sticky cards hung in the trees in late May. Examine the wing bands to distinguish the species.
- *Chemical Control:* Sprays need to target the adults before egg laying begins. Adults should be controlled 5 to 6 days after they emerge.

PEACH SILVER MITE

- *Damage:* Peach silver mite sucks sap and causes peach leaves to turn silver in appearance.
- *Appearance*: This is an extremely small mite but may be seen with a hand lens. It is whitish and somewhat triangular and may be found on both surfaces of the leaves, usually near the veins.
- *Life History and Habits:* The adults hibernate under bud scales and in the bud axils. They become active in spring as the buds begin to open and foliage appears. Several generations are completed during the growing season.
- *Chemical Control*: In general, sprays applied to control other mite species will control peach silver mite as well.

GREEN JUNE BEETLE

- *Damage:* Because green June beetle is a direct feeder on peach fruit, a single adult can cause significant damage. Like the Japanese beetle, initial feeding by a few green June beetles often attracts additional beetles. These beetles can attack both green and ripening fruit, and their damage contributes to Japanese beetle fruit feeding activity.
- *Appearance:* Green June beetles are 3/4- to 1-inch long and are emerald green except for a tan border on the sides of their wing covers. The grubs have stout, grayish to white bodies with brown

heads. When mature, they are 1 1/2- to 2-inches long. Unlike other white grubs, green June beetle grubs are more straight-bodied and have the curious habit of crawling on their backs.

- *Life History and Habits*: Green June beetle grubs feed more on compost, manure, or decomposing organic matter in the soil. However, they throw up mounds of soil, and their burrowing activity can disturb the root system of grasses. In October or November when soil temperatures begin to cool, the grubs cease feeding and move more deeply into the soil where they spend the winter. The following March they begin migrating back to the sod zone where they continue to feed. They stop feeding in late May and move back down into the soil to transform into adult beetles. Green June beetle adults emerge in early July and are usually obvious as they feed on trees, shrubs, and flowers of many plants. Eggs are laid in soil with decaying vegetation.
- *Monitoring and Thresholds:* There are few threshold guidelines relative to when peaches need to be treated for green June beetle. However, the first colonizers in early July will attract others into the orchard, so early control can reduce later infestations.
- *Chemical Control:* Carbaryl is the most effective insecticide used in managing green June beetle. Repeated insecticide applications may be necessary at 7- to10-day intervals to prevent reinfestation during the adult flight period, or after heavy rains. Use of a spreader/sticker in the spray mix can increase the duration of effectiveness.

4

Wood Material Utilization in Forest

While it is not usual for those responsible for the growing and harvesting of timber crops to be directly concerned with their conversion and manufacture into usable products, there should be a close liaison between the two operations of forest production and wood utilization. Wood has inherent qualities that make it particularly suitable for many purposes—good strength in relation to weight, pleasing appearance, insulation against heat and sound, ease of fastening with nails, dowels, screws, or glue and case of working with relatively simple tools. However, wood is by no means a homogeneous material: it differs from one species to another in structure and physical properties and hence in the use for which it is best suited.

Within environmental limitations it is desirable to grow the species that will best meet the requirements of local industry and available export markets. Furthermore, to make the fullest use of the timber crop at all stages of development and to minimize logging and manufacturing waste, it is important to co-ordinate as closely as possible forest operations and the wood-using industries and also to bring about a closer integration of the wood-using industries themselves — as, for example, the sawmill and pulp industries.

FOREST POLICY

In many countries forests are of vital importance to the over-all economy, not only in relation to industrial development, but also in providing benefits which may be difficult to evaluate in monetary terms. For this reason and also because of the long-term nature of the forest enterprise, most countries seek to ensure the perpetuation of their forests and to safeguard the interests of present and future generations by placing the forests under some degree of government control.

In some countries the government retains ownership of a large proportion of the forested land and leases it to the wood-using industries for timber extraction: provision may be made for some degree of forest management by regulation or through agreement. Other countries have granted or sold much of their forest land to private interests, though considerable areas have sometimes been retained by the state to provide for timber reserves, recreational facilities and the protection of drainage areas; in some cases there is considerable public control of forest management on the freehold lands.

Elsewhere long-term forest management is entirely neglected and forest policies, if any exist, may be concerned only with the method and intensity of exploitation. Progress in the development and practice of forestry varies widely throughout the world. While forestry is well advanced in much of Europe and in certain tropical regions, only a small proportion of the accessible forests of the world is under intensive management today.

In some regions man still depends on the forests for his food and shelter and contributes little toward their development; in many countries man long ago destroyed the forests and has done little to restore them; elsewhere uncontrolled forest exploitation is still in evidence and proper protection and management lie in the future. In North America we are still in an early stage of forest conservation. In recent years, however, we have begun to see the forests as an integral part of our economic structure and to take steps against the destructive exploitation that has threatened to exhaust them.

RESISTANCE TO DAMAGE

Trees, like other organisms, are damaged and destroyed by various agencies, including fire, insects, fungi, grazing animals, industrial fumes and such climatic factors as wind, frost and snow. The capacity to resist such damage depends upon the vitality and the protective devices of the species. A knowledge of the relative resistance of different tree species is of importance in forest management.

DISTRIBUTION

Apart from the chance results of geological history, the continental distribution of tree species and the limits of tree growth, both in altitude and latitude, depend primarily on heat; the east-west distribution, on the other hand, is more often determined by moisture conditions—the rainfall in the interior of a continent may be too scanty to support tree growth. Thus each species of tree has a range of natural distribution determined primarily by climate. Obviously there is a great deal of overlapping and many species grow under similar conditions — in fact it is common to have a number of species sharing the same area in a mixed stand.

Pure stands of a single species do not commonly occur in the natural forest, though stands approaching this condition may be found where a species is well adapted to extreme conditions of climate or soil moisture, or where fire has brought about a situation particularly favorable to one species. The most heterogeneous and complex forest communities are found in tropical regions of high rainfall.

The distribution of a species within its natural range is determined mainly by soil, topography and competing vegetation, as well as by local variations in climate, which also affect the long-term development of many species: climatic races or strains develop, more or less adapted to local variations of climate within the range as a whole. This fact accounts, in part at least, for the wide variation in results obtained by using seed of other than local origin in establishing forests.

Scots pine, for example, has developed a number of climatic races and experience in both Europe and North America shows that efforts to introduce a "foreign" race of this species without a proper correlation of climatic conditions at seed source and planting site will almost certainly prove unsatisfactory, particularly in the susceptibility of the trees to damage by snow and sleet and the incidence of disease. This, of course, also holds true for attempts to establish any tree as an exotic outside the boundaries of its natural range. However, where the climatic conditions of two regions are similar, the introduction of tree species from the one to the other may be undertaken successfully, as, for example, in the introduction of conifers from the Pacific coast of North America into Great Britain.

An understanding of the various biological factors is of fundamental importance in the development of forestry practice, particularly in the fields of silviculture and protection. It must be emphasized, however, that practical forest management must depend also on the knowledge and application of sound business principles and on the development of suitable methods and techniques to be used in collecting, analyzing and organizing the basic data to be obtained from the forest.

Branches of Forestry

As is implied by what has been said above, forestry has a number of subdivisions, each of which is concerned with some particular aspect or phase of the subject. Hence, an understanding of forestry itself is dependent on knowledge of its parts, the extent to which they are interdependent and how best they may be co-ordinated.

BIOTECHNOLOGIES BASED ON VEGETATIVE PROPAGATION

Strategies supporting large-scale utilization of genetic material with a narrow genetic base must be appropriately integrated into tree improvement programmes. Vegetative propagation within such programmes allows for a fast release of new materials and for appropriate matching of clones to different local conditions. It also allows continued cultivation of given clones and to efficiently change the mixture of clones used in a given programme. Vegetative propagation also supports other currently available biotechnologies (*in vitro* storage and cryopreservation; *in vitro* selection).

IN VITRO STORAGE AND CRYOPRESERVATION

In vitro storage refers to the storage of germplasm in aseptic culture under laboratory conditions, while cryopreservation refers to the storage of cells, tissues, seeds, etc. at temperatures of liquid nitrogen (-196°C). The two techniques do not seem to be widely used in genetic conservation activities for forest trees, but they may serve as back-up strategies for species with seed storage problems.

Micropropagation

For crop and horticultural species, micropropagation (*in vitro* vegetative propagation of plants) is now the basis of a large commercial industry involving hundreds of laboratories around the world. Successful protocols now exist for a large number of forest tree species and the number of species for which successful use of somatic embryogenesis has been reported is increasing (somatic embryogenesis is a step in micropropagation where somatic cells are differentiated into somatic embryos). So, in the future, it is likely that micropropagation in the forestry sector will become commercially more important. Compared to vegetative propagation through cuttings, the higher multiplication rates available through micropropagation seem to offer a quicker capture of genetic gains obtained in clonal forestry strategies.

One major factor impeding early application of micropropagation in many large-scale forest plantation programmes, is that breeding and selection of desired clones are not sufficiently advanced for clonal forestry to be contemplated. Current high costs will also be an impediment to the direct use of micropropagation in many programmes. Technologies resembling those used commercially in horticulture are most likely to be affordable for a limited number of high-value forest tree species, particularly those for which propagation by cuttings is difficult. Micropropagation is unlikely to be used for the production of planting stock of non-industrial forest tree species.

In Vitro Selection

In vitro selection refers to the selection of germplasm based on test results using tissue culture under laboratory conditions. Many recent publications for crop plants have reported useful correlations between *in vitro* responses and the expression of desirable field traits, most commonly disease resistance. Positive results are available also for tolerance to herbicides, metals, salt and low temperatures. For the selection criteria of major general importance in forest trees (in particular vigour, stem form and wood quality), poor correlations with field responses will limit the usefulness of *in vitro* selection. However, *in vitro* selection may be of possible interest in forestry programmes for screening disease resistance and tolerance to salt, frost and drought.

WOOD-CONVERSION INDUSTRIES

Wood may be converted into other useful products by mechanical, chemical, or biological methods, or combinations of methods. Some wood pulp is made mechanically, some chemically and some by a combination of the two. Sugar and some other chemicals are made from wood by chemical methods, yeast and alcohol by chemical conversion followed by biological action. Other organic products can also be made by fermentation of wood sugar.

Charcoal and other wood-distillation products are the result of chemical change brought about by heating without the use of additional chemicals. Lactic, butyric and acetic acids have been made experimentally in substantial yields by direct fermentation of sawdust without intervening chemical treatment, although this conversion method is not known to be in commercial use.

From the standpoint of volume of wood used, number of employees, value of products and influence upon the world's economy, the most important of the wood-conversion industries is the pulp and paper industry..

The pulp and paper industry is really two industries which overlap to such a degree that neither one is entirely distinct from the other. In addition, the fiberboard or building-board industry is closely related to the pulp industry. Some mills produce pulp only and sell it to other manufacturers. Some produce pulp for their own use only, in the manufacture of paper or other pulp products. Other mills produce pulp that they convert into building boards. Many mills produce paper only, from purchased pulp. Paper, in turn, is produced in a multitude of grades and qualities and converted into an almost unlimited number of end products.

THE WOOD-PULP INDUSTRY

Wood pulp may be made by strictly mechanical means, by cooking wood with chemicals, or by combinations of chemical and mechanical methods.

Mechanical Pulp. This is made almost entirely by pressing blocks of wood against huge grindstones revolving in the presence of hot water. This is the cheapest method of producing pulp and gives yields of 90 per cent or more, based on the dry weight of the wood.

Approximately one third of the world's pulp is made in this manner. Most of it is used in the manufacture of newspapers, but substantial amounts go into other kinds of paper. Cheap power is a basic requirement in the profitable manufacture of mechanical or "groundwood" pulp, for power is the major product consumed, in addition to the wood itself. Groundwood pulp mills are usually located, therefore, where both wood and water power are abundant and relatively cheap.

The amount of power used per ton of pulp and the quality of the pulp produced are influenced by the species of wood used, the condition of the grindstone surface and the operating conditions. Since paper made from groundwood pulp alone is too weak for most uses, stronger pulp must be added. Newsprint paper, for example, commonly contains between 15 and 20 per cent chemical pulp and the rest groundwood pulp.

Chemical Pulp. This is made by cooking wood chips under pressure in solutions of chemicals which dissolve out the lignin and some of the other soluble materials in the wood, leaving the cellulose fibers in a more or less pure state. The yield of pulp is less than half of the original weight of the wood. The quality and usefulness of the pulp depend upon the species of wood, the cooking chemical used and other cooking and processing details.

In the *sulfite* process, the cooking liquor consists of a solution of calcium, magnesium, ammonium, or sodium bisulfite plus sulfurous acid. This results in a strong, relatively light-coloured pulp, which is adaptable to the widest variety of uses of any of the commercial wood pulps. One of its important uses is in mixture with groundwood pulp for the manufacture of newsprint paper.

When especially purified, sulfite pulp is used in the manufacture of viscose rayon and other cellulose derivatives. The sulfite process is used mainly with light-coloured, long-fibered, nonresinous softwoods like spruce, hemlock and balsam, but some birch, aspen and southern yellow pine are also employed. The *sulfate* process uses an alkaline cooking liquor containing sodium sulfide and sodium hydroxide. It is adaptable to both resinous and nonresinous species and, in fact, can be used with almost any wood.

In the United States the sulfate process is used principally in the manufacture of pulp from pine, but its use for other species is growing. Most grades of sulfate pulp are used for wrapping paper, bags, fibre shipping containers and other purposes where high strength is important. The pulp can also be bleached and used for the manufacture of high-grade papers for many purposes. In the *soda* process, also, an alkaline cooking liquor is employed, consisting of a solution of caustic soda. This method is used mainly for hardwoods and the pulp produces a bulky paper with low strength but good opacity, when used alone. Mixed with longer-fibered pulps added for strength, soda pulp is used extensively for high-quality printing papers.

Semichemical Pulps. These are newer than the others and differ from them in being produced partly by chemical and partly by mechanical means. The wood chips are softened by partial cooking in chemical solution and then pulped by passing them between revolving discs or plates. The chief cooking chemical used is neutral sodium sulfite solution, but both alkaline sulfate and acid sulfite solutions may also be used.

The yield of pulp depends upon the extent of the cooking and the severity of the processing, but is usually within the range of 70 to 80 per cent because much of the lignin is left with the fibers, in contrast with the strictly chemical processes that take out practically all of the lignin and, as pointed out, yield less than 50 per cent.

The principal use for semichemical pulps is in the production of corrugating board for fibre boxes. Bleached semichemical pulps are being used also for high-quality white papers, including glassine, bond, book, magazine and other grades. The semichemical process is especially suitable for pulping miscellaneous hardwoods, a characteristic which, together with the high yield and wide utility of the pulp, favors its increasing use.

A variation of the semichemical process consists in heating the wood in steam or water, followed by mechanical fiberization. These methods produce, with yields of 85 to 95 per cent, coarse pulps used mainly in the production of fiberboards, roofing felts and similar products. Dissolving Pulp Products. A relatively small but growing use for chemical wood pulps is for the production of highly purified pulps called dissolving pulps or high alpha-cellulose pulps. These are used in the manufacture of such products as rayon fabrics and tire cord, cellophane, absorbent tissue, lacquers, smokeless powder, photographic film, plastics and a variety of cellulose chemicals.

It is estimated that more than 700 thousand tons of dissolving wood pulp were produced in the United States in 1952 and more than 420 thousand tons in Canada. Other important producers are Sweden, Norway, Finland, Western Germany, Austria, Italy and Japan.

Production and Consumption

The world production of wood pulp in 1952 was estimated by FAO at 37.7 million metric tons. Of the world production reported to FAO (not including U.S.S.R.), about 35 per cent was mechanical, 29 per cent sulfite and 36 per cent all others. The relative production of the principal pulp-producing countries and an indication of the growth of the industry are evidenced. From year to year pulp production is strongly influenced by world economic conditions; hence the relative proportions of the total contributed by different countries as well as the proportions of the different kinds of pulp vary considerably. The United States is not only the greatest producer but also the greatest consumer of wood pulp, for its pulp imports greatly exceed its exports. Canada and the Scandinavian countries produce much more pulp than they consume and depend heavily on the export markets. The United States consumes about one half of the total world production of pulp.

BUILDING BOARD OR FIBERBOARD

This classification includes coarse fibre products such as hardboards, insulation boards and similar products, largely used in building construction but also to an increasing extent for industrial products of various kinds. The estimated world production in 1953 was 2.62 million metric tons, of which hardboards, including semihardboards, constituted 48 per cent. In Europe the production of hardboards was twice as great as that of insulation boards.

The reverse was true in the United States and Canada, where insulationboard production was more than twice that of hardboards. About one third of the reported production of hardboard and two thirds of the insulation board were manufactured in the United States.

Hardboards are made in large dense sheets from various kinds of wood pulp by the use of high temperatures and pressures in hot-plate presses. Insulation boards are much lighter in weight and less dense and are dried with hot air in tunnel dryers, without pressure.

CHARCOAL AND WOOD-DISTILLATION INDUSTRY

When wood is heated to high temperatures in air it burns, but in the absence of air it turns to charcoal, giving off much gas in the process. The gas consists mostly of water, but it includes acetic acid, methyl alcohol, various light oils and tars, the proportions varying with the species of wood used and the temperatures maintained. Methyl alcohol and acetic acid made in other ways compete strongly with the wood products and the commercial value of wood tars and other distillation products is limited. As a result, most of the value of the products of wood distillation lies in the charcoal and most of the world's charcoal is produced without saving the by-products.

When the gaseous by-products are not to be saved, charcoal can be produced very simply by stacking the wood together, covering it with earth, igniting it in one or two places and maintaining a slow heat by controlling the admission of air. After the wood reaches a temperature of about 500° F. an exothermic chemical reaction begins, which furnishes much heat and reduces the amount of air required to finish the carbonization.

The time required to complete the charring process and cool the charge subsequently so that the charcoal may be removed without igniting when exposed to the air varies from three or four days to two or three weeks, according to the size of the charge and the character of the pit. Simple kilns made of brick, concrete blocks, or even metal, are often used instead of the earth covering. An ordinary five-cord concrete-block kiln can complete a charge in about eight days. When the gases are to be condensed and saved, much more expensive equipment is required, such as steel ovens, tram cars, condensers, water pumps and refining apparatus, which make the investment much higher than when the gases are wasted.

Ovens of this kind in common use can carbonize a charge in about one day, but generally the preheating requires three days and the cooling three days in addition, making a total of seven days. Statistics on charcoal production are incomplete and inadequate. Wood used for charcoal is sometimes reported with that used for fuel and the amounts of charcoal produced are seldom reported. India is also a heavy producer, but no data on Indian production are available. Italy used about an eighth as much wood for charcoal as did Argentina.

The U.S. charcoal production is estimated at about 335,000 tons. Charcoal is used mainly for domestic fuel and carbon disulfide manufacture. In Japan and certain other countries it is used to some extent for truck and bus fuel. In the United States much charcoal is used for recreational cooking at picnics and barbecues and on similar occasions. There are, of course, innumerable industrial uses for charcoal in metallurgical, chemical and other manufacturing operations.

WOOD-HYDROLYSIS INDUSTRY

When wood is heated with acid under properly controlled conditions the celluloses and hemicelluloses are converted into sugars which can be separated and purified to produce glucose and xylose; or else the solution can be neutralized and fermented to produce alcohol, glycerine, yeast, citric acid and a number of other products.

The production of sugars from wood has been experimented with extensively in the United States and high yields have been obtained, but the process is not yet commercially successful in this country. It has found commercial use in Europe for the production of sugars, alcohol and yeast — especially in Germany, France and Switzerland. Production statistics, however, are not available.

Other Forest-Products Industries

A multitude of products other than wood comes from the forests, including naval stores, tannins and cork, which are important articles of world commerce. Rubber could be included as a forest product of great economic importance, but most of the natural rubber is now being produced from plantations instead of from wild trees in the forest. Oils, gums, waxes and resins in great variety are obtained from the forests of the world.

Many fruits, nuts, spices and other edible products originally obtained from wild trees are now produced from plantation trees of improved varieties and thus have become agricultural rather than forest products, but others are still wild-forest crops.Christmas trees have become an important wild-forest product in the United States and Canada although gradually coming in larger quantities from plantations. In parts of the United States and Canada where hard maple is plentiful, maple syrup and maple sugar are products of local importance and esteem. The list could be extended to include hundreds of products, although many of them have local rather than international importance.

THE NAVAL-STORES INDUSTRY

Turpentine and rosin are the two principal products now included in the term "naval stores," a term that dates from the time when all ships were made of wood. Turpentine and rosin are obtained by the distillation of the oleoresin or "pitch" that exudes from living pine trees when they are "tapped" or scarified.

At present these products are also obtained by steam distillation and solvent extraction from chips of resinous wood, mainly pine stumps, that have been left in the ground until the nonresinous portions have rotted away. Substantial amounts of turpentine are also produced as a by-product in the sulfate pulping of pine chips and a much smaller amount as a by-product from the destructive distillation of pine wood. Of the 1951 production of turpentine in the United States, it was estimated that 36 per cent was "gum" turpentine, 33.6 per cent wood turpentine (from stumps), 29.7 per cent from sulfate pulping and 0.7 per cent from destructive distillation. Ten years earlier the percentages were 52.0, 34.6, 11.8 and 1.6. Of the 1951 U.S. rosin production, 39.4 per cent was from gum and 60.6 per cent mainly from solvent extraction. World statistics on turpentine and rosin production, which, however, does not include data from the U.S.S.R. and a number of other turpentine-producing countries.

Of the total production, the United States produced more than did the rest of the world combined. A large percentage of the production of each major producing country is exported. In addition to turpentine and rosin, pine oil, pine tar and other navalstores items are produced and an increasing number of chemicals are being synthesized from the primary naval-stores products.

Table. World Production of Turpentine and Rosin

Country	Turpentine (Thousand U.S. gal.)	Rosin (Thousand lb.)
World	53,309	1,511,797
United States	34,856	1,086,888
Other countries	18,453	424,909
France	4,685	123,457
Portugal	5,122	147,691
Sweden	3,000	6,000
Spain	2,410	57,143
Mexico	1,600	52,000
Greece	1,058	27,122
Japan	428	10,496
Others	150	1,000

CORK

Commercial cork is the outer bark of a species of oak that grows mainly in countries bordering the western half of the Mediterranean Sea, particularly France, Portugal, Spain and North Africa. Cork trees also grow in other countries in small quantities and have even been introduced into the United States, but are not industrially important there. Every eight or ten years the outer bark, usually one or two inches thick, is carefully stripped from the trees without damaging them or interfering with their growth or production of subsequent crops.

The bark is air-dried for a few days and then taken to market and sold for manufacture into insulation, bottle stoppers and cork compositions of various kinds. Portugal, the world's largest producer of cork, accounts for about half of the world's estimated production of 310,000 tons and cork and cork products are important items of export from that country. The United States uses almost one half of the world's cork crop.

TANNIN EXTRACTS

The production of tannin extracts for leather tanning amounts to a considerable industry throughout the world. The quebracho wood of South America produces more tannin extract than does any other single source. This wood, which is very heavy, hard and high in tannin content, is chipped and leached to yield an extract that is highly valued by tanners. More than 250,000 tons of it were exported from Argentina and Paraguay in 1950.

On a smaller scale, chestnut wood chips are extracted for tannin in the United States, Italy and France. Commercial tannins are also derived from barks, fruits and leaves as well as from the wood of plants. Wattle bark from Australia and South and East Africa is an important source. Many of the forest species of the United States contain substantial quantities of tannin in their barks, but in general only the barks of eastern hemlock, chestnut oak and tanbark oak have found extensive commercial use in this country. Improvements in by-product utilization may in time bring into commercial use other barks that are available as waste in large amounts at sawmills, pulp mills, veneer mills and other factories. Tannin production in the United States has been decreasing for years and importation increasing. In 1927 nearly 56 per cent of the vegetable tannins used in the United States were from domestic sources, but in 1950 domestic tannins constituted only about 15 per cent of the total and in 1952 the ratio was still lower. Most of the domestic tannin was obtained from chestnut wood, with much smaller amounts coming from oak and hemlock barks. Nearly 70 per cent of the tannins imported in 1950 were obtained from quebracho wood and about 19 per cent from wattle bark.

Waste Utilization

In the conversion of trees into wood products, on the average much less than half of the tree gets to market in the form of finished products. The waste is usually greater in countries that are heavily timbered or where labour costs are high than where the reverse is true. The U.S. Forest Service in 1946 estimated that of the wood cut in or imported into the United States in 1944, 43 per cent reached the hands of the user, 22.5 per cent was used under the boilers of mills and factories that processed the wood and 34.5 per cent found no use at all. This estimate did not include any bark, most of which finds no use or is burned as fuel.

The waste in producing fuel wood, mine timbers and fence posts is relatively low in comparison with the waste in the production of lumber, hewed ties and cooperage, for which the estimated losses are from 68 to 72 per cent. In the manufacture of chemical pulp, the lignin and other constituents dissolved out by the pulp liquor and thus lost, amount to 50 to 60 per cent of the original pulpwood, although fuel values are received from a part of these liquors. Altogether, the total amount of lignin, bark and wood in the United States that finds no use may exceed 100 million tons annually.

As the Forest Service has pointed out, such waste could be considered equivalent to a neglected forest, a source of raw material of great potential value. The extent to which it can be used economically, however, is limited by its accessibility to markets and the cost at which it can be made available to prospective users. Manufacturers can be expected to use wood residues, instead of roundwood or lumber, as their raw material only where it is financially profitable to do so. The wood-waste problem is not one of wilful wastefulness but one of economics and technical developments.

Since 1944 much progress has been made in the use of logging, veneermill and sawmill wastes in the manufacture of pulp and building boards and these uses seem to be growing steadily. More and more sawmills are removing the bark from their logs before sawing, which increases the acceptability of the slabs and edgings from the logs for pulp manufacture. Matson estimated that the equivalent of nearly five million cords of logging and mill residues is being used annually in the production of pulp chips, hardboards and soft boards, chiefly in the Pacific Coast states.

There is also a newer and relatively much smaller use for sawdust, planer shavings and other fine material: they are combined with resins and then, by means of heat and pressure, the mixture is converted into hard, smooth sheets that can be used like plywood, hardboard and similar products, in the manufacture of doors, cabinets, toys and numerous other things. The production of such "particle boards" is expanding.

Sawdust, shavings and "hogged" sawmill waste are suitable for chemical processing and it is known that a great variety of useful chemicals, such as sugars, acids, alcohols, yeast and furfural can be obtained from them in good yields. Many of these products may in time be made profitably from wood residues but, for the moment, they do not offer attractive commercial possibilities in North America.

Many useful products can be made from waste bark. Douglas-fir bark, for example, is being ground up and segregated into a variety of products to which proprietary names have been given and which, through aggressive merchandising, find many uses. Douglas-fir bark also yields useful tannins and waxes, but not yet in commercial production. The bark of California redwood is being processed commercially into a fibrous insulating material and a by-product dust. It is improbable, however, that more than a small fraction of all the bark available can be used in such ways.

Wood flour, of which possibly 80,000 tons are produced annually in the United States, is made by grinding sawdust or shavings from acceptable species and sifting it to the degrees of fineness needed. It is used in the manufacture of linoleum, adhesives, certain molded plastic products, dynamite and numerous other products where an absorbent filler is needed.

Small pieces of wood in standardized sizes and shapes are produced in substantial quantities from sawmill slabs and low-grade lumber and are sold to manufacturers of various products. Such material from softwoods is commonly called "cut stock," and that from hardwoods "dimension stock." Much skill and efficiency are required to produce these products from wood residues with the degree of accuracy and uniformity of dryness desired by the purchaser and at a profit to the producer. Many other potential uses exist for products that can be made from the residues of sawmills and woodworking factories. The problem for the plants producing the residues is first to find markets for which their residues or the products made from them are suitable and then to deliver the residues or the products to the user at a cost that is attractive to him and profitable to the producing plant. Although innumerable products can be made from wood residues, only a limited number can be made profitably.

Large owners of timberland who are operating on a sustained-yield basis and running their own manufacturing plants can make a variety of products and locate their plants so that the waste from one may be used conveniently as the raw material for another. In a number of notable instances, by integrating their industries in this way and by good management, large lumber manufacturers have eliminated their wood losses or reduced them to very low figures. Although the small sawmill operator is much more limited in his ability to use his residues, he can improve his machinery and methods so as to reduce the waste to the lowest practicable minimum. It may also be practicable in some localities for small, separately owned plants making different products so to integrate or group themselves that most of the residues from the primary breakdown of the log will find profitable use.

THE DEMAND FOR FUEL WOOD

The first forest product used by man in quantity was obviously firewood. The gathering of naturally dead and fallen wood need not reduce the area of the forest, but the practice of killing and felling trees by girdling and the use of fire must have been invented or discovered at a very early time. The method has been recorded from so many peoples that it might almost be called universal. Primitive hunters and food gatherers annually set fire to the vegetation, for several reasons: to facilitate hunting by clearing away underbrush, to attract game by the salt content in the ashes, to make it easier to dig up roots and tubers or to find nuts fallen on the ground and sometimes to drive game with fire in organized group hunts. Besides, trees and polewood killed by fire became more readily available for fuel. It must be assumed that regular burning, repeated for thousands of years, destroyed trees over wide areas.

Henry Youle Hind, while exploring the Saskatchewan River valley in 1858, observed that Indian fires had driven the forest back 180 miles from its former limit. On their expedition through the northern Rocky Mountains in 1803, Lewis and Clark found extensive prairies in valleys where trees can grow very well. Alexander Henry in 1814 and Charles Wilkes in 1841 described large open grasslands in the Puget-Willamette valley of Washington and Oregon. Wilkes, as did Lewis and Clark, concluded that these prairies owed their origin to Indian fires. Evidence supporting the conclusion is seen in the fact that the forest naturally began to encroach upon the prairies again after the American settlers had occupied the land and the Indian burning had ceased. Similar evidence is known from many other parts of the world. The full effect of the ancient hunters on the vegetation can probably never be measured exactly, but it is certain that the process of reducing the forest area began long before agriculture was known.

The larger populations that became possible when farming had been introduced increased the demand for fuel wood; the inventions of the pottery kiln, fired bricks, metallurgy and other industrial arts made further heavy demands on the forest; and today the need of fuel wood continues unabated in many lands.

In some countries deforested regions have been replanted because of the demand for fuel wood. Up to 1941, according to B. H. Hunnicutt, the Paulista Railroad in southern Brazil had harvested more than 70 million cubic feet of firewood from eucalyptus plantations started in 1904. M.D. Chaturvedi reminds us of the present emphasis on fuel-wood plantations in India, some of which have been remarkably successful. Troup, average annual increments of 500 cubic feet per acre have been obtained on plantations in southern India by the coppice or sprout method, a rate of growth that can be called phenomenal.

The consensus of all reports is that destructive fuel-wood cutting in the tropical forests is far more common than conservative use. William Vogt says that at least 75 million people in Latin America depend on the forest to provide firewood or charcoal for all their cooking and heating -if heating can be afforded--and that large areas have been so stripped that some people are practically without wood and use anything that will burn. Hunnicutt describes clear-cutting on a large scale in parts of Brazil and the necessity of hauling wood for long distances to the railroads because wide sections along the tracks have been deforested.

In The Amazing Amazon, Willard Price speaks of having seen hundreds of cords of mahogany and rosewood brought aboard river steamers for firewood.

H. F. Mooney informs us that charcoal burning has been the greatest single factor in forest destruction in southwestern Asia in recent years. Reports in a similar vein from other countries indicate that the oldest cause of forest reduction, the demand for fuel wood, is still operating. For this reason alone the area of many of the low-latitude forests continues to shrink steadily.

To my knowledge, no practical solution of the fuel problem in most of the tropical countries is in sight, except perhaps the coppice method. It is not likely that coal, petroleum, natural gas, or electricity will become economically available to most of the people, at least not in the near future. But many tropical and subtropical trees coppice vigorously and if this method could be widely and systematically applied it would not only tend to save the forest but would help to solve the fuel problem where it is most pressing.

An anomalous situation has come about in some countries. The thinning of growing timber and the removal of cull trees and other undesirable wood is good forestry practice, but if a market for the thinnings is lacking it may be economically impossible to perform the improvement operation. Not all of this removed material is suitable for industrial purposes, but practically all of it can be used as fuel. Much of the low-value fuel wood from the French forests is difficult to sell.

Thorsten Streyffert points out that one of the serious forestry problems in Scandinavia is the finding of a greater market for firewood and a similar problem exists in the United States, particularly in the eastern hardwood forests. The Stanford Report and the other forecasts predict a sharp decline in the use of fuel wood, not because such wood is lacking, nor for reasons of good silviculture, but because people increasingly prefer to cook their food and heat their houses with other fuels.

THE DEMAND FOR CONSTRUCTION MATERIAL

The second cause of deforestation, the demand for construction material or industrial wood in the widest meaning of the term, was hardly a factor of great significance among the ancient hunters, but in the later history of settled regions this requirement put an even greater pressure on the forest than did the cutting of fuel wood. Old London, which burned down several times, was built of wood, as were hundreds of other European cities. A large part of the forests that were clear-cut in Michigan, Wisconsin and Minnesota during the latter part of the nineteenth century went into millions of homes built in the prairie states and the old logged-over areas in the Pacific Northwest provided construction material for early Seattle, Tacoma, Portland, San Francisco, Los Angeles and many other communities.

The demand for mine timber was one of the significant reasons for the early deforestation in southwestern Asia, on Cyprus, in the Sierra Nevada and Sierra Morena of Spain, in the Erzgebirge of Germany and in many other mining regions. The need of wood in the mining industry was so critical that it led to some of the earliest forestry regulations; for example, volume allotments were instituted during the late Middle Ages in Germany and Austria in order to assure the supply of mine timber. Deforestation near the mines in Mexico had become so serious by 1550 that the native people petitioned for protection of their land and wood supply and a forestry regulation, probably the oldest in the New World, was promulgated by the Viceroy in 1550. The effect of mining on the forest is strikingly expressed by Dan DeQuille, who says in The Big Bonanza -- perhaps with a trace of Western overstatement -- that the Comstock Lode can in truth be called the tomb of the forest of the Sierras.

The demand for ship timber put another heavy pressure on the ancient forests, particularly in western Europe and most of all around the Mediterranean Sea. In this region, because of steep slopes, generally thin soil and a frequently violent type of rainfall, a single clear-cutting may result in such serious soil denudation that the forest cannot come back and a brushy vegetation takes the land instead. It may be that the primary reason for deforestation in this region was the clearing of land for agriculture and the effect of grazing livestock, particularly the browsing of free-ranging goats, but surely the second most important cause was the cutting of ship timber.

The forest was one of the casualties in the naval wars fought by the Phoenicians, Etruscans, Greeks, Romans, Carthaginians, Arabs, Genoese, Venetians and all that followed after. The mountains of Greece were nearly stripped of trees by the fifth century B.C. Thucydides informs us that one of the purposes of the Sicilian expedition of Athens in 415 B.C. was to gain control of the abundant supply of ship timber in the forests of Italy.

A century later Theophrastus observed that ship timber was scarce everywhere in the eastern Mediterranean, but he praised the fine forests of Italy and other lands around the western Mediterranean. These western forests did not last long. In the first century B.C. Strabo was impressed by the superb timber in the Atlas Mountains, but in the first century after Christ Pliny observed that some of these forests were already exhausted. The demand for ship timber continued and spread northward. The Dutch fleets of the early seventeenth century were built mostly with timber from the oak forests of western Germany, which were heavily logged to raise money for indemnities after the Thirty Years War.

In the opinion of Henry Cavaillés, the three main reasons for the greatly accelerated deforestation in the French Pyrenees during the seventeenth century were the consumption of wood by the Catalan forges, the clearing and burning of woodland to make pasture for sheep and goats and the cutting of ship timber, particularly for the construction in the 1660's of a French navy consisting of 60 ships-of-the-line and 40 frigates. In Britain, as H. C. Darby says, the naval wars with the Dutch in the seventeenth century and with the

French in the eighteenth and the Napoleonic wars in the early nineteenth put such a strain on the English oak forests that they have never recovered.

When “the far-flung navies melt away” they take the forests with them. Ship timber is no longer one of the major requirements, although wood in large amount is still needed in ships and shipping. It is estimated that 350,000 board feet of lumber were used in the construction of every Liberty ship built in the United States during World War II and the wood that goes into boxes, crating and dunnage on a modern general cargo ship runs into 100,000 board feet or more.

Today the heaviest demand for industrial wood products -- lumber, pulp, plywood and the like -- is on the midlatitude forests of North America and Europe. There is a chance, however, as far as the pressure of this demand is concerned, that improved forestry methods may check any further significant reduction of the forest area in these regions. If the area is reduced in the coming years the reason is more likely to be the need of more agricultural land, or the need of space for industrial and urban expansion.

The old need for local construction material and the relatively new demand for industrial products are also felt in the tropical and subtropical forests and one of the great uncertainties ahead is what will happen to them if and when they become fully exploited for industrial purposes. If problems of tropical silviculture, management and economics -- including aboriginal economy -- should be successfully solved, these forests may become a very great source of wood products, but if the problems are not solved the exploitation may result in an enormous disaster. One of the complicating factors is that the most potent pressure on the tropical forest comes from the third of the ancient demands, namely the need of agricultural land.

THE DEMAND FOR CROPLAND AND PASTURE

Exactly how large a part of the forest reduction in the world has been caused respectively by the requirements for fuel wood, construction material and agricultural land cannot be determined, but it seems safe to say that the forest has yielded far more ground to farms and pasture than to any other demand. Much of the world's cropland has been gained without infringing on forested areas -- for example, in the interior plains and prairies of Canada and the United States, the pampas of Uruguay and Argentina, the steppes of southern Russia and the wheat region of Australia-but even more has been made by clearing forest land. About one third of the original forest in the United States has been cleared, mostly in the eastern woodland and much of this once timbered land is now under cultivation.

Brazil has probably lost 40 per cent of her forest area, most of it to agriculture. Pollen analysis has shown that virtually all of western and central Europe was covered with forest at the beginning of Neolithic time and it is reasonably certain that India and China were also predominantly forested when agriculture was first introduced. At the present time it is estimated that only about 9 per cent of China is forested and 18 per cent of India and the figure for the average of western and central Europe is also 18 per cent.

The principal reason for the clearing of these vast forests was undoubtedly the need of food-producing land. In like manner the world over, farmers and herdsmen have won most of their land at the expense of the forest and there is good reason to think that the forest will continue to lose ground.

The tradition of making farms from forest land may be 10,000 years old. Like the primitive hunters, the ancient farmers knew tree girdling and the use of fire in clearing land, a method still widely employed in the shifting agriculture that prevails in many tropical and subtropical countries from Middle America around the world to New Guinea.

André Aubréville says that in the deforestation now going on in central Africa because of shifting native agriculture we are witnessing the finishing touches to a process that began long ago. H. L. Shantz estimates that the original area of the tropical forest in Africa has already been reduced by two thirds; and P. W. Richards believes that almost the whole of the tropical virgin forest may disappear within our lifetime unless determined efforts are made to change the farming system. In the highlands of New Guinea, according to O. H. K. Spate, shifting native cultivation is eating into the main forest at an accelerating rate and a serious crisis is likely to develop within a few decades if the system continues unchanged.

Besides giving ground to aboriginal farming, the tropical forest has yielded space to commercial plantations and there is no indication that the pressure from this type of agriculture will be lessened in the future. On the contrary, many are looking to the tropics for more land.

The amount of what is defined as unused but potentially cultivable land in the world is variously estimated from 1 billion acres by the FAO to more than 6 billion acres by C. B. Fawcett and L. Dudley Stamp. In these estimates more than half of the new land is supposed to be in tropical regions. According to the calculations of Robert M. Salter, one of the requirements for providing the people of the world with an adequate diet in 1960 is the opening of 1 billion acres of new tropical soil. It is not made clear, however, precisely where all this new soil is.

In a thought-provoking article published in the Agronomy Journal Robert L. Pendleton calls attention to the fallacies in estimates such as that of Salter and cautions against overestimating the potential of tropical forestry and farming, particularly large-scale commercial grain farming. It is often overlooked, as Pendleton and many others who have studied the tropics point out, that rich alluvial soils are not abundant and that most of the tropical soils are poor in nutrients.

These poorer tropical lands have been cultivated for several thousand years by farmers who have depended on forest soil and one wonders what would happen to the remaining forest if another billion acres were to be plowed up and made into modern farms on the European or North American pattern. In all probability, the long continued aboriginal cultivation has been possible only because it is forest soil that has been used. The method developed by the ancient farmers makes use of the forest, which is an essential part in the cycle of shifting agriculture: the forest is the fallow and cover crop that restores fertility and maintains the soil properties and it may well be that in the long run this method is the only sure way of producing food from the poorer tropical soils.

The system is destructive of timber and has therefore frequently been condemned, particularly by foresters, but it is not destructive of soil if the rotation period is long enough. There may be more practical wisdom in the old way than in trying to change it. At any rate, whether one thinks of aboriginal or of modern commercial agriculture, it is clear that the tropical forest is under increasing pressure and will not long remain intact.

Attempts to check further agricultural encroachment upon the forest are being made. India's New Forest Policy of May 1952, as reported in Unasylva changes the previous conception that the forest has no right to a site if it can be used for agriculture and declares that land must be allotted to farms and forest according to a system of land use under which each site will yield most of the product for which it is best suited; but, the report continues, it is also recognized that the need of agricultural land is greater than ever. It may be difficult to implement this worthy policy in the future, for always and wherever agricultural need has pressed on the forest, it is the forest that has yielded.

The midlatitude forests in southern Chile, South Africa, extratropical Australia and New Zealand, which never were very large in comparison to those in the northern midlatitudes, have in the past lost much ground to logging, fuel-wood cutting and agriculture. At the present time it might be said that the forest area is more or less stabilized in New Zealand, Australia and South Africa and is even increasing somewhat because of much planting of exotic trees, especially Monterey pine. Whether the forest area of Chile can be called stabilized is more doubtful, for reports of recent heavy logging in the rough terrain of southern Chile suggest that deforestation may be the result.

But in Chile, too, Monterey pine has been introduced and is said to be growing very well. The frontier between farm and forest in the mid-latitudes of the northern hemisphere is more or less stationary for the present; indeed, the forest has regained some ground in recent time. The clearing of land in western Europe continued until the forest was reduced to a minimum size at a time in the past that is difficult to determine for lack of comprehensive records, but was perhaps a century or two ago; and then the forest gradually began to increase because of reforestation projects. In 1750, according to authorities quoted by George P. Marsh, forests covered about 32 per cent of the area of France and a century later only half that much.

It is scarcely possible to say exactly how much the forest area has increased since 1850, for national boundaries have changed somewhat and the definitions of what constitutes forest land are not uniform. The reforestation continues at the present time. After visiting Germany in 1953, Leo A. Isaac reports that he was greatly impressed by the widespread planting of trees, including North American species; he speaks of one nursery firm which alone produces and contracts with other growers for a total of 600 million seedlings a year, or about the equal of the entire annual forest planting in the United States.

Table. Forest Land as Percentage of Total Area, 1850, 1939, 1953

	1850[a]	1939[b]	1953[c]
Germany	26.7%	26.80%	
West Germany			28.1%
East Germany			25.6
France	16.8	19.1	20.7
Great Britain	5.0	5.0	6.5

More than 400 forests have been established in Great Britain since 1920 and in connection with them a number of new and interesting forest villages have been built, mostly in Wales, northern England and Scotland. The Forestry Commission of the United Kingdom hopes to afforest some 3 million acres over a 50-year period.

The 2-million-acre pine forest of Les Landes in southwestern France has been built up during the last century, reforestation is under way in the French Pyrenees and a project of planting trees on some 500,000 acres of deforested land in the southern French Alps was initiated in 1928 and is approaching completion. Whether or not the forest area in western Europe can be permanently maintained at its present size is essentially an economic question.

The highly industrialized west-European countries have long drawn upon the resources of other continents for food, fertilizers, livestock feed and raw materials for industries, in return for manufactured goods; and the demand for imports is rising rather than falling off. A Hamburg lumber importer, interviewed by The New York Times on January 2, 1955, predicted that Europe will buy more and more wood products abroad. One of the reports made at the annual meeting of the American Paper and Pulp Association in February 1955 dealt with the present "immense" demand for paper in Europe and another speaker said that the most significant development in the industry during 1954 was the emergence of the United States as a net exporter of pulp, much of which is going to Europe. These rising demands obviously imply that Europe's own forests are under heavy industrial pressure.

The agricultural land of Europe is likewise under heavy pressure. Georg Borgström expresses the opinion, which everyone shares, that Europe will be forced to develop its agriculture to the utmost and makes the suggestion, which is not endorsed by all, that the aim should be selfsufficiency in food, or as close to it as possible. Borgström also believes and again not everyone agrees, that a general soil depletion and deterioration in western Europe is concealed by heavy use of fertilizers and that the high yields cannot be expected to continue to rise very much higher. He suspects that the Netherlands has almost reached the possible limit of yield at reasonable cost.

Whether Borgström is right or not, it is clear that the success of the whole economic experiment depends on Europe's ability to pay for the increasing imports, including all the fertilizers; and the fate of the forests rests on this ability. If the time should ever come when Europe can no longer draw on the resources of other continents at the same rate as now, then self-sufficiency in food would become an urgent necessity and agriculture would surely gain priority over forestry. There will no doubt always be forests in western Europe, for some of them grow on land that is not suitable for other crops, but parts of the forests occupy soil that can be farmed if necessary. It must be remembered that nearly all of the cropland in western Europe has been made from original forest soil, improved and built up through the centuries and the areal limit of that sort of improvement has not been reached.

Another fact to reckon with in western Europe is the increasing amount of ground needed for urban and industrial expansion, highways, airfields and the like. Much of this ground will doubtless have to be yielded by the forest. The large forests in the higher latitudes of North America and Eurasia -- Alaska, Canada, the Scandinavian peninsula, Finland, northern Russia and Siberia -- are in no great danger of agricultural encroachment at the present time and probably never will be, mainly because they grow in regions where neither soil nor climate is favorable to crop farming.

To be sure, a certain amount of forest clearance for the making of farms is still going on, for example in Alaska, along the margin of the pioneer fringe in the Canadian prairie provinces, in the clay belt of northern Ontario and Quebec and on the "cold farms" of northern Finland and presumably clearing is still in progress here and there on the southern margin of the forests in Russia and Siberia. In view of the great extent of the northern timberlands, however, these scattered clearings can hardly be regarded as a threat to the forest; and, besides, the area of timberland thus lost to farms is more than likely balanced by reforestation of other areas.

The forest situation in the United States partly resembles that in the high latitudes and partly that in western Europe. Much of the timber, particularly in the West, occupies land that probably never can be used for agriculture; but some of it, especially on farm wood lots in the East, grows on soil that could be used for crop farming if necessary. As in western Europe, the forest area of the United States has increased somewhat in recent time.

Many sections of the cutover lands in the upper Great Lakes region have been reforested; the timbered area in the South has increased substantially because of widespread planting of pines; and a rather large amount of farm land in various regions has been abandoned, turned into pasture, or naturally reverted to forest. For example, the Alabama Black Belt, which was at least two-thirds forested 150 years ago, was largely cleared for agriculture during the nineteenth century, but more recently much of the cropland has gone back into forest. In 1936, according to the Forest Service, 45 per cent of the area of the counties in which the Black Belt lies was timbered, but the corresponding figure for 1953 was 54 per cent, showing an increase in forest acreage of about 20 per cent in 17 years.

While the forest area in the United States has increased somewhat in the recent past and the farm acreage has decreased, these trends have been characteristic only of certain regions and are not likely to affect the national balance in land use for very long. On the contrary, in a national sense a reversal of the trends seems to be coming. It is clear that timberland is already yielding ground to other forms of land use and probable that the crop acreage will soon have to be increased. All who have given thought to the future agree that the demand for farm products in the United States will rise by 35 or 40 per cent before 1975 and that this demand must be met either by higher yields per acre, or by increasing the crop acreage, or by both and most think that both will be required. The amount of additional land needed is variously estimated.

The general consensus of the participants in the section dealing with agricultural land at the Mid-Century Conference on Resources for the Future in December 1953 was that from 25 to 30 million acres of new cropland would be needed in 1975, even though most of the demand is expected to be met by greater productivity per acre.

Samuel H. Ordway, Jr., makes a point that does not seem to have received sufficient attention in some of the other forecasts, namely the loss of cropland to highways, airfields, cities and industries. If the trend of recent decades continues, this loss, in Ordway's estimate, will amount to 48 million acres by 1975 and therefore, merely to keep the crop acreage the same, 48 million acres of new land must be found; and to this must be added the new land required to meet the demand in 1975. Ordway uses the estimate of the Agricultural Research Administration, 68 million acres and concludes that a total of 116 million acres of additional cropland will be needed.

At the other extreme is the 1975 forecast of the Paley Report: a 40 per cent increase in the demand for farm products is foreseen, most of which can be met by wider application of technological improvement, which is counted on to raise the average yield per acre by 33 per cent; hence, the report considers it unlikely that any large amount of additional cropland will be required. However, some 15 million acres of new land will be needed to replace farmland lost to urban and industrial expansion and most of this replacement is expected to come from forest land on farms.

Not everyone feels confident that the average yield per acre will increase by 33 per cent in the near future. In fact, some people will be happy if the present yields can be maintained. In the words of Ordway, there is no acceptable evidence, only hope, that we shall not need additional productive acres in 1975.

The needed farm land, whatever the exact amount may be and the increasing space required for other types of land use add up to a total acreage that will not be small. Much of this land, perhaps most of it, will be taken from the forest and a substantial reduction in the forest area is therefore to be expected.

Richard E. McArdle, Chief of the U.S. Forest Service, summarizes his expectations for the future by saying that we shall probably have enough timber to meet our needs, but not our wants; and that we may encounter some shortage of timberland if high-quality forest land continues to go out of timber production for crop farming, highways, reservoir developments, expansion of urban areas and other uses.

To sum up: the ancient human pressure on the forests of the world continues at the present time and in all probability will become more severe in the future. There are regional differences in the reasons for the pressure, but all have the same effect: the forests are losing ground and it is generally the best and most accessible timberland that is being lost. Hence, the now inaccessible forests may become increasingly important and the question is how much we can count on them in the foreseeable future.

The Inaccessible Forests

The term "inaccessible" as used in the inventories has meaning only in the economy of modern industrial culture: the term implies the economic impossibility of exploiting certain forests for industrial purposes because of difficult terrain, great distance to markets, or lack of means of transportation. But these forests are not physically inaccessible to man. Aboriginal peoples-hunters, food gatherers, fisher folk, cultivators-have lived in and affected them for a very long time. Travelers crossing the Congo basin by airplane in recent years have been astonished by the numerous openings in the forest, large clearings under cultivation or abandoned after having been farmed; and similar openings, though probably not so many, are reported from the Amazon forest.

Approximately 50 or 60 million acres of forest land in the Yukon drainage area of interior Alaska, most of it inaccessible except for local use, has been swept by so many devastating fires that in the opinion of Governor Frank Heintzleman, 4 formerly regional forester for Alaska, not more than 20 per cent of this forest remains intact. These fires apparently became more severe after the white settlers came, but they were not unknown before that time. On his exploration of the interior Labrador peninsula in 1862, Henry Youle Hind 5 found that fire had destroyed the vegetation and converted huge areas into what he called desert and that stories of extensive conflagrations were traditional among the Indians; the "dark days" of Canada in October 1785 and July 1814, Hind suggested, were most likely caused by smoke from enormous forest fires in the northland.

Large sections of the forests in Finland and northern interior Sweden were burned over in the early days by people practicing shifting agriculture -- svedjebruk -- a system that continued well into the nineteenth century; and from Raphael Zon's account of forestry in the U.S.S.R. we learn that forest fires, set by natives, are still common in remote parts of Russia. It would be most unrealistic to think of the world's inaccessible timberlands as a reserve consisting of 5 billion acres of unbroken virgin forest ready to be opened whenever we decide to do so.

No universally adopted definition of what constitutes forest land exists and in all fairness it must be said that the FAO faced a well-nigh impossible task in deciding what to include in the term. Whatever the reason may have been, the category of inaccessible forest was made so broad that it came to include a large amount of land that cannot possibly be regarded as potentially productive timberland. An example is found in North America. The 770 million acres of accessible forest shown in the table is probably just about right. But the inaccessible, truly potentially useful timberland in the United States today scarcely amounts to 50 million acres, if that much; in the interior of Alaska, 50 million; and in Canada, about 200 million; a total of 300 million acres.

Hence, in order to account for the 850 million acres of inaccessible forest shown in the FAO inventory, it would be necessary to include some 150 million acres in the United States that are covered with chaparral, piñon, juniper, mesquite and sparse forests at high elevations and in addition about 400 million acres of taiga land in northern Canada. None of this is regarded by American foresters as potentially productive timberland. Thus, more than half of the 850 million acres is not a true forest reserve, whether inaccessible or not. Likewise, more than half of the 785 million acres of inaccessible forest ascribed to the U.S.S.R. is probably nothing but small-stick taiga forest.

Africa provides another example. The FAO inventory shows a total of more than 1.9 billion acres of forest land on this continent, of which 695 million acres are classed as accessible. In striking contrast are the figures of André Aubréville, who says that the total area of the rain forests in the Guinea coast region, Congo basin, Abyssinia and East Africa is about 570 million acres and that only some 60 per cent, or 340 million acres, are covered with actual forest, the remainder being given over to shifting agriculture. The other woodland in Africa, described as "dry forest" by Aubréville, consists of savanna with scattered bush and trees, yielding wood only for local needs and offering virtually no opportunity for industrial forestry. These lands are annually burned over and the forests that remain, in Aubréville's opinion, seem to be destined for degradation and complete conversion into savanna. It is obviously this type of land that constitutes the billion and a quarter acres of inaccessible "forest" in the FAO inventory, a most dubious reserve of timberland. Even the 695 million acres classed as accessible forest become questionable if weighed against the data of Aubréville. One cannot help suspecting that a large part of the 2? billion acres of inaccessible forest assigned to Latin America, Asia and the Pacific Area may in fact be no better than the alleged reserve in Africa.

Another misconception comes from expressing the forests in areal units, a method that is particularly misleading when applied to the forests in the far north. A better expression of the forest resource is the annual growth of wood per unit area, but unfortunately this measure could not very well have been employed in the inventory, for observed annual growth rates are not available from all forest regions.

But enough data are at hand for making a rough comparison between the forests in the far north and those in the midlatitudes of Europe and North America. Unasylva reports annual growth in cubic feet per acre as follows: Denmark, up to average in Europe, average in the U.S.S.R., average in the United States, average in Canada.

For Scandinavia, Thorsten Streyffert indicates: southern Sweden and Denmark, central Sweden, northern Sweden, southern Norway, central Norway, northern Norway. The high yield in intensively managed forests, such as those of Denmark or Germany, cannot be taken as representative of the natural growth rate in the midlatitudes, but it seems safe to say that the annual natural increment in the forests of western Europe and the United States, certainly in the southern and Pacific coast states, runs from 40 to 60 cubic feet per acre.

The conspicuous fact is the decrease northward: per unit area, the forests in the midlatitudes produce 4 or 5 times as much wood annually as the forests in northern Scandinavia or northern Canada and the difference would be much greater if a comparison with the true taiga forest could be made. If there actually are, as the FAO inventory says, about 1,500 million acres of inaccessible forest land in the far north, it must be remembered that trees do not grow on all of those acres. Anyone who has made his way through the northern forests knows that the obstacle to travel is not the density of the forest but the seemingly innumerable lakes, swamps, wet bogs and other morasses bearing no timber; and, furthermore, many scattered areas of permafrost extending well to the south of latitude 55° in the Canadian Shield and Siberia preclude the growth of real forest trees. Besides, the trees that do exist in the taiga are small and slow of growth. It may well be that midlatitude forest land annually produces 10 or 20 times as much wood per acre as the taiga forest and perhaps the difference is even greater.

The upshot is that the 1,500 million acres of inaccessible forest in the far north is not equivalent in productive capacity to more than some 200 or possibly 300 million acres of forest land in the midlatitudes.

Not many records of annual growth are available from the low-latitude regions, but there would seem to be no reason for any significant difference between the inaccessible and accessible tropical forests, for, by and large, both types occur in lands of high temperature and long growing season.

Both the total stand of all timber and the rate of annual growth are high, but not markedly higher than in the midlatitudes and, in fact, perhaps not so high as in the best of the extratropical forests.

Furthermore, the latter are composed of relatively few species, nearly all of which have a market, whereas the tropical forest consists of very many species, only a few of which have commercial value at the present time and trees of such species are scattered far and wide in the forest.

According to André Aubréville, the volume of the total stand of all trees in the Guinea Forest of Africa may be as high as 300 cubic meters per hectare, of which 100 cubic meters might be considered as commercial timber, but the stands that can be utilized at the present time range only from 10 to 30 cubic meters per hectare. In comparison, the volume of standing timber in the Douglas-fir region of the Pacific Northwest is conservatively estimated to be about 38,000 board feet per acre-some estimates run as high as 50,000 board feet per acre-a value that must equal at least 400 cubic meters per hectare, practically all of which is merchantable timber. The tropical forest potential is high, but it is well not to exaggerate it.

It is not possible at the present time to gauge exactly the reserve in the inaccessible forests, whether expressed in acres or in annual growth per acre, for most of these forests have not been surveyed. Unasylva warns that the "bald summary" of the inventory exaggerates the world's potential forest resources, for some of the forests have primarily a protective function and for that reason should not be counted on as a future source of timber. After leaving the protection forests out of the reckoning, Unasylva concludes that the world still has a reserve of some 3.4 billion acres (1,400 million hectares) of potentially productive forest, or considerably more than the total forest area exploited today. I suspect that the reserve is greatly overestimated.

As part of the task in evaluating the outlook, it has seemed necessary to review the reports on the inaccessible forests, the alleged large reserves of which appear to have given some of the lookouts a comfortable, but I think false, sense of security. It does not really matter how large the reserves are if they are not accessible. It is reasonable to expect that a certain amount of new timberland will become available in the near future, but nothing in the evidence we now have suggests that we shall be able to open up a really large amount, say, a billion acres or more, in the next generation or two.

The important point is that for all practical purposes the forestry problem of the world is the problem of making the best use of the forest land within reach. We are deluding ourselves if we expect the solution to come from the inaccessible forests. The chance of achieving the best use of the forest land depends on trends in silviculture, wood technology, management and other technical knowledge and on our ability to make effective use of such knowledge.

The Half-Promise of Technology

Like a log, the wood industry has two ends, one in the forest where the trees grow, the other at the mill where the timber is cut up and made into different products. It has often been said that roughly one quarter of the wood is wasted in the forests and another quarter at the mill, so that only one half reaches the market. The saying is misleading, for the slash, "left to rot in the forest," as one report puts it with misplaced indignation, is not really wasted but decays and helps to maintain the fertility of the soil; and much of the mill waste is utilized as fuel, providing energy that drives the mill. However, much wood is still wasted by some operators and one of the main problems of silviculture in the forest, technology at the mill and management at both ends is the elimination of needless waste.

Resulting from forest-products research that has been dealt with in a preceding chapter -- have eliminated practically all waste at some of the mills. At the modern integrated plants, some of which can be seen, for example, in the Pacific Northwest, the visitor is impressed by the smooth efficiency of new machinery and methods of operation that convert the log into lumber, plywood, pulp, paper and a long list of by-products made from material that formerly was wasted. Everything in the log is utilized and almost everything can be made from it. Technology is no doubt one of the brightest promises for the future, but we need to be clear about the distinction between promise and fulfillment.

Not all the plants are integrated: Again the Pacific coast can serve as example. The Forest Products Directory of Western North America lists about 3,400 sawmills, most of which are situated in the Pacific coast states. In 1952, according to The Lumberman, approximately one third of the lumber produced came from 100 large plants and the rest from small or medium-sized sawmills.

Complete use of the log is made only at the large mills, which are integrated with pulp, paper and other wood-products factories. At more than 3,000 sawmills in the West the most conspicuous structure in the landscape is the big, black waste burner, in which a never-dying fire consumes bark, sawdust, slabs, trimmings-in short everything that is not merchantable lumber.

It might be thought that this appalling waste is needless and could easily be eliminated, but the meaning of the term "needless" depends on the definition of need and it certainly would not be easy to change the system. The small mills, which account for two thirds of the lumber production, are needed in order to meet the market demand and as long as they are needed and continue to operate they must dispose of the mill waste in some manner and the easiest way is to burn it. An integrated plant capable of using everything in the log requires more capital than the small operators have and the waste can be sold only if there is a market for it. Pulp mills, utilizing chips and the new hardboard industry in the West have begun to provide such a market, but we have no assurance that it will ever be able to absorb all the residual from some 3,000 sawmills.

There is hope that a substantial part of the waste can be eliminated, but not much hope that all of it will be. The reason for the big incinerators lies at least partly in ingrained custom: mill waste has always been burned, so one continues to burn it; but the main reason is economic and not lack of technical knowledge, or wasteful habits in the operators -- nobody throws away material if it can be turned into money. This waste might be completely eliminated if the entire wood industry were organized into large combines financially able to build a few hundred integrated plants, but such a reorganization would require profound economic and cultural changes.

Deeply rooted in American tradition is the belief that there is and ought to be a place for the little man engaged in small, independent enterprise: farmer on the land, lumberman in the forest and mill operator in the wood industry. Technology runs into the dilemma of small ownership, a dilemma that cannot be resolved by technicians, for it is not a technical problem. Perhaps the nation will have to choose some day between preserving the old tradition or going all out for technology. Meanwhile technology is a half-promise, because it can effectively reach only half or less than half of the industry.

The situation in western North America is not representative of the world; there is less waste in some regions, but more in others. In a report to the British Wood Preserving Association in 1952, T. J. Price says that estimates of forest officers in the Gold Coast indicate that only about one fortieth of the timber felled in the colony is shipped as exportable logs or lumber. This is perhaps an extreme case, but it is probably true that on the average much less than half of the timber cut for industrial purposes in the tropical forests ever reaches the market.

One of the reasons for the great loss of tropical wood is the attack by fungi and insects on felled timber, a problem that can be solved only by the wood-preserving industry. The significance of wood preservatives deserves wider general recognition, for the future success of the wood industry, particularly in tropical regions, will in large measure rest on the development and widespread use of chemical solutions that will not only save the timber but prolong the useful life of forestry products.

Railroad ties, pilings, poles, fence posts and mine timber treated with creosote last more than twice as long as formerly when they were untreated and in some instances have lasted more than five times as long. Mechanical wear, not decay, is now the main reason for the replacement of railroad ties. Virtually colourless and odorless chemical preservatives that act as fungicides, insect repellants and fire retardants and help to prevent warping and splitting have appeared on the market of late and are coming into wider use as applications on finished wood, such as window sash, doors, partitions, flooring and the like. In recent years some 300 million cubic feet of wood have been treated with preservatives annually in the United States, or about 5 per cent of the total amount of wood used, not counting pulpwood and firewood.

There is no reason to think that all wood will ever be so treated, but the use of preservatives will doubtless increase and therefore, because wood products will last longer, a commensurate easing of the demand on the forest might be expected.

Preservatives are even more important in the tropical regions, where they might well be called indispensable. T. J. Price reports that tropical woods rarely arrive in Britain free from fungus stains or wormholes and for that reason several desirable species have already lost their market; preservation in the tropics, he suggests, must begin in the forest the moment the tree is felled, for it is the log that must be preserved. J. Benoit, Chief of the Wood Preservation Service at the Centre Technique Forestier Tropical in France, makes the same suggestion. Reporting on African field experiments before the British Wood Preserving Convention in 1953, he describes how insects (including ambrosia or pinhole borer beetle, the most troublesome of all) swarm over a tree as soon as it is felled and he states that in most cases the only effective way of preserving the log is to spray it directly after felling.

It has been found that ordinary preservatives and insecticides, including DDT, are ineffective against ambrosia; in fact, under certain conditions this beetle is attracted by creosote derivatives. However, mixtures have been achieved that promise good results (benzene hexachloride, known in Britain as BHC, in France as HCH); and on the whole, Benoit concludes, the development of preservatives for use in the tropics is showing a substantial advance. Judging from other reports, Benoit's conclusion seems to be a fair statement of the present trend, although many technical problems besides that of ambrosia remain to be solved.

The outlook for preservatives in the tropics can be evaluated only by experts in this field, but even a layman can see that utilization of tropical woods is bound to be highly wasteful unless logging and wood preserving go hand in hand in the forest. Whether they will successfully join hands is another question and one that cannot be answered by technology. Discussions at the British Wood Preserving Convention revealed that tropical woods allegedly treated with preservatives before they are exported arrive in England full of fungus and wormholes. The question is not only what they may have been treated with, but whether the treatment may not have been more fictitious than real. Technologists may provide preservatives, but someone else must apply them.

A moot question, which needs more attention than seems to have been given to it, is whether technological improvements in the long run will have the effect of decreasing or increasing the demand on the forest. According to the Paley Report, technical developments in the United States have had two opposite effects on materials: they have increased our efficiency of use, but have also greatly increased the drain upon the resources.

The Stanford Report, which is primarily an analysis of the demand, does not go into the question of whether the forest can provide the supply; but the Weyerhaeuser Timber Company, at whose request the study was made, introduces the summary of the report with some conclusions of its own, one of which is that technical developments will make it possible to manufacture a larger volume of products from the same amount of wood, so that timber supplies will stretch farther than in the past. It is true that many new wood products have appeared on the market and that the consumption of them does not increase the demand on the forest, provided they are made from parts of the log that were formerly wasted. Some of them are, but others are not.

Hardboard, for example, which is made from millwaste, can replace lumber for certain purposes and if it were only a question of hardboard and lumber the timber supply would indeed stretch farther. But plywood and veneer, which cannot be made from waste, compete with lumber for saw logs and increase the demand for timber. Some of the plywood companies have found it necessary to purchase timberland in order to assure their supply of peeler logs. And the demand for plywood and veneer in 1975, according to the Stanford Report, will be at least ten times as great as the demand for hardboard. It is perfectly clear that technological improvement has resulted in a much more efficient use of the raw material and the production of many new articles, but it is also true that we are consuming more different wood products than ever and in larger amounts.

It may be that technology has, so to say, whetted the appetite of the consumer for new products and stimulated the desire for rapidly replacing them as soon as new models appear on the market. The Stanford Report expects the lumber requirements in the near future to increase more slowly than does the population; but the anticipated demand for veneer, plywood and pulp and all its derivatives, typical products of the "new technology," will increase much faster than does the population, in some instances more than twice as fast.

A vast undertaking of research in forest products and wood use is being carried on at the present time: at Dehra Dun in India, Nogent-sur-Marne in France, Madison in Wisconsin, Princes Risborough in England and at hundreds of other institutions throughout the world. In recent years, according to the American Forestry Association (1950), 1,600 organizations in the United States alone have been engaged in more than 12,000 projects of wood research. It would seem that more of this energy, money and intellect might well be concentrated on the question of how technology will affect the supply in the forest. It can scarcely be said today that we really know what the answer is.

Management and Forest Genetics

Forestry is in one sense a very ancient practice. It is related in Genesis that Abraham planted a grove at Beersheba; according to Plato, the Greeks of the fifth century B.C. raised forest trees; Cato says that Roman farmers began at an early time to cultivate plantations of trees to provide timber for local needs; and Pliny describes at length the system of coppicing woods in regular rotation to obtain firewood, vineyard stakes and the like.

But forestry and silviculture in the modern sense are not very old, dating from the work of Heinrich Cotta and G. L. Hartig in Germany in the 1780's and from that of Bernard Lorentz and Adolphe Parade in France shortly thereafter. The age of modern silviculture, it might be said, is about the same as three or four generations of forest trees and in terms of certain species as only two or three generations. A plant breeder working with annuals would hardly feel confident about the outcome of his experiment after only two or three generations of the plant he is studying.

He can observe many successive generations and is able to correct a mistake after only a few years have elapsed, but the forester, as R. S. Troup puts it, rarely lives long enough to see the result of his mistake. It would be captious not to recognize the great contribution of silviculture and the remarkable improvements in forestry that have been achieved in recent years, but we must keep in view the fact that modern silviculture is still in its early stages; the outcome of the great experiment is not yet certain. New systems are still being tried out: Mark L. Anderson of England has recently described the result of "spaced-group planting"; the method of "area selection" or "patch-cutting," now prevalent in the Douglas-fir forest of the Pacific Northwest, is scarcely twenty-five years old in that region; and in the mixed coniferous forest of the Sierra Nevada in California, after analysis of forty years of research and experience, a new system known as "unit-area control" is to be initiated.

There are two major types of silvicultural systems: coppice methods, probably the oldest, depending on vegetative or sprout reproduction; and what are known in Europe as high-forest systems, which depend on seed reproduction. The use of coppice is declining, except perhaps for firewood production in some regions and industrial forestry in the future will no doubt depend increasingly on seed reproduction. About a dozen high-forest systems are in use, most of them developed in western Europe during the nineteenth century and they can also be divided into two major groups: clear-cutting types and selection-cutting types.

A number of variant forms have come into existence, such as "irregular shelterwood," "strip system," "coulisse method," "wedge-cutting," "two-storied high forest," "high forest with standard," and several others. Some of them are essentially clear-cutting in different patterns, others are fundamentally selection-cutting. On the whole, regeneration after clear-cutting produces even-aged stands, often dominated by a single species, while a forest under selection-cutting remains mixed both as to age classes and species. It is not necessary to discuss these systems in detail; it will suffice to say that each has advantages and disadvantages and that none can be called the best system for all regions. Clear-cutting has sometimes been uncritically condemned, but under certain circumstances it is undoubtedly the most suitable method and, in fact, the only system that can be used in some regions. It is not so much the method as the manner of employing it that may be dubious.

Another grouping can be made according to two different trends in recent silviculture, or two different attitudes toward the forest and these attitudes will surely be of more critical importance in the future than any particular system of forestry. The attitudes are not always well expressed, but in their clearest form, one of them can be described as faith in artificial forest, the other as faith in natural forest.

On the one hand, there are those who believe that the forest should be treated as a plantation producing a maximum amount of the type of wood that is most profitable to man, the pace of production being forced by human action; on the other hand, those who think that the best result in the long run will be obtained if natural conditions in the forest are disturbed as little as possible and the forest, while encouraged by the co-operating hand of the silviculturist, sets the pace of timber production.

The difference some what resembles the contrast in agriculture between those who believe in "organic farming" and those who have what Sir Albert Howard calls the "N K P mentality" and place their faith in forced feeding with commercially produced chemical fertilizers. These attitudes toward the forest are found in every country, but the history of German forestry provides the clearest examples and the longest record.

Coppice, clear-cutting and selection-cutting have been practiced ever since the Middle Ages in the forests of Germany and western Europe generally, locally more or less regulated at certain times but more commonly not. Unregulated selective cutting in high forest was widespread, but it was the wrong kind of selection: the best trees were cut out, leaving the poorer specimens to regenerate the forest.

Some of the forests were excessively grazed by livestock; the custom of gathering fallen material for fuel and the raking up of leaves for bedding the livestock in winter denuded many forests of their litter; and yet other forests, because of the passion for hunting among the nobility, became overstocked with game, causing great damage to the forest seedlings. The result was a general deterioration of the forest, which by the end of the eighteenth century had become so serious that strong remedial measures were initiated. Heinrich Cotta, 6 Director of the Forest Academy at Tharandt, suggested that the only way of rehabilitating the forest was by clear-cutting and making a fresh start, helping the regeneration by the planting of trees, if necessary. This method became common in the forests of Saxony and was adopted in many other regions as well. It should be noted that the proposal was remedial in purpose, not intended as a permanent system and Cotta himself warned against the danger of monoculture. But that is what the method turned out to be.

It became standard practice in the nineteenth century and early part of the twentieth to clear-cut, plant and harvest on short rotations and the upshot was that many forests became dominated by even-aged, nearly pure stands of spruce or pine, while fir and beech, oak and other deciduous trees well-nigh disappeared. The dominant motive was profit: spruce or pine, worked on short rotations for lumber, pulp, mine timber and the like, gave-for a while-the highest and quickest economic returns. This was the earliest and one of the clearest expressions of the first of the two attitudes, faith in artificial forest. As Franz Heske says, wood changed from a carefully rationed essential material to an ordinary commodity, the production of which was governed primarily by financial considerations.

But signs that all was not well began to appear: stunted growth of trees, yellowish needles, increasing growth of lichens and peat, decrease in grass, formation of hardpan in some soils; and it also became apparent that the trees suffered increasing damage from wind, frost, fungus and insects. A series of thorough investigations of forest conditions in Saxony, undertaken by E. Wiedemann in the 1920's, revealed marked soil deterioration and a loss in the average annual growth rate, which according to estimates quoted by R. S. Troup amounted to more than 9 cubic feet per acre over a 100-year period, or a total loss of 3 1/2 billion cubic feet over the whole of the state forests in Saxony. Unsatisfactory results in managed forests have also been observed in other regions.

Reports from the large planted pine forest of the Landes in southwestern France indicate that the heavy losses of recent years have resulted not only from catastrophic fires, windfalls and attack by beetles, but also from the more fundamental cause of soil deterioration. After about 100 years of intensive management, described by G. Roux as "absolute monoculture," the trees are less healthy, the annual increment has dropped, the yield of turpentine has decreased and natural regeneration is almost impossible to obtain; seed cones have long been imported. J. L. Arend estimates that the area underlain with hardpan in the region has increased by approximately one half. One of the proposed remedies is the introduction of broadleaf species to mix with the pines.

Anders Holmgren, after fifty years of work in the forests of Swedish Norrland, during which time he has given much of his attention to the problems of regeneration and methods of cutting, recommends greater reliance on natural forest and natural reproduction, advises against clearcutting with the coulisse method in the mountains and the far north and suggests that, if the method must be used, the areas cut should not be larger than 25 to 35 acres.

The experience of Saxony does not prove, as many foresters have pointed out, that clear-cutting alone was responsible for the deterioration, nor that pure stands are necessarily bad: some forests naturally consist of nearly pure stands and adverse results have also been observed in mixed forests managed with different systems. The reasons for the unsatisfactory outcome were many and complex, but certainly the fundamental cause was the forced pace dictated by financial considerations. It is not likely that any forest soil, whether supporting pure or mixed stands, can long stand up under intensive short-rotation cropping.

Not all foresters approved of the system. Strong protests were made, particularly by Karl Gayer, who advocated the reestablishment of forestry operations based upon and not contrary to, natural laws.

Many followed Gayer's lead and a reaction set in against clear-cutting and monoculture, reaching an opposite extreme in the idea of Dauerwald, or forest with continuous cover and selective-cutting in such a manner that "the forest hardly notices it" Clear-cutting and artificial regeneration are still practiced in many regions, but the present trend in European silviculture is in another direction. E. W. Jones summarizes his survey of present-day ideas among European foresters by saying that opinions preponderantly favour mixture of species, natural regeneration, constant preservation of the forest canopy and flexible cutting methods and stress particularly the importance of giving careful attention to soil and site conditions in reaching decisions on silvicultural policy for a given region.

That forestry is essentially an economic operation is not overlooked; on the contrary, the belief is that these methods of cooperating with the bent of nature will in the long run give the highest returns and at the same time preserve the health and beauty of the forest. It is interesting to note, as an indication of the trend, that Joseph Köstler concludes a recent major work by reminding us that "forestry is also landscape care," and leaves the reader with a quotation from Goethe: "We must eavesdrop on Nature to learn her ways, lest we coerce her into obstinacy by procedures against her will" Here is the clearest expression of the second attitude, faith in natural forest.

The trend of European silviculture may be toward natural forest, but whether the industry as a whole will follow suit is another question. There is no doubt about the persistence of strong faith in artificial forest and forced pace in countries outside of Europe. The new fast-growing pine forests in Chile, New Zealand, Australia, Kenya and South Africa are essentially monoculture plantations, with some of the rotations as short as thirty years. The technique developed in South Africa, as described by W. E. Hiley, consists of early and heavy thinning according to mathematical rules and "has the object of producing the greatest value of timber in as short a time as possible and the prescriptions are based on strictly economic principles" No better definition of nineteenth-century forestry in Saxony could be found, but whether the result will be the same only future generations will know. Hiley concludes that it is too early to pronounce judgment on the principles of this "daring experiment"

The large industrial forests in the southeastern part of the United States, where pines have always been abundant, are becoming dominated by softwoods at the expense of hardwoods because of the emphasis on planting pine for industrial purposes. Southern forestry, in fact, seems to be destined for monoculture.

The prevailing attitude is indicated in speeches made at the meeting of the Southern Pulpwood Conservation Association in February 1954. As reported in American Forests, one of the speakers said: "Put every acre to work -- that's good management, make every acre work harder -- that's applied research, make every acre grow better trees faster-that's basic research" The speakers also questioned the wisdom of growing high-grade saw timber on 80-year rotations in the national forests of the South, where there are so many pulpmills. The planting of trees is certainly a commendable practice and no one can object to improvements in management and yield, but it is obvious that here we are again on the old Saxon road. The intention is to make the acres do what we bid. It will also be observed that the statement of the speaker reflects the widespread confusion over what constitutes basic research. Not only is the task of basic research assigned, but the expected result is laid down.

One can imagine the feeling of a research scientist: produce better trees, or else! In the rugged terrain of the dense Douglas fir forest in the Pacific Northwest the method now almost universally employed and approved by virtually every forester familiar with the region, is clear-cutting of staggered blocks, sometimes called "area selection" There is no standard size of the blocks to be logged, for conditions vary from place to place, but in conversation with foresters having lifetime experience with Douglas fir one hears the opinion that, by and large, the blocks ought not to exceed 30 or 40 acres. It may be recalled that Holmgren made the same suggestion for northern Sweden. But the economic advantage of cutting larger blocks is great and clear-cut areas of 300 acres are common. It can hardly be said that the long-run effect on the forest of such large blocks is known, but many foresters suspect that it will not be good. The size of the blocks is indicative of an attitude that is determined more by financial interest than by silviculture. We are clearly moving in a direction opposite to that suggested by Goethe. We intend to coerce nature, hoping that she will not respond with obstinacy, but there is plenty of evidence that she will and none that she will not. The claim that good management can cause-or force -- the American forest soil to yield twice as much wood as now on a sustained basis has almost become sacrosanct doctrine. Some speak of increasing the yield even more. We know that forestry has improved remarkably in recent years, we know that good management can raise the annual growth, but we do not know, we only hope, that the total annual production can be doubled and permanently maintained at that level.

The science of genetics has come to the fore during the last half century and may conceivably have a tremendous effect on the forests of the future. The renewed interest in natural forest during Karl Gayer's time was accompanied by a reawakening curiosity about the nature of the forest, the ironic result of which may be the end of both the concept and the existence of natural forest. Like silviculture, modern forest genetics has its precursors.

Duffield points out that as early as the 1780's a German forester by the name of Wangenheim was aware of hereditary strains in forest trees and recommended the planting of seeds from superior individuals and that Patrick Matthew, a British forester, in 1831 proposed a theory of natural selection derived from his experience with raising trees. In more recent years it has been discovered that there are local races within the species of trees and that a number of traits are hereditary, such as form of trunk, type of branching, rate of growth, vigour of seeding. Applied forest genetics is based on these discoveries. It is no longer a question merely of planting seeds but of planting seeds from the best trees.

Several thousand "elite" trees have been registered in the forests of Scandinavia and Finland, seeds are regularly collected from these protected individuals -- with the aid of specially constructed steel ladders erected from movable platforms, making it possible to harvest the seeds without harming the tree -- and careful records are kept of the progeny. C. Syrach-Larsen began to establish seed tree orchards in Denmark during the 1930's for the purpose of raising scions of superior trees and to carry out cross-pollination experiments. In Sweden pines from the north, characterized by slow growth but highquality wood, have been crossed with faster growing southern stock in the hope of combining the best qualities of both parents.

Experiments with X-ray treatment to achieve polyploid varieties have produced a "giant aspen" having twice the ordinary number of chromosomes and maturing in 30 years. J. W. Duffield and F. I. Righter report about two dozen new pine varieties produced at the experiment station near Placerville in California, many of which show decided hybrid vigour, grow faster and are more resistant to cold, insects and diseases than either of the parent species. A cross between eastern white pine and Himalayan white pine is watched with particular interest, for this hybrid not only outgrows eastern white pine but has shown itself resistant to blister rust in a rigorous exposure test that now has been in progress for nearly a decade.

Perhaps we are on the threshold of a period that future historians will refer to as the time when forest trees were domesticated and natural forests passed away. It is to be expected, if the domestication succeeds, that the character of the future forests will depend on present trends in applied forest genetics. While all geneticists work in the same general direction with controlled breeding experiments, there seem to be different opinions about the advantage of exotics over native species, some placing most of the faith in elite indigenous trees, others expecting more from introduced species, or from a combination of exotics and superior native stock.

The principal trend is no doubt in the direction of combination. On the whole, European experience with introduced species has not been particularly happy; and J. S. Boyce, in a thought-provoking article, describes the risks of spreading diseases and insect pests by introducing foreign trees, warns against placing too much faith in them and concludes by saying that "exotics are not all foredoomed to failure, but for every exotic the chance of failure appears to be much greater than the chance for success" The "combination" idea is expressed by Syrach-Larsen in the following statement, although, if anything, he seems to favour exotics:

"Broadly speaking, for improvement of the forest stock intensive silviculture places most reliance on the introduction of exotics and on selection of the best indigenous trees" He also says: "To succeed to any appreciable extent we must soon approach the stage when it becomes obvious how much more profitable it will be to create rapid-growing pure plantations on favorably situated areas"

This is artificial forest with a vengeance. The image conjured up resembles a field of hybrid corn: super-trees, single cropping and fast rotations. Perhaps the forest operator of the future, like the farmer now, will have to spend a large part of the income for fertilizers. In line with this trend a representative of the fertilizer industry attended the Fourth American Forest Congress in 1953 and suggested a programme of spreading fertilizers from airplanes on forest soils, thereby "permitting the cutting of trees perhaps ten years sooner"

The problems in nature facing the forester are many and complex: fire, windfalls, frost damage, insects and fungi, but surely the most critical long-run problem in any silvicultural system, whether of the traditional type or the new, is the maintenance of a fertile and healthy forest soil; for if the soil fails, everything fails. Fortunately the foresters, at least in some regions, are well aware of the problem. C. W. Scott stresses the importance of watching the soil in the new Chilean plantations of Monterey pine and André Aubréville has given the same advice to the managers of eucalyptus groves in Brazil. Most of the larger companies in the forest industries of the Pacific Northwest have had soils experts on their staffs for several years. Watching the soil is one thing, but providing a remedy if it should turn out to be deteriorating is another. It would hardly be economically feasible, even if it were physically possible, to raise forest trees with the aid of fertilizers, as we raise many of our crops. My own belief is that the best promise for the future lies in another direction, namely natural forest, but we do not seem to be going that way.

In a purely technical sense, silviculture in the forest is perhaps as bright a promise as technology at the mill, but both run into the dilemma of the small-ownership and economic problems not of their making. Excellent results have been obtained by application of the Dauerwald idea in some of the European forests, but Dauerwald, which requires, so to say, Dauerpflege or continuous and careful attention to details and much hand labour, is, for economic reasons, impossible in most of the world and certainly impossible on both small and large forest holdings in North America.

Every forest operator knows that he cannot always follow even the best advice of the silviculturist, for he would go broke if he tried. Clear-cut patches in the Pacific Northwest on which Douglas-fir seedlings should be coming up are in many regions taken over by alder or choked with brushy vegetation; some of the young stands need thinning; and many trees should be pruned for best results.

Only the larger companies are able to perform improvement operations of this type, for money spent on pruning, thinning, or fighting brush might not be returned until 75 or 100 years have elapsed and under present circumstances most of the operators are unable to finance such long-term investments.

The result is that much of the forest is growing up unattended and the land is not producing so much good timber as it might. According to the Annual Report of the Forest Experiment Station at Portland in 1952, the use of pulp chips obtained from mill waste in Oregon and Washington has meant an annual salvaging of material in recent years equivalent to several hundred million board feet of logs, a very good saving, but at the same time the practice has had a detrimental effect on the market for pulpwood obtained by thinning, making good forest management more difficult. Thus, what is good economy at the mill is not always good silviculture in the forest.

WOOD AND OTHER PRODUCTS

Tropical forests provide many valuable products including rubber, fruits and nuts, meat, rattan, medicinal herbs, floral greenery, lumber, firewood, and charcoal. Such forests are used by local people for subsistence hunting and fishing. They provide income and jobs for hundreds of millions of people in small, medium, and large industries. Tropical forests are noted for their beautiful woods. Four important commercial woods are mahogany, teak, melina, and okoume. Honduras mahogany (Swietenia macrophylla), grows in the Americas from Mexico to Bolivia. A strong wood of medium density, mahogany is easy to work, is long lasting, and has good colour and grain. It is commonly used for furniture, molding, paneling, and trim. Because of its resistance to decay, it is a popular wood used in boats.

Teak (Tectona grandis) is native to India and Southeast Asia. Its wood has medium density, is strong, polishes well, and has a warm yellow-brown colour. Also prized for resistance to insects and rot, teak is commonly used in cabinets, trim, flooring, furniture, and boats figure. Melina (Gmelina arborea) grows naturally from India through Vietnam. Noted for fast growth, melina has light colored wood that is used mainly for pulp and particleboard, matches and carpentry. Okoume (Aucoumea klaineana) is native to Gabon an the Congo in west Africa. A large fast-growing tree, the wood has mod erately low density, good strength-to density ratio, and low shrinkage during drying. It is commonly use(for plywood, paneling, interior fur niture parts, and light construction.

Other Economic Values

Tropical forests are home for tribal hunter-gatherers whose way of life has been relatively unchanged for centuries. These people depend on the forests for their livelihood. More than 2.5 million people also live in areas adjacent to tropical forests. They rely on the forests for their water, fuelwood, and other resources and on its shrinking land base for their shifting agriculture. For urban dwellers, tropical forests provide water for domestic use and hydroelectric power. Their scenic beauty, educational value, and opportunities for outdoor recreation support tourist industries. Many medicines and drugs come from plants found only in tropical rain forests. Some of the best known are quinine, an ancient drug used for malaria; curare, an anesthetic and muscle relaxant used in surgery; and rosy periwinkle, a treatment for Hodgkin's disease and leukemia. Research has identified other potential drugs that may have value as contraceptives or in treating a multitude of maladies such as arthritis, hepatitis, insect bites, fever, coughs, and colds. Many more may be found. In all, only a few thousand species have been evaluated for their medicinal value. In addition, many plants of tropical forests find uses in homes and gardens: ferns and palms, the hardy split-leaf philodendron, marantas, bromeliads, and orchids, to name just a few.

Environmental Benefits

Tropical forests do more than respond to local climatic conditions; they actually influence the climate. Through transpiration, the enormous number of plants found in rain forests return huge amounts of water to the atmosphere, increasing humidity and rainfall, and cooling the air for miles around. In addition, tropical forests replenish the air by utilizing carbon dioxide and giving off oxygen. By fixing carbon they help maintain the atmospheric carbon dioxide levels low and counteract the global "greenhouse" effect.

Forests also moderate stream flow. Trees slow the onslaught of tropical downpours, use and store vast quantities of water, and help hold the soil in place. When trees are cleared, rainfall runs off more quickly, contributing to floods and erosion.

DEFORESTATION

Before the dawn of agriculture approximately 10,000 years ago, forests and open woodland covered about 15.3 billion acres (6.2 billion ha) of the globe. Over the centuries, however, about one-third of these natural forests has been destroyed. According to a 1982 study by FAO, about 27.9 million acres (11.3 million ha) of tropical forests are cut each year-an area about the size of the States of Ohio or Virginia. Between 1985 and 1990, an estimated 210 million acres (85 million ha) of tropical forests were cut or cleared. In India, Malaysia, and the Philippines, the best commercial forests are gone, and cutting is increasing in South America. If deforestation is not stopped soon, the world will lose most of its tropical forests in the next several decades.

Reasons for Deforestation

Several factors are responsible for deforestation in the Tropics: clearing for agriculture, fuelwood cutting, and harvesting of wood products. By far the most important of these is clearing for agriculture. In the Tropics, the age-old practice of shifting, sometimes called "slash-and-burn," agriculture has been used for centuries. In this primitive system, local people cut a small patch of forest to make way for subsistence farming. After a few years, soil fertility declines and people move on, usually to cut another patch of trees and begin another garden. In the abandoned garden plot, the degraded soil at first supports only weeds and shrubby trees. Later, soil fertility and trees return, but that may take decades. As population pressure increases, the fallow (rest) period between cycles of gardening is shortened, agricultural yields decrease, and the forest region is further degraded to small trees, brush, or eroded savanna.

Conversion to sedentary agriculture is an even greater threat to tropical forests. Vast areas that once supported tropical forests are now permanently occupied by subsistence farmers and ranchers and by commercial farmers who produce sugar, cocoa, palm oil, and other products. In many tropical countries there is a critical shortage of firewood. For millions of rural poor, survival depends on finding enough wood to cook the evening meal. Every year more of the forest is destroyed, and the distance from home to the forest increases. Not only do people suffer by having to spend much of their time in the search for wood, but so does the land. Damage is greatest in dry tropical forests where firewood cutting converts forests to savannas and grasslands.

The global demand for tropical hardwoods, an $8-billion-a-year industry, also contributes to forest loss. Tropical forests are usually selectively logged rather than clear-cut. Selective logging leaves the forest cover intact but usually reduces its commercial value because the biggest and best trees are removed. Selective logging also damages remaining trees and soil, increases the likelihood of fire, and degrades the habitat for wildlife species that require large, old trees-the ones usually cut. In addition, logging roads open up the forests to shifting cultivation and permanent settlement. In the past, logging was done primarily by primitive means-trees were cut with axes and logs were moved with animals such as oxen. Today the use of modern machinery—chain saws, tractors, and trucks -makes logging easier, faster, and potentially more destructive.

Endangered Wildlife

Forests are biological communities-complex associations of trees with other plants and animals that have evolved together over millions of years. Because of the worldwide loss of tropical forests, thousands of species of birds and animals are threatened with extinction. The list includes many unique and fascinating animals, among them the orangutan, mountain gorilla, manatee, jaguar, and Puerto Rican parrot. Although diverse and widely separated around the globe, these specles have one important thing in common. They, along with many other endangered species, rely on tropical forests for all or part of their habitat.

Orangutans (Pongo pygmaeus) are totally dependent on small and isolated patches of tropical forests remaining in Borneo and Sumatra, Indonesia. Orangutans spend most of their time in the forest canopy where they feed on leaves, figs and other fruit, bark, nuts, and insects. Large trees of the old-growth forests support woody vines that serve as aerial ladders, enabling the animals to move about, build their nests, and forage for food. When the old forests are cut, orangutans disappear.

The largest of all primates, the gorilla, is one of man's closest relatives in the animal kingdom. Too large and clumsy to move about in the forest canopy, the gorilla lives on the forest floor where it forages for a variety of plant materials. Loss of tropical forests in central and west Africa is a major reason for the decreasing numbers of mountain gorillas (Gorilla gorilla). Some habitat has been secured, but the future of this gentle giant is in grave danger as a result of habitat loss and poaching.

The jaguar (Leo onca), a resident of the Southwestern United States and Central and South America, is closely associated with forests. Its endangered status is the result of hunting and habitat loss. The Puerto Rican parrot (Amazona vittata), a medium-sized, green bird with blue wing feathers, once inhabited the entire island of Puerto Rico and the neighboring islands of Mona and Culebra. Forest destruction is the principal reason for the decline of this species. Hunting also contributed. Today, only a few Puerto Rican parrots remain in the wild and their survival may depend on the success of a captive breeding programme.

In addition to species that reside in tropical forests year round, others depend on such forests for part of the year. Many species of migrant birds journey 1,000 miles or more between their summer breeding grounds in the north and their tropical wintering grounds. These birds are also threatened by tropical forest destruction.

THE PRACTICE OF FORESTRY

Forestry-loosely defined as the systematic management and use of forests and their natural resources for human benefit-has been practiced for centuries. Most often, forestry efforts have been initiated in response to indiscriminate timber cutting that denuded the land and caused erosion, floods, or a shortage of wood products.

Ancient Forestry Practices

In ancient Persia (now Iran), forest protection and nature conservation laws were in effect as early as 1,700 B.C. Two thousand years ago the Chinese practiced what they called "four sides" forestry-trees were planted on house side, village side, road side, and water side. More than 1,000 years ago, Javanese maharajahs brought in teak and began to cultivate it. In the African Tropics, agroforestry (growing of food crops in association with trees) has been practiced for hundreds of years. In the Yucatan Peninsula of southern Mexico, the ancient Mayas cultivated fruit and nut trees along with such staples as corn, beans, and squash. Bark, fibers, and resin were obtained from plants grown in fields, kitchen gardens, and orchards. Early in their civilization, the Mayas practiced slash-and-burn agriculture. As their population grew, they found more efficient methods of growing crops. They terraced hillsides, learned how to decrease the time between "rotations" of agricultural land with native forests, dug drainage channels and canals to move water to and from cultivated areas, and filled in swampland to plant crops.

The agricultural sophistication of the Mayas enabled their civilization to grow and flourish. What brought about their decline about A.D. 820 is not fully known, but some believe that as their society developed, the Mayas made unsustainable demands on their environment. Relatively little is known about tropical forestry before the mid1800's in most places. At that time, the European colonial empiresnotably the Dutch, English, and Spanish-brought modern forest management practices to Indonesia, India, Africa, and the Caribbean. Centers for forestry and forestry research were established, and more careful records were kept.

Sustainable Forestry

Modern forestry has its basis in 18th-century Germany. Like the Chinese and the Mayan forest practices, German forestry is essentially agricultural. Trees are managed as a crop. Two concepts are important: renewability and sustainability. Renewability means that trees can be replanted and seeded and harvested over and over again on the same tract of land in what are known as crop "rotations." Sustainability means that forest harvest can be sustained over the long term. How far into the future were foresters expected to plan? As long as there were vast acres of virgin (original) forests remaining, this question was somewhat academic. Today, however, sustainability is a vital issue in forestry. Most of the world's virgin forests are gone, and people must rely more and more on second- growth or managed forests. Perhaps we now face, as never before, the limits to long-term productivity. In the German forest model, forestry is viewed as a continual process of harvest and regeneration. Harvest of wood products is a goal, but a forester's principal tasks are to assure long-term productivity. That is achieved by cutting the older, mature, and slow-growing timber to make way for a new crop of young, fast-growing trees.

Harvest-Regeneration Methods

Three examples of timber harvest-regeneration methods (silvicultural systems) illustrate how foresters manage stands to produce timber on a sustained basis.

Selection

Individual trees or small groups of trees are harvested as they become mature. Numerous small openings in the forest are created in which saplings or new seedlings can grow. The resulting forest has a continuous forest canopy and trees of all ages. Such systems favor slow-growing species that are shade tolerant.

Clearcutting

In clearcutting, an entire stand of trees is removed in one operation. From the forester's point of view, clearcutting is the easiest way to manage a forest-and the most economical. Regeneration may come from sprouts on stumps, from seedlings that survive the logging operation, or from seeds that germinate after the harvest. If natural regeneration is delayed longer than desired, the area is planted or seeded.

Clearcutting systems are often used to manage fast-growing species that require a lot of light. Resulting stands are even aged because all the trees in an area are cut-and regenerated-at the same time. Clearcutting has become controversial in recent years because it has the potential to damage watersheds and because it tends to eliminate species of wildlife dependent on old growth trees. If clearcuts are kept small and the cutting interval is long enough, however, biological diversity may not be impaired.

Shelterwood

In shelterwood systems, the forest canopy is removed over a period of years, usually in two cuttings. After the first harvest, natural regeneration begins in the understory. By the time the second harvest is made, enough young trees have grown to assure adequate regeneration. Shelterwood systems favor species that are intermediate in tolerance to shade. Such systems are difficult to use successfully and are the least used of the three silvicultural methods described.

Multiple-Use Forestry

Gifford Pinchot, the first Chief of the U.S. Forest Service, was also this country's first professional forester. Pinchot advocated the use of forest resources-all resources, not just timber-for human benefit. Pinchot was a strong and charismatic leader, and his ideas helped shape the course of forestry in the United States. Pinchot had a vocal opponent in John Muir, a young naturalist from California who believed that public lands should be preserved rather than used. Eventually Muir and Pinchot became rivals for public approval. Oddly enough, there was no

loser in this early conservation battle. Muir's preservation ethic became embodied in the philosophy of the National Parks, and Pinchot's concept of wise use became the guiding principle of the National Forests. National Forests are still managed under the concepts of multiple use and sustained yield. The dominant uses of National Forests are considered to be wood, water, wildlife, forage (for domestic cattle and wildlife), and recreation. Extraction of minerals and other valuable products is also considered a legitimate use of National Forests. Because Pinchot's philosophy left room for the "highest and best use" of a given area, the U.S. National Forests now include a wilderness system of more than 32 million acres (13 million ha) in which timber harvest is not allowed. Today it is generally recognized that most, if not all, nondestructive uses of forest are valid. Some areas may be set aside as parks; others for wildlife habitat or as wilderness. Still others will be managed for timber harvest or multiple benefits. Today, conflicts arise primarily over where these different uses will be dominant. In the National Forests, such decisions are made through a land-use planning process in which the public has ample opportunities for input and involvement.

FORESTRY RESEARCH

At the turn of the century, very little was known about the world's native forests or how to manage them. In the United States, foresters were quick to recognize the value of information about forests and a branch of research was established in the Forest Service in 1915. Early research was done primarily in support of reforestation efforts, but, as forestry grew in size and complexity, so did the research. Today, the USDA Forest Service has six regional experiment stations located in important forest regions. Each experiment station has several field laboratories generally with specialized assignments for a geographic region or a specific subject area, and numerous sites for field research. In addition, the Forest Products Laboratory in Madison, WI, serves as a nationwide centre for research and development of new technology relating to wood, including tropical woods. Two laboratories are dedicated exclusively to tropical forest research: the International Institute of Tropical Forestry in Puerto Rico and the Institute of Pacific Islands Forestry in Hawaii.

Research is vital for modern forest management, which is information intensive. Today's foresters require vast quantities of data and a knowledge of ecology: they must understand not only the parts of ecosystems but how different parts of the environment interact. Scientific investigations are conducted in support of all kinds of forestry activities: silviculture, forest insect and disease control, wildlife habitat management, fire prevention and control, range and watershed management, forest products utilization, forest survey, reforestation, ecology, and economics.

5

Commercial Deployment of New Biotechnology-Based Forest

Commercial deployment of new biotechnology-based forest tree varieties of a limited number of taxa (*Eucalyptus*, *Pinus taeda*, *P. radiata*, *P. pinaster*, *Populus*) can be expected in the near future. The impact of biotechnology applications on global wood supply may, however, take longer to materialize (not before 2020, according to Seppälä 2003). In this context, genetic modification appears to be the most significant and controversial sector of forest biotechnology. It is noteworthy that in some countries the terms 'biotechnology' and 'genetic modification' are used as synonymously, adding to the confusion.

During the last decade, several collaborative groups or consortia have been created by public or private partners for applying specific forest biotechnologies in a number of countries, including:

- ArborGen, created in 1999, with the aim of enabling research on genetic modification, has been working on herbicide resistance, growth and fibre quality for paper pulp in *Eucalyptus, Pinus radiata*, *Populus, Pinus taeda* and *Liquidambar* sp.
- GenForSA, created in 1999 with the main objective of modifying and testing disease resistance of *Pinus radiata*, while improving wood quality and formation.
- Monfori Nusantra Indonesia, created in 1996, has been involved in mass production of tissue cultures of *Tectona grandis, Acacia* and *Eucalyptus* for field trial establishment and commercialization.
- The Oregon State University's Tree Genetic Engineering Research Cooperative (TGERC) has carried out research on the use of GM poplar in plantations.
- The Poplar Molecular Genetics Cooperative (PMGC) has been established with the aim of improving knowledge of genetic and molecular mechanisms responsible for productivity and quality trait variations in hybrid poplars.
- The North Carolina State University Forest Biotechnology Industrial

Research Consortium (FORBIRC) mission is to integrate genome technology, metabolic engineering, traditional tree breeding and wood and paper science into a research organization directed to the creation of superior wood as a raw material and as a product.

The interest induced by the first genetic modifications of tree species at the end of the 1990s has since declined. Regulatory frameworks have been established in many countries for GM organism (GMO) field experiment or commercial deployment, with increased testing costs. In a number of countries, hostile reaction from the public or/and environmental groups has also limited deployment of GM trees. Some authors have questioned the profitability or return on investment of MAS techniques, at least in the short term and the medium term.

FORESTRY RESEARCH PRIORITIES

In Europe, forestry research priorities now focus on stronger synergies between biotechnology and conventional selection and breeding programmes mainly for wood improvement. The aim is to fill the gap between forest trees and agricultural crops in terms of the level of domestication. As a result, tree improvement tends to use genomic and proteomic approaches for achieving trait selection more rapidly. This is done by taking advantage of the advances obtained in model plants such as *Arabidopsis*, and of the considerable increase of information exchange resulting from the latest bioinformatics progress. More and more effort is devoted to research aimed at better understanding biotic and abiotic stress responses. Significant investments have been made in genomics research on *Populus* and *Pinus* within the framework of the Poplar Genomics Initiative and the Pine Mapping Project. In the tropics, wood production in volume per hectare is significantly higher than in the temperate zone.

Several large-scale companies have recently invested in tree planting in tropical and subtropical areas. Some paper companies tend to fund their own research activities and keep them confidential. More emphasis seems to be put in genomics and proteomics, and particularly sequencing, thanks to developments on *Arabidopsis thaliana*, *Oryza* and *Populus*. At the same time, there is increasing information coming from genome banks on wood quality, growth and stress resistance. Good examples of EST development include *Eucalyptus* in the framework of the Genolyptus project in Brazil, *Populus*, *Pinus taeda*, and *P. pinaster* and *Quercus*, although applications for practical field selection are very unlikely to be operational for at least a decade. Some predict that the next few years are more likely to see significant field applications of forest genomics research than of genetic transformation research, whose application is seriously hindered by strong environmental opposition and stringent regulations. Developments in agricultural crops are very meaningful is this respect. Genomics and proteomics outcomes are expected to boost conventional tree breeding and improve the efficiency of existing programmes.

The production costs of a new forest tree variety using conventional selection (well-known species, with a short juvenile phase) is roughly estimated at about US$400 000 and requires 15 to 20 years to be developed. The calculation does not include the investment required for the selection of the suitable species. On the other hand, genomics, supported by bioinformatics in the framework of MAS, requires substantial equipment and running budget (as an indication, a supportive biotechnological platform costs several million dollars). In turn, such biotechnologies are expected to enable breeders to access more accurate and numerous traits of interest for greater improvements, while reducing the new variety production costs thanks to bioinformatics.

As regards developing countries and countries with economies in transition, there are few references available on their involvement in forestry biotechnology. The limited literature mainly refers to micropropagation in Vietnam, Malaysia, Indonesia and India. Malaysia has a reported strong oil palm molecular biology programme, including genetic modification.

However, some emerging countries with advanced financial, institutional and human capacities (including Brazil, India, and China) have made significant breakthroughs in advanced forest biotechnology. Brazil has been actively involved in research on *Eucalyptus* through the Genolyptus project that constitutes at the moment the cutting edge in research on the genus worldwide. China is very active in poplar genetic modification. It can be reasonably assumed that after a period of time, results from research in technically advanced countries will benefit an increasing number of developing countries and countries in transition. African countries, with the exception of Kenya and South Africa, are far less advanced in forest biotechnology. However, notable development of agricultural crop biotechnology in countries like Nigeria, Uganda, Côte d'Ivoire, Morocco and Algeria (particularly through AAB) and even in the forestry sector (ISAAA in Kenya), constitute positive indicators suggesting that the situation is improving. Adoption of well-targeted forest biotechnology, possibly for drought-resistance, seems a crucial issue for the future of such activities.

Micropropagation has for the past several years been the most striking example of forest biotechnology applications, at least in terms of numbers of plants produced and acclimatized. Applications have been reported mainly in Asian countries, although there is still a lack of reliable information as far as operational field planting is concerned.

GENETICS OF CHROMOSOMES

It might be best to consider the nature of the genotype at this point so that genetic principles are understood before we consider the complexities of the effects due to environmental fluctuations. As noted before, the physical bases of heredity are the genes that are borne on the chromosomes carried in cell nuclei. These chromosomes can be observed during mitosis (when the cell divides and reproduces an exact replica of its genetic content, and in meiosis (including reduction division), where the chromosome number is halved to produce the gametes (spermatozoan or ovum) involved in sexual reproduction.

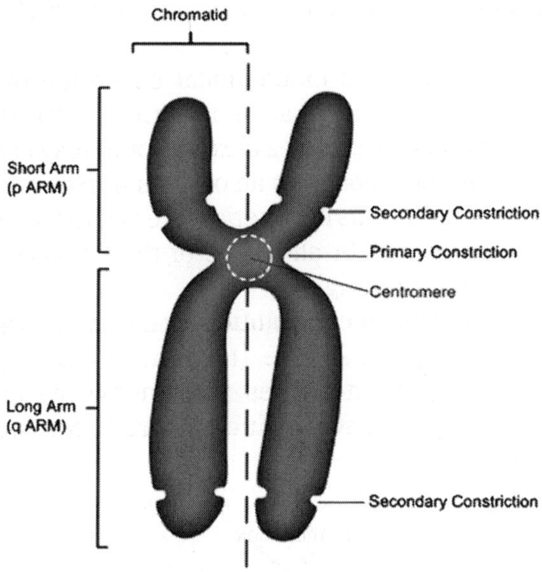

Fig. Human Chromosomes

The human species' 46 chromosomes are arranged in 23 pairs; in females all 23 pairs are homologous (correspond structurally and derivatively); in males, there are 22 homologous pairs plus a nonhomologous *X* and *Y* chromosome. The chromosome pairs, when properly stained, differ in length and appearance so that they can be individually recognized, much as one might recognize the familiar face of a friend. During meiosis, when the gametes are formed and the chromosome number is halved, all the chromosomes in a single gamete differ because each gamete has only one member of each chromosome pair. At fertilization, two gametes unite, each containing a set of 23 chromosomes to form the fertilized cell (zygote) with 23 pairs or 46 chromosomes once again. The number of chromosomes contained in gametes (23) is referred to as the haploid number and that of zygotes ($2 \times 23 = 46$) is the diploid number written as n and 2n, respectively. The process of meiosis affords a significant opportunity for increasing genetic variability. The most important consistent mechanism of change is crossing over, which takes place during the early phases of meiosis. Homologous chromosomes, lying next to each other, break, exchange equivalent portions of their genes, and reanneal to produce combinations that differ in arrangement from the genetic endowment provided by parents. This phenomenon is omnipresent in the normal meiotic process; when it fails to occur in at least one parent, the resulting organism is often sterile. Crossing over is the most consistent of a number of processes that provide the raw material upon which natural selection may act.

A coarser and more sporadic mechanism is individualized chromosome rearrangement, of which there are four possible types: deletion, duplication, inversion, and translocation. Deletion is the sometimes lethal removal of a gene or sequence of genes.

Death may occur because a biochemical sequence has been excised, resulting in an intolerable structural or functional gap in the organism. Duplication of a chromosome or part of a chromosome may cause an imbalance of gene activity, reducing the viability of an organism. However, because some organisms can tolerate duplications of genetic material, these duplications might play an evolutionary role if part or all of one of the duplications mutated and functioned differently from the original.

Indeed, this is one way in which evolutionary change is postulated to occur. Greater genetic variability gives rise (in minute quanta, to be sure) to physiologic and morphologic differences that endow the organism with an enlarged Behavioural repertoire so that it can more successfully cope with an ever changing environment.

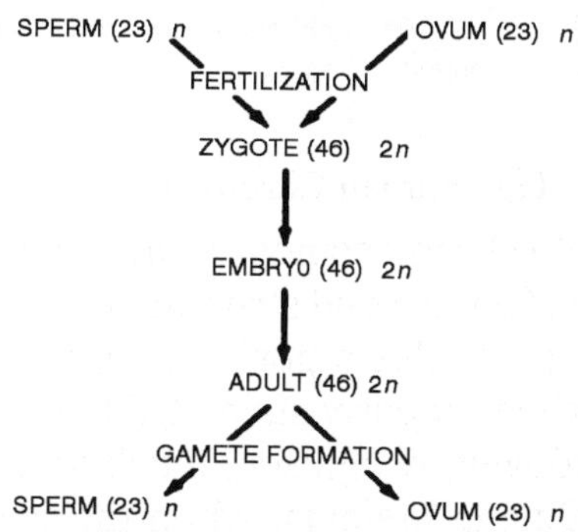

Fig. Human Chromosomes

This can be generalized for sexually reproducing organisms, which will be designated by n, where n is the haploid number and $2n$ the diploid number. An inversion occurs when a chromosome breaks in one or two places and the segment between the break rotates 180° leading to a reversal of gene order and a resulting change in biochemical product. A translocation occurs when two chromosomes from different pairs break simultaneously and exchange segments.

According to current knowledge, not all these chromo-somal and gene changes are equally important in their effects on Behaviour. To date, inversions and chromosome number changes appear to form the main categories of importance. But, crucial to the meiotic process (which reproductive systems evolved to support) is the shuffling and mingling of genetic material into infinite numbers of combinations.

MUTATION AND MUTANTS

Until now this discussion has centred on the shuffling of genetic material in sexually reproducing species. Genetic variability within a species has been discussed with emphasis on the variety of gene combinations attributable to meiosis, crossing over, duplication, and translocation. All these simply involve the reassortment of a set of variable genes indigenous to a population.

The essence of creating novelty is the process of mutation. These are alterations in the chemical structure of a gene that lead to changes in gene products that may ramify throughout the organism. Mutations at some loci occur spontaneously on the average of one in a million newly arisen gametes. The frequency is increased by radiation and by some chemicals known as mutagens.

A veritable storehouse of genetic mutations in mice is found at the Jackson Memorial Laboratory in Bar Harbor, Maine. Through selective inbreeding, when a mutant trait of interest is found, a strain of mice each of whose members possess the trait is established and bred for research uses. Many of the known mutant genes affect the nervous system during its development and produce a range of deficits in Behaviour from minor disturbances to severe defects. One of these found in the mouse is a mutation on chromosome five, an altered gene called *fidget*.

When this gene is present on both members of the fifth chromosome, it produces a Behavioural phenotype in which the mouse's head moves from side to side continuously. The gene is written *fi* and the mouse is said to have an *fift* genotype. In most mice the fidget gene is not present at this locus but the normal alternative gene, written as +, is. If the mouse is genotypically either *fi*+ or ++, the Behavioural alteration is not seen. Only if there are two *fi* genes present does the fidget phenotype become visible, as the wild type (+) gene is dominant to the recessive *fi*). However, dominance and recessivity are not necessarily complete, because heterozygotes (individuals with non-identical alleles) are often distinguishable from both homozygotes (individuals with identical alleles). Even if at first sight dominance appears to be complete, detailed biochemical tests may reveal differences between the heterozygote and the homozygote.

Untreated phenylketonuria in man provides one such example. Superficially viewed, the disease is controlled by a recessive gene *p*. with afflicted individuals being *pp* and normals *p*+ or ++. However, at the biochemical level, *p*+ and ++ may be distinguishable because the *p*+ individual tends to have more phenylalanine in blood serum. Dominance, in this case is incomplete. So, depending on which phenotype component is being examined, different conclusions as to the dominance may be drawn.

If a simple dominant—recessive relationship is assumed with two pairs of alleles at two loci (i.e., *A* and *B* are dominant to *a* and *b*), what might ensue? If a double heterozygote (*AaBb*) is crossed with a double homozygote (*aabb*) what are the statistical expectations? The double recessive homozygote produces only *ab* gametes. From the double heterozygote each separate locus yields equal numbers of *AB, Ab, aB* and *ab* gametes according to Mendel's principles of segregation and of independent assortment of alleles during gamete formation. If these gametes are fertilized with the recessive *ab* gametes, the results, assuming two independent pairs of alleles and simple dominant recessive relationships, will yield a 1:1:1:1 phenotype ratio.

The backcross to a recessive parent (one with an *aabb* genotype) is often used as a testcross in practical breeding programmes because it can determine the actual genotype of an organism whose recessive alleles may be obscured by dominant ones. The use of this wholly recessive dihybrid parent unmasks recessive alleles in the alternate parental genotype. If we return to Mendel's garden peas and specify physical traits in a dihybrid cross (*A* is the gene for yellow and *a* the gene for green; *B* is the gene for a smooth surface and *b* the gene for a wrinkled one), we emerge with four types of gametes (*AB, Ab, aB, ab*) produced in approximately equal numbers, which yield the basic Mendelian ratio of 9:3:3:1.

RECOMBINATION AND GENETIC MAPS

There is an important exception to Mendel's second law. Assortment of genes is not generally independent if gene loci are on the same chromosome. The closer genes are to each other along a given chromosome, the more closely linked they are during gamete formation. Although genes on the same chromosome may recombine with each other during meiosis, the frequency of the recombination depends on the spatial distance between the relevant loci. Generally, the further apart the loci are, the greater the possibility of recombination.

From the frequency of these recombinational events, chromosome maps have been constructed for individual chromosomes. Genes located on the same chromosome are said to belong to the same linkage group. Thus, in humans we have 23 linkage groups (one for each pair of chromosomes, i.e., the haploid number of chromosomes [*n*] corresponds to the number of linkage groups).

The house mouse, possessing 40 chromosomes, has 20 linkage groups; in *Drosophila melanogaster,* an organism of historical importance in Behaviour genetics, the numbers are 8 and 4 respectively. During the formation of human eggs and sperm (gametes), the chromosome number is halved (meiosis), leaving the egg with 22 autosomes and an X chromosome, and the sperm with 22 autosomes and an X or a Y sex chromosome. Males produce four sperm from each specialized cell in the testes and X- and Y-bearing sperm occur in about equal numbers.

GAMETES AND FERTILIZATION

In ovaries, meiosis also produces four products, but three of these become polar bodies, whereas the fourth becomes the chromosome containing nucleus of the developing egg. Human females are born with their entire reproductive quotient already present and the limits are set for the number of eggs they can produce (about 400 during a reproductive life span). Males, however, continue to produce sperm from sexual maturation until senescence.

Differences in gametic size and function are extreme. The head of the sperm is about 3 to 5 microns long and 2 to 3 microns wide. The tail, which propels it, is 10 times the length of the head. Essentially, it is a pared down packet of miraculously organized DNA (genetic material) with a launching and traveling apparatus to respond to and fertilize ova.

The human egg, in contrast, is one of the largest cells in the body—a sphere about 130 microns in diametre. It moves only passively when it is essentially shoved along, taking a few days to progress from ovary to uterus. As noted by Hartl (1977, cited in Probber and Ehrman,: "All the sperm that gave rise to all the people that ever lived could be carried in a teaspoon the eggs, would require a small bucket." The egg is comparatively enormous because in addition to its complement of chromosomes, its cytoplasm provides the nourishment for, and biochemical assistance to the developing embryo for translation of the information carried in its genes.

Unlike the female gamete the function of the sperm as a cell is complete when the sperm contacts and penetrates the egg (having shed its tail), and sperm and egg fuse. With fertilization of the egg, formation of the embryo begins, but it is not until 6 weeks afterward that sex chromosomes become operative. At this point the *Y* chromosome, if present, will cause the inner layer of the gonads to develop into testes. If, however, there are two X chromosomes, at about 12 weeks the outer layer of the gonads will elaborate into ovaries.

X-CHROMOSOME INACTIVATION

A further complication relates to sex. Although every human embryo is predisposed to develop as a female, its eventual sex phenotype will be determined by the types of chromosomes borne by the sperm. Our species has 23 pairs of chromosomes, each of which carries thousands of genes that compose the genotype that develops and unfolds to become the visible organism, the phenotype. In humans, 22 of the total number of chromosomes are autosomes; that is, two copies of these chromosomes are present in both sexes. The 23rd pair is made up of the unique sex chromosomes; the female has two of the X chromosomes; the male has one X and one Y chromosome. These were originally called X from X the unknown factor, and Y because it follows X alphabetically. The X chromosome was first described in 1891 from insect material as "a peculiar chromatin body" but its function was not known then or for a long time afterward. Although the female had two doses of all the genes on the X chromosome and the male only one, there seemed to be no significant differences in X chromosome activity between the sexes.

It was not until 1962 that a basis for dosage compensation emerged. This has been named the Lyon hypothesis after Mary F. Lyon, who first presented detailed studies on the subject. She hypothesized that in the normal female, one X chromosome is inactivated in each somatic body cell early in embryonic development. Whether the maternal or paternal X is rendered inert is a matter of chance and the result is that patches of cells with X-chromosomal lineages emerge.

The deactivated X becomes a tightly packed mass usually found close to the inner surface of the nuclear membrane of somatic cells in normal female mammals. It is called a Barr body after its discoverer. Accordingly, each cell in a male or a female has only one operative X chromosome. Because a female's cells are of two types (maternal X and paternal X), she is a genetic mosaic. A male, however, has only his maternal X chromosome to express. What are the implications of this double X-chromosome dose for the female of our species? At the molecular level the X chromosome is a much more information packed and therefore biochemically active chromosome than the Y, which is much smaller and, except for male sex determination and fertility, inert. Recent work by McKusick and Ruddle cited more than 100 genes in the human X-chromosome map. The Y chromosome bears only two known genes, one controlling the TDF (Testis Determining Factor) and the other coding for the H—Y antigen.

The H—Y antigen plays an important role in the expression of male physiological and morphological characteristics, specifically for the differentiation of the testis. "Further male differentiation is imposed on the embryo by the action of testicular hormones against the inherent tendency toward the female phenotype." It is also known that males with their X and Y chromosomal complement are relatively unprotected from the expression of unpaired deleterious genes on their single X chromosome. Every thing on their maternal X chromosome is expressed if not modified by genes on the autosomes. This is demonstrated by the much greater percentage of males afflicted by X-linked defects such as colour blindness and hemophilia. Genes on the X chromosome are said to be sex linked. In females the principles of heterozygosity and homozygosity apply to genes on the sex chromosomes just as they do for the genes on autosomes. In males, loci on the X are not matched by loci on the nearly inert Y chromosome; this condition is known as hemizygosity.

EPIGENESIS

The process of development involves constant interaction and coaction of an organism's genes and its environment. Given identical genotypes as in monozygotic human twins or highly inbred mice and *Drosophila,* phenotypes are usually very similar. Even under optimal circumstances however, phenotypic variation occurs because of perturbations in the environment. In a given situation one can speak of the average phenotype of individuals with the same genotype as a *norm of reaction*. The *range of reaction* covers the spread of phenotypes that could potentially develop from a given fertilized ovum. Clearly, the range of reaction increases if the environment is not carefully controlled. The *epigenetic* view of development postulates that genotype and environment are involved at every stage from fertilization to birth and beyond. At fertilization, these actions begin at a molecular level. Enzyme activities and changes in the intracellular environment are initiated. The egg provides a source of energy for the newly formed zygote and a well primed machinery for protein synthesis, which initiates the primary series of cleavages and invaginations that produce the embryo, fetus, and finally, the neonate. The control mechanisms fostering maturation are encoded within the genes, in the internal and external environment of the DNA, and in the interrelations and interactions between them. At every level molecular, cellular, and organismic there are functional gene environment interactions and products.

GENOME RELATIONSHIPS AND WIDE-HYBRIDIZATION

Domestication of crops is likely to have captured only part of the allele diversity present in the wild progenitor gene pools, leaving many desirable traits absent from the crop. In some cases, such as barley 3 *Hordeum spontaneum* or maize 3 teosinte, crops can be crossed easily with wild relatives and there are few barriers to the incorporation of novel alleles since the genomes are essentially homologous. In contrast, recently evolved polyploids such as hexaploid wheat will have had few opportunities to enlarge the gene pool. This is reflected by RFLP studies of wheat, barley and maize which show that wheat is less polymorphic. The frequency with which allopolyploid species occur in nature shows that the processes of pollen germination, correct pollen tube growth, fertilization of the egg cell, and embryo development are often conserved in related species. This can be exploited to synthesize new combinations of genomes. Such interspecific or intergeneric hybrids are often referred to as wide-hybrids or wide-crosses. Wide-hybrids can be used to enlarge crop gene pools for any character but are most valuable where suitable genetic variation does not exist within a species, or where screening for a desired character would be difficult, time consuming or expensive. Disease resistance or abiotic stress tolerances are the characters most frequently targeted for transfer.

Barriers to Wide-hybridization

The ease of producing wide-hybrids, and the ease with which genetic material may be transferred by backcrossing to a crop parent, varies greatly from cross to cross. Successful gene transfer may be restricted by prefertilization, postfertilization or chromosome instability barriers. These are not necessarily directly related to the degree of genetic divergence between two species and the genetical control of hybridization barriers may be simple.

Prefertilization barriers usually involve failure of pollen germination or inhibition of pollen tube growth, which may be at the surface of the stigma, in the stigma, or in the transmitting tissue of the ovary wall. The best studied incompatibility reactions affecting cereal wide-hybrids are due to the *Kr* genes of wheat where dominant alleles inhibit the growth of pollen tubes of other species at the base of the style. These genes are best studied in crosses with rye and *Hordeum bulbosum*. No method is known for overcoming the effect of these genes, so wide-hybrids involving bread or durum wheat usually use genotypes with recessive, noninhibiting alleles. In grasses there are, of course, well-characterized genes for self-incompatibility (the *S* and *Z* loci) but it is not known whether there is homology between these and the *Kr* loci or whether a similar mechanism of inhibition operates.

Postfertilization barriers leading to seed abortion are common in widehybrids. In cereals, this is frequently due to defective development and subsequent collapse of the endosperm. Embryos, however, are often viable, and hybrid plants can be recovered if the embryos are excised from developing seeds and cultured on suitable nutrient agar. In the *Triticeae*, using these techniques, a huge catalogue of wide-hybrids have been produced.

For example, hexaploid, tetraploid and diploid wheats and their wild relatives can be intercrossed but embryo rescue is usually unavoidable, especially if different ploidy levels are crossed. Similarly, ovule or embryo culture has been used to recover hybrid plants from *Brassica* crosses. In some crosses, for example between bread and durum wheats, hybrid necrosis kills the developing seedling. This is due to the epistatic interaction of independent genes in the respective species. Crosses within the separate gene pools are unaffected, probably because selection has eliminated necrotic combinations of alleles. Incompatibilities between nuclear genomes and the cytoplasm donated by the female parent may also contribute to lethality or sterility in certain crosses.

In vitro fertilization using isolated gametes, and the subsequent recovery of plants, is a means by which prefertilization and some postfertilization barriers might be overcome. Currently, *in vitro* fertilization has only been used successfully in maize, but such systems have potential for wide-crosses.

A further barrier to the production of wide-hybrids is the phenomenon of chromosome elimination. In such cases fertilization results in the formation of a hybrid zygote, but the complete chromosome complement of one of the parents is lost during subsequent mitotic divisions, leading to the production of a haploid embryo. The best characterized system is the cross between barley and its wild relative *Hordeum bulbosum*. However, the phenomenon occurs in a wide range of other cross combinations in the *Triticeae* including wheat 3 *H. bulbosum*, wheat 3 barley and wheat 3 maize, sorghum or pearl millet.

The mechanism of chromosome elimination is not understood but involves the inability of the chromosomes of one genome to successfully attach to the mitotic spindle. In some crosses, the occurrence of chromosome elimination is affected by the parental genomes since some barley 3 *H. bulbosum* crosses produce a high frequency of barley haploids while others give a high frequency of relatively stable hybrids. The genetical control of this regulation is probably fairly simple. Although chromosome elimination is detrimental to gene transfer, efficient haploid production methods benefit breeding and genetic analysis programmes and are extensively used by workers in these areas.

Utilization of Hybrid Plants

Once hybrid plants have been produced there are usually problems of sterility due to the inability of the hybrid to form viable gametes, particularly male gametes. Usually, this is due to a lack of chromosome pairing and chiasma formation at meiosis between the respective genomes and the subsequent formation of unbalanced gametes from random combinations of univalents. This can occur even in cases where close homoeology between the genomes can be shown by comparative chromosome and genetic analysis. Clearly the mechanisms controlling the pairing of chromosomes can be so precise as to prevent homoeologous pairing between different genomes. Such mechanisms may be genetically simple, as in wheat where pairing between the A, B and D genomes of bread wheat is restricted by a few pairing control genes of which the most powerful is *Ph1* on chromosome 5B.

Fertility can be partially or fully restored in many hybrids by doubling the chromosome number with colchicine or an equivalent spindle inhibitor as this provides identical pairing partners for each chromosome. In such cases new allopolyploid 'species' can be established. The major motivation for creating wide-hybrids is the introgression of useful genes into cultivated species. Generally, there are three levels at which the transfer of genetic material can be considered. Firstly, a whole genome can be introduced into an existing crop species to create a new allopolyploid species.

For example, the introduction of a diploid rye genome into durum or bread wheat to create hexaploid or octaploid triticales, respectively, has been highly successful. New genome combinations such as *Tritordeum* (the amphiploid between wheat 3 *Hordeum chilense*) are also being investigated as crops for specific environments. Secondly, whole chromosomes can be introduced to create substitution or addition lines. Thirdly, translocations can be induced between homoeologous or nonhomoeologous chromosomes so that only individual arms or small segments are introduced. The starting point for introgression is to backcross a wide-hybrid to the cultivated species. In many cases this is possible using the hybrid as the female parent, even if the hybrid is highly sterile, since a proportion of egg cells contain a nonreduced or otherwise viable chromosome complement. If there is homoeologous pairing between genomes, translocation products can be identified using *in situ* hybridization or molecular markers and stabilized by further backcrossing. Backcrosses can also be tested for the presence of the desired trait or traits in the 'alien' chromosome segment.

Combinations of cytological and marker techniques can then be used to characterize suitable lines. In some hybrids pairing between homoeologous genomes does not occur, or occurs too infrequently to be useful. In wheat, this can be overcome by using the *ph1* mutant or plants lacking chromosome 5B (carrying the *Ph1* pairing control gene). This allows higher levels of homoeologous chromosome pairing. Alternatively radiation treatments can be used to induce translocations. In such circumstances it is usually more effective to develop single chromosome addition or substitution lines prior to selecting recombinants, particularly as such stocks can confirm that the desired trait is carried by a single chromosome.

Plant breeders are generally interested in maintaining the characteristics of the cultivated species while adding one or more useful genes from the alien source. Generally the aim is to reduce the size of the alien chromosome segment as much as possible since donor species may have many alleles that are undesirable in the crop. This is greatly assisted by the use of molecular markers which define the alien segment. *In situ* hybridization using total genomic DNA or species-specific repeats is also valuable because it reveals the number and size of the alien segments. *In situ* methods are also increasingly able to locate low-copy sequences on plant chromosomes, thereby linking the molecular and physical mapping.

Comparative mapping information is important for the manipulation of alien chromosome segments because of the importance of producing plants with genetically balanced genomes. For example, substitution of rye chromosome 1 for a wheat group 1 chromosome is successful because these chromosomes show conserved synteny throughout their length. In contrast, rye chromosome 7 is translocated with respect to the wheat chromosomes of groups 2, 4 and 5 and substitution of rye chromosome 7 for a wheat group 7 chromosome produces a plant with reduced doses of group 7 segments and extra doses of group 2, group 4 and group 5 segments. Comparative mapping therefore provides a framework for the assembly of new chromosome segments in genetically balanced combinations. This principle also applies to other crop groups such as *Brassica*.

BIOTECHNOLOGIES AND TREE IMPROVEMENT

The author has reviewed the current status of biotechnology, and applications in tree improvement in detail in a work soon to be published (FAO, in press). The following paragraphs summarize the most important biotechnologies.

CRYOPRESERVATION AND IN VITRO STORAGE

This comprises the maintenance of cells, tissues or organs in cultures where growth is slowed (e.g. by the reduction of light, temperature or nutrients) or suspended (by immersion in liquid nitrogen). Many technical difficulties are involved, particularly in the subsequent regeneration of plants from the cultures, but recent results are generally encouraging. Regeneration from cryopreserved tissues has been induced for more than 70 plant species, including coconut, rubber, cocoa and coffee, and for several forest tree species. These results have led to hopes that the technologies may have a number of applications in tree improvement.

Gene conservation. Although increasingly used for the storage of threatened germplasm of agricultural species (Engelmann, 1991), *in vitro* storage and cryo-preservation have little to offer for this purpose with regard to forest trees. Gene pools of most of the established industrial species are reasonably well preserved in stands, both *in situ* and *ex situ,* and in seed stores. Undoubtedly, the gene pool of many tree species is threatened, particularly among the tropical hardwoods and non-industrial species. Distributions of these species are poorly known, as are their biological characteristics. Major impediments to the preservation of forest tree germplasm are: the inadequacy of resources for the survey and collection work that would be required before any germplasm could be stored; and the unreliability of many existing seed storage facilities. Even for the recalcitrant (hard-to-store) species, priority would be better directed to the establishment of *ex situ* plantings, which should facilitate urgently needed evaluations of the material. In the longer term, cryo-preservation and *in vitro* storage may have some application as a backup conservation strategy, but only for populations of well-surveyed, recalcitrant species.

Maintenance of juvenility. Suspension of the growth processes also implies the maintenance of the maturation state previously attained in the tissues - without any of the uncertainty associated with alternative strategies such as long-term hedging or serial propagation. Cryopreservation therefore warrants much more attention as a means of maintaining juvenility during simultaneous clonal testing and thus capturing genetic gains offered by clonal forestry with industrial species. The technology is therefore applicable mainly in cases where good breeding programmes are in place, where clonal forestry is a realistic goal and where "rejuvenation" is difficult -particularly for the conifers.

Use of molecular markers

The use of molecular markers involves the examination, using sophisticated biochemical techniques, of variations in cellular molecules such as DNA and proteins. As an alternative to traditionally measured features such as vigour, stem quality and various morphological aspects, molecular markers offer the advantages of being unaffected by the environment or the developmental stage of the plant while also being very numerous. These characteristics have led to a number of potential applications in tree improvement.

Genetic fingerprinting. The inherent characteristics of molecular markers render them much more useful than morphological traits in establishing the identity of a particular tree or tracing its genetic relationship to other trees. For example, using molecular markers, it was possible to identify each of 39 peach cultivars individually (Ballard *et al.,* 1992). Markers have important immediate applications in supportive research for advanced breeding programmes with industrial species mainly for quality control, e.g. checking clonal identification, orchard contamination and within-orchard mating patterns by "fingerprinting". Markers also have important immediate applications in supportive research for tropical hardwoods and non-industrial species, in particular for essential taxonomic studies and investigations of mating systems.

Quantification of genetic variation. Molecular markers are potentially more useful for quantifying genetic variation than are traits such as vigour and stem form, for which environmentally induced variation is frequently a confounding factor (i.e. it is not clear if traits are produced genetically or by external factors). Markers have been used to compare within- and inter-population variation in several tree species (Muller-Starck, Baradat and Bergmann, 1992). The quantification of genetic variation to aid in sampling strategies for the development of gene conservation and breeding populations of new industrial and non-industrial species is a potentially useful application of molecular markers. Markers may, however, provide underestimates of genetic variation with respect to traits (e.g. vigour and stem quality) that are more subject to evolutionary pressures and, therefore, will need to be used with caution.

Marker-assisted selection. This refers to indirect selection on the basis of markers shown to be associated with commercially important genes. Unaffected by the environment or developmental stage, markers offer the possibility of highly effective and early selection, long the hope of forest tree breeders (e.g. selection for wood quality at the young seedling stage). Although the possibilities are very attractive, there are limitations that will prohibit application in the short or medium term (Strauss, Lande and Namkoong, 1992): i) marker analysis is currently too expensive to permit the screening of large populations of seedlings; ii) associations between markers and economically important traits have to be established separately for different families, thus, even when cheaper markers are available, marker-assisted selection will apply mainly to advanced and sophisticated breeding programmes - those for which the creation and maintenance of the appropriate pedigree structures can be afforded and where clonal forestry is achievable. For most species, current resources would be far better directed towards moving breeding programmes to this stage of advancement, rather than to the development of marker-assisted selection.

The major current value of molecular markers lies in long-term strategic research; marker studies are making great contributions to advances in the understanding of basic genetic mechanisms and genome organization at the molecular level. An important emphasis of this work in coming years will be the study of quantitative traits of forest trees, of which a few model species will receive most attention, for example loblolly pine *(Pinus taeda).*

Genetic engineering

This comprises the insertion of novel genes into a plant or else the modification of existing genes through manipulation of the DNA molecule. Crops to which genes for insect, virus and selected herbicide resistance has been added are being or are near to being applied commercially. A tree crop into which these genes have been inserted is the poplar. Many projects are under way for forest trees, the reduction of lignin biosynthesis, for instance, but numerous technical difficulties remain to be solved. The insertion of currently available insect- or herbicide-resistant genes into a new species would constitute a major research undertaking, and successful application would be dependent on being able to regenerate from the transformed cells. The manipulation of more complex traits would be an even more formidable undertaking and much research remains to be done. An often overlooked research component is the extensive testing that would be required before a responsible recommendation for the large-scale deployment of transgenic plants could be made. Research projects of this type are necessarily intensive and must be regarded as long-term with only a modest expectation of success.

Insect resistance is of potential value, for example in poplars and some tropical hardwoods. However, the work involved in introducing several different resistance genes, sufficient to ensure that insects do not acquire tolerance during the rotation, should not be underestimated. The reduction of lignin biosynthesis is a very valuable objective for the pulp species. The introduction of herbicide-tolerant genes is of some interest but, in many programmes, the advantage of practicing unguarded herbicide applications may not be sufficient to pay for the research programme. Cold-tolerant genes are likely to be of some commercial value for many species, in particular the eucalypts. Much remains to be done, however, to establish that sufficient tolerance can be conferred using antifreeze proteins and to extend the work to tree species.

Prevention of the escape of genes into wild populations is likely to become an important concern, and sterility should be an early target of genetic engineering work with forest tree species. The major factor limiting application of genetic engineering in forest trees is the state of knowledge of molecular control of the traits that are of most interest - those relating to growth, adaptation and stem and wood quality. Genetic engineering of these traits remains a distant prospect. It is important that genetically engineered genotypes be of high quality with respect to other traits as well. The clonal test is the most logical basis for the integration of genetic engineering into traditional tree improvement programmes. For these reasons, genetic engineering is most appropriately conducted with species for which breeding programmes are advanced and clonal forestry can be realistically contemplated. Research on this subject should not assume a high priority with species for which natural variation available within the taxon remains poorly investigated.

Micropropagation

This refers to in vitro plant propagation methods. The principal approaches are axillary budding (actually a miniaturization of propagation with cuttings): the induction of adventitious buds on non-meristematic tissue (i.e. inducing a shoot where one would not normally develop): and somatic embryogenesis (where individual cultured cells or small groups of cells undergo development resembling that of the zygotic embryo). As an alternative to other vegetative propagation methods, the attraction of micropropagation lies in its ability to multiply elite clonal material very rapidly. More than 1 000 plant species have been micropropagated, including more than 100 forest tree species (Bajaj. 1991; Thorpe, Harry and Kumar, 1991). Successful experimental practices probably could be developed for most tree species. For most industrial forest plantation species, the costs of planting stock and insufficient data regarding field performance remain major obstacles to be overcome before a broader use of micropropagules as direct planting stock may be contemplated (Haines, 1992). Micropropagation has an immediate application, however, in integrated clonal propagation systems featuring the commercial planting of cuttings harvested from rapidly multiplied, micropropagated stool plants of the selected clones. This approach is of value only in very advanced breeding programmes incorporating the identification of outstanding clones currently only a few programmes are at this level.

Appropriate integration into breeding programmes is essential. Where clonal testing on a relatively large scale is possible and affordable, the currentapplicability of techniques mainly to juvenile material is not necessarily an impediment to the capture of good gains through clonal forestry. This conclusion, however, is dependent on the ability to store juvenile material for the period of a clonal test. Genetic variation in response, often substantial, is not likely to be a major problem where clonal testing can be preceded by screening for responsive genotypes, although demonstrating the absence of adverse correlations with economic traits is important. Breeding programmes new industrial species and non-industrials are not sufficiently advanced to warrant much use of micropropagation in the short term.

Micropropagation may have a wider application in the multiplication of stool plants of industrial species as breeding programmes become more advanced and other limitations to clonal forestry (e.g. maturation problems) are overcome.

For some non-industrial tree species, micropropagation may ultimately have a role in the multiplication of selected varieties prior to release. Development of simple micropropagation techniques for those species for which such methods are not already available is therefore a useful research objective but should not take priority over issues such as advancement of the breeding programme.

Work done with some crop species indicates the possibility of encapsulating somatic embryos to form artificial seeds which can then be handled like conventional seeds. With considerable research, developments in this area may overcome the constraint of planting stock costs (discussed above) and enable the direct use of such propagules in forest plantation establishment.

For industrial species, therefore, the development of these technologies is a useful long-term research objective but one which is best pursued with one or two model species, for example *Picea abies* and *Pinus taeda.*

In vitro control of the maturation state

There have been several reports of cultured mature buds displaying a reversion to a more juvenile state in response to the culture techniques and conditions. This has led to hopes that in vitro rejuvenation may be the solution to the poor rooting and vigour displayed by shoots collected from trees of selectable age for many forest plantation species.

The major limitation to this approach is that there is little evidence of complete, permanent and reliable rejuvenation. In fact, some studies have clearly demonstrated the effect to be a temporary response to the culture condition.

Further empirical work with this objective has a low probability of success. An understanding of the molecular basis of maturation (e.g. Hutchison and Greenwood, 1991) is much more likely to lead to practical manipulation, but this work is in its infancy and the reversal or promotion of maturation to precise levels remains a distant prospect.

For clonal forestry with industrial species, the maintenance of juvenility is about as useful as rejuvenation for many purposes (Haines, 1992) and it is probably able to be achieved using technologies such as cryopreservation or coppicing.

Nevertheless, a more fundamental control of the maturation state remains one of the most valuable objectives of long-term strategic research in forest tree improvement with industrial species. Rejuvenation is most applicable to efforts where good breeding programmes are in place and where other limitations to clonal forestry do not exist.

RESEARCH PRIORITIES

Supportive research

The preceding analysis suggests that the short-term possibilities for applying biotechnology in supportive research in forestry are:

- The use of molecular markers for quality control in advanced breeding programmes with established industrial species, e.g. for checking clonal identification, orchard contamination and within-orchard mating patterns by fingerprinting;
- The use of markers in essential taxonomic studies and investigations of mating systems;
- The use of markers for the quantification of genetic variation to aid in the design of sampling strategies for gene conservation and breeding population collections for breeding programmes with "new" industrial and non-industrial species.

Strategic research

Strategic research priorities relating to the application of biotechnology in tree improvement can be grouped into three broad areas:

Long-term generic research. This is most efficient if conducted collaboratively with a small number of model species, thus avoiding the diffusion of resources and efforts. High priority should be accorded to:

- Genetic engineering for sterility - this will underlie many of the eventual applications of genetic engineering;
- The use of molecular markers and DNA transformation techniques to investigate genetic processes at the molecular level, in particular those relating to complex traits such as growth, adaptation and stem and wood quality, is of particular relevance to industrial species, but will also pave the way for the application of biotechnology to non-industrial trees;
- Molecular studies of the maturation state for industrial plantation species.

A somewhat lower priority should be given to the development of somatic embryogenesis in combination with artificial seed technology as an inexpensive method of clonal propagation.

Long-term specific research. Two high priority areas are:

- Genetic engineering of useful traits, including lignin reduction in pulp species; cold tolerance, particularly in eucalypts; and insect resistance, e.g. in poplars and perhaps Meliaceae (when appropriate breeding programmes are in place). Transformation with appropriate genes (the introduction of several genes in the case of insect resistance) may be achieved within the short to medium term (the next five to ten years) but must be followed by perhaps ten years of field testing before responsible commercial deployment may be recommended;

- Marker-assisted selection, for species where breeding is advanced and where the creation and maintenance of the appropriate population structures are feasible and affordable - it will probably be ten years before this is possible on an operational scale.

Short- to medium-term research. Areas that warrant attention include:

- The examination of genetic correlations between regenerative competence and commercially important field traits (high priority);
- The development of cryopreservation methods as a means of maintaining juvenility in advanced breeding programmes with industrial species (high priority);
- The development of cryopreservation as a backup measure for gene conservation in proven species for which breeding programmes are in existence and for which seed recalcitrance has been demonstrated (moderate priority);
- The development of simple micropropagation techniques for species where none is yet available (low to moderate priority).

EQUITY OF ACCESS TO GENETIC MODIFICATION BIOTECHNOLOGIES

Private investors have taken the lead for most investment in modern biotechnology, and in so doing have also had to manage the associated economic risks. In many situations such investment risks are protected by patents. An obvious concern is, therefore, that access to GM crops and trees will be controlled almost entirely by private corporations.

Agreements for the use of techniques or material could be prohibitively expensive, preventing the use of the technology where it would be of value. In the past, governments were the main investors in breeding and molecular biology research. Their role may have to expand once again in this regard, in order to provide a flow of material and information that can be developed and shared by both private and public institutions (Santos and Lewontin, 1997), as opposed to being controlled largely by private companies. Nevertheless, financial constraints, rather than available technical knowledge, are probably the largest challenge ahead for applying modern biotechnology in less developed countries, although it has been shown, e.g. by the Agricultural Biotechnology Support Project (www.iia.msu.edu/absp/), that arrangements through donations are possible.

Even if financial considerations can be addressed in the research phases of biotechnology, moving the technology from the development stage to operational reality provides a new set of technical challenges with additional expenses (Polonenko, 1999). It is likely that only a few economically important species will warrant such additional investments, and it is on these economic situations and realities that the debate should focus.

Public acceptance of genetic modification

As has been alluded to previously, in terms of biotechnology research, one of the most important economic considerations separating forestry from agriculture is that trees require substantially more time for proper evaluation under both laboratory and field conditions. Rapid generation turnover in crop species has allowed genetic modification technology to develop quickly. Although this has been desirable from an economic point of view, the ecological ramifications have been far more difficult to study, as complex experiments must be carried out over several years. Public and government acceptance of GM plants is now probably as dependent on biological risk assessment and risk issue management (Leiss, 1999) as it is on technical developments or economics alone.

For GM trees, there is clearly a need to provide scientifically sound, neutral and intelligible information to the public, so that when and if GM trees become commercially available, informed decisions can be made about their release into the environment. Foresters are generally very aware of the need for public involvement in many other areas of forest management (e.g. chemical applications in forests, game management, felling patterns and plans, allowable annual harvesting levels), so decisions concerning GM trees may simply be an additional issue to be managed in the process.

CONCLUSIONS AND FUTURE CHALLENGES

The main issues that should remain relevant to the use of any modern biotechnology in forestry, now and in the future. The following are some summary points and conclusions:

- The development time, from the laboratory to the field, for all new types of tissue culture or GM trees will be substantial, usually in the order of a decade or more. While this may provide forest scientists and managers with time to evaluate many of the issues being faced in agriculture, the economic realities of relatively long generations will continue to be a major challenge to investors in biotechnology in forest trees. It appears that genetic modification will therefore become a reality only for particularly novel and valuable traits in short-rotation species in intensively managed plantations. Herbicide resistance is likely to be the first trait for which genetic modification will become economically viable, but even this may be several years away from large-scale use.
- All modern biotechnologies require large research and development investments. The allocation of funds, through either private or public

agencies, needs to achieve a balance between building scientific capabilities and knowledge and supporting more applied, well proven forestry technologies (Burdon, 1994). In this regard, the investment and use of any biotechnology needs to be assessed on a case-by-case basis, considering the specific technology in the specific ecological, political and economic environment.

- The public must be confident that regulatory frameworks are in place so that commercial development of biotechnology does not take precedence over biosafety issues. However, the development of useful biotechnology in forestry should not be put at risk with extraordinary regulatory compliance costs that are driven by unwarranted concerns.
- Government participation in the development of biotechnology appears necessary to keep some flow of material and knowledge in the public domain. Without this, the costs of using privately owned genes could seriously limit further interest and development of genetic modification technology in trees. Concerns about monopolies arise with any new products, but governments can step in to deal with substantial inequities or corporate actions that are not in the interest of society as a whole.
- Even if GM trees are not used on a wide scale in the future, research and development in GM technology will provide a great deal of information about gene action and regulation, which could be of long-term value to conventional breeding programmes.
- Other forest management decisions with potentially more serious ecological consequences, including large-scale species introductions or inappropriate use of provenances or improved trees from even conventional breeding, need to be evaluated by foresters, managers and regulatory agencies in the same way as the products of modern biotechnology are.
- Geneticists have always attempted to create genetic options for the future. If we look carefully at where these current and future biotechnologies are heading in forestry, it is obvious that the research and development will largely be used in support of advancing clonal forestry. Therefore, it is in the context of appropriate clonal forestry that most of the issues need to be examined and evaluated.

From a genetic perspective, concerns that biotechnology is “unnatural” ignore the dynamic changes in the genetic code that occur within and across species genomes through modification of transposable genes or elements by virus vectors and through mutation. This article has attempted to point out that humans have already had a huge effect on the genetic structure of many plants and animals, using many different technologies which will continue to develop and change over time. However, while those in the forefront of any technology will promote its potential benefits, in the end it will be the economic and regulatory systems of governing bodies at the national and global levels that must evaluate the technology's relevance and appropriateness.

CHANGES OUTSIDE THE FOREST SECTOR

The example in the last paragraph, which would be applicable to many other developing countries, indicates the unpredictability of changes and their impact on forestry. Even with the best forecasting tools that were available three decades ago, very few of the changes described could have been visualized. Sector analysis focused on measurable values and on estimating demand and supply, largely based on projecting historical trends. The key parameters considered in assessing the direction of forestry development were limited to population, urbanization and changes in income, supply and prices of substitutes. Fundamental system-wide changes in the economy as a whole could not be captured by such forecasting techniques; thus many changes visualized then were rather off the mark. Many intentional efforts to bring about changes in forestry had little impact, while most changes were unintentional and not necessarily the outcome of planned efforts.

Since changes in the rest of the system can alter development scenarios drastically, it is important to understand how system-wide changes take place and what may be done by foresters to take advantage of the emerging opportunities. For convenience, the changes could be categorized as economic, institutional, environmental or technological.

ECONOMIC CHANGES

Among the major driving forces of change is the increasing economic integration and interdependency of countries. The rejection of centralized planning and the adoption of liberal economic policies have opened up economies, with the market mechanism becoming the most dominant determinant of change. Some of the impacts that are relevant to forestry include:

- Opening up of local markets for imported forest products;
- Increased transnational investment in forestry, especially logging, plantation development and wood processing, largely based on perceived comparative advantages;
- Improved opportunities for marketing unique products (e.g. products from medicinal and aromatic plants, which were earlier used only locally; ethnic foods) and services (e.g. ecotourism);
- probable adverse impacts on local industries that fail to adapt to the changes and thus lose the comparative advantages that they had in a protected environment.

The impacts of globalization, which results in the easy movement of capital, technology, goods and services across national boundaries, are sometimes difficult to assess. The set of criteria previously used for measuring comparative advantage for investment in forestry (e.g. nearness to markets and raw material supplies; quality and quantity of raw material) is expanding to include very different criteria (e.g. need to reduce pollution; degree of openness of economies; barriers to trade).

Institutional changes

While globalization is increasing the interdependency of countries and societies, more pluralistic institutional arrangements are emerging (FAO, 1999). Improved access to information is changing the centre of power and authority and, as part of a larger process of devolution of administrative responsibilities with wider emphasis on participatory approaches, local communities are playing a lead role in resources management decisions. The involvement of the private sector in forestry, including forestry research, is increasing (Enters, Nair and Kaosa-ard, 1998).

Furthermore, increasing awareness of forestry issues has brought civil society to the forefront of influencing forestry decision-making (FAO, 1998a). Improved access to information, the media's new consciousness of the public and the emergence of democratically functioning and transparent institutions, as well as impartial and just mechanisms to redress grievances, have all enhanced the ability of civil society to intervene in critical issues of public interest.

Increased concern for environmental protection

Undoubtedly, the most important development affecting forestry in recent years is the increasing awareness of the environmental issues relating to forest resources management, with initiatives at all levels (including the global-level Convention on Biological Diversity, Convention to Combat Desertification and Framework Convention for Climate Change, which were initiated by the United Nations Conference on Environment and Development [UNCED], and country-level efforts to revise forest policies) giving emphasis to environmental benefits. Outcomes of this awareness include ongoing efforts to develop and refine criteria and indicators for sustainable forest management, certification and labelling, codes of logging practices and the extension of protected areas.

Technological changes

Developments in science and technology, both within and outside the forestry sector - more particularly the latter - have had a significant impact on forestry (FAO, 1998b). Advances have included enhanced utilization potential for several species, reduced raw material requirements through improved efficiency in processing, substitution and lowered emission of pollutants.

Recycling has increased and, as in the case of paper, could drastically reduce the demand for wood (Abramovitz and Mattoon, 1999). Research on biological aspects of forestry, including ecosystem processes, have enhanced the tools and techniques available for sustainable management of plantations and natural forests. Genetic selection, tree breeding and rapid multiplication techniques, coupled with refined site management practices, have enhanced the potential to realize higher productivity from tree plantations.

Developments in information and communication technologies are unleashing unprecedented changes. Resource monitoring capabilities, including imaging resolution and interpretation, have improved considerably. Forest cover changes, pest and disease infestations and outbreaks of fire are more easily monitored, and further improvements that will enable wider easy real-time access to resource information are a distinct possibility in the near future. The spread of e-mail and Internet access is improving the ability to communicate and thus to share information and experience. Distance is no longer a critical barrier for interaction among people. This is altering the structure and functions of organizations, with purpose-oriented networks becoming a dominant institutional arrangement (Nair and Dykstra, 1998), while the traditional vertically structured organizations are slowly fading out (EIU, 1997).

EXPLAINING THE CHANGES: LOOKING BELOW THE TIP OF THE ICEBERG

Given the potential of the economic, institutional and technological changes, why has the forestry situation failed to improve, or even worsened, in several countries during the past few years? Why is change rapid in some countries but painstakingly slow in others? In some cases efforts to bring about economic, institutional and technological changes are not possible to sustain, as the system as a whole in some way rejects them. The adoption of change seems to be largely dependent on system-wide developments and, more particularly, structural shifts in economies.

Structural changes in economies

Historically, agriculture (including animal husbandry) has been the main source of livelihood for most people in the early stages of economic development. Economic progress is largely dependent on the production of surpluses over and above consumption, and their transformation into other goods and services through either trade or investment. As populations grow, production is increased by bringing more land under cultivation or by intensifying agriculture through yield-enhancing technologies. Under the extensification scenario, forest cover tends to decline, as most agricultural expansion takes place on forest land. In many countries, agriculture remains the mainstay of the economy, and consequently population increase implies additional forest clearance for cultivation. Shifts away from land-based activities occur largely with the emergence of other sectors.

In several countries, the discovery of oil and natural gas has led to structural shifts in the economy (Wunder, 2000), with the extractive sector starting to contribute most of the gross domestic product and foreign exchange and accounting for most of the new job opportunities. In some countries, the growth of the manufacturing and services sectors has facilitated a shift from agriculture, reducing the pressure on land, thus slowing deforestation and, in some cases, even reversing the process. Dependence on land for the production of goods declines drastically as materials-based production improves or is replaced by knowledge-based industries. The attendant impact on forest cover is evident from the historical experience of most developed countries, and recently also from that of some developing countries.

Limits to structural shifts

If structural shifts in the economy can alter forest resource use patterns fundamentally, a key question is: What is the probability of such shifts taking place in most countries in the foreseeable future? Alternatively, what are the chances that such changes may not take place at all in some countries? In a highly interlinked and competitive global economy, the economic niches are increasingly being redefined and redrawn, largely on the basis of comparative advantages. This may limit significant structural changes in the economies. For example, development of the manufacturing sector in many developing countries remains constrained by low investments, small markets or the lack of competitiveness of local production. Manufacturing in developing countries is often agriculture-based and its expansion tends to have a negative effect on forests because of increased raw material requirements. In several developing countries, the indigenous manufacturing sector has relied on highly protected internal markets. Economic liberalization is, however, resulting in major changes, and removal of import restrictions is often undermining local industries because of increased competition from cheaper, and at times better-quality, imported goods.

Although a number of countries are hoping to expand their services sectors to enhance income and employment, there are inherent limitations to such an option. Most of the high value-added service sectors, including mature sectors such as banking, commerce and shipping as well as the newer sectors such as entertainment, information, technology and finance, are centred in the developed countries, whereas most developing countries focus on services at the low end of the value-added spectrum.[1] Even in the case of the "brain-power industries", which a number of countries are hoping to use for fast-track development, most of the services occupy low value-added niches - e.g. processing of credit card bills, "screwdriver technology" such as electronics assembly, and similar labour-intensive operations that take advantage of cheap labour. The potential for generating surplus from these activities is limited and there is very little scope for significant structural shifts.

Windows of opportunity provided by the development of the extractive sectors (e.g. oil and natural gas) have often been lost through private appropriation of the benefits or through the tendency to become too dependent on a single sector (Dale, 2000). When oil or natural gas (or for that matter any booming sector that generates a significant inflow of foreign exchange) becomes an important source of income, it pushes up the exchange rates, makes other sectors non-competitive and discourages diversification of the economy (a phenomenon sometimes called "Dutch disease" because it was first noted in the Netherlands in the 1960s following the discovery of large reserves of natural gas).

BEYOND STRUCTURAL CHANGES: SOME POSSIBLE FUTURE SCENARIOS

If major structural shifts are unlikely to take place in the foreseeable future, what will be the direction of changes and what are the implications for forestry? Will the future bring a situation of stable equilibrium that will enable most countries to bring about the necessary changes within the framework of the market mechanism? There are reasons to believe that changes could take place in new directions, largely because of the emergence of new fault lines in the global social and economic system.

Fault lines, disequilibrium and new scenarios

Predictable changes take place under equilibrium conditions in which it is easy to assess and understand the changes, whereas system-wide changes occur under disequilibrium conditions. As society develops and evolves, there are always new fault lines along which major changes take place. During most of the latter half of the twentieth century, the political-economic divide between the centrally planned and the market economies was the major fault line at the global level (Thurow, 1996). This divide tended to overshadow other fault lines. The collapse of the Soviet Union and the emerging dominance of the market mechanism were the major changes during the past decade.

However, the end of the cold war has brought to the surface some dormant or unnoticed fault lines. Many of the issues, such as authoritarianism, violation of human rights and corruption, that were overlooked (or sometimes tacitly approved or tolerated) to keep countries within a given political bloc have become unacceptable. Emphasis on democratic decision-making is percolating to the subnational level, resulting in efforts to devolve administrative responsibility to local levels. The emphasis on human rights is lending support to ethnic, linguistic and religious groups, often fuelling new conflicts and sometimes undermining the concept of nation states. Market-oriented development, especially increased flow of funds, technology and goods and services, is blurring national borders (Giddens, 1998). Alliances or interest groups that cut across countries are becoming important players in the economic, social and political fields. In a less polarized world with weaker nation states, new tensions are already surfacing which will have direct and indirect impacts on natural resources management, including forestry. These tensions include:

- The increasing economic and political power of transnational companies, the emergence of a large number of interest groups based on religion, ethnicity, language, etc., and consequent emerging conflicts;
- High material consumption and deteriorating environments;
- Widening inequalities and worsening distribution of income emerging from unequal access to skills, resources, technologies and markets.

Ruptures or tensions along the above fault lines are already evident. The large multinational companies, which are now consolidating their position through mergers, acquisitions and improved technologies, will face increasing pressure. In many cases, future conflicts will be between very active, highly networked organizations, representing the interests of a multitude of diverse groups, and economically powerful multinational corporations.[2] The rapid growth of information and communication technology is accelerating the process of change along the fault lines, with the increasing emergence of groups and communities that have a strong local base but interact in a transnational environment. There may also be strong local groups that are excluded from or unable to take advantage of the market mechanism. While the market mechanism has triumphed in recent years, it will have to adapt to a situation where new non-market systems are becoming an integral part of the economic and institutional framework.

PREPARING FORESTRY FOR CHANGE

Since there is uncertainty about the precise nature of future changes, it is difficult to indicate how forestry today should gear itself to adapt to those changes. Although in many situations the traditional problems will persist for some time, it is imperative that foresters and forestry organizations foresee changes beyond the immediate future and are prepared to adapt to emerging situations. Some of the most important changes that could be envisaged are outlined in the following sections.

Changing role of organizations

National-level objectives, programmes and plans will become less relevant as local communities, subnational entities and the private sector become the dominant forces in forest resources management decision-making. It is quite likely that forestry as a distinct sector or profession will cease to be relevant and will be integrated into a much broader framework encompassing all natural resources sectors. The role of national entities, especially hierarchically structured organizations such as forest departments, will diminish considerably. The ability of traditional forestry organizations to influence forest management - in terms of knowledge, resources, acceptance by stakeholders, etc. - has already declined considerably. Possible future roles for such organizations will be as impartial arbitrators to facilitate resolution of conflicts among the large number of emerging players, and as apex bodies to develop standards for the various practices that are acceptable to the diverse players. Since most of the productive functions will be performed by farmers, communities and private enterprises, the public sector is unlikely to have any major role in wood production. At most its domain will be limited to managing public goods and services, often under contractual arrangements.

New skills and approaches

Environmental issues, especially those relating to protection of watersheds and biodiversity, are becoming more prominent. Although ensuring the sustainability of public goods supply will still be a public domain responsibility, the approach will have to be very different from what it is today. Those who have to refrain from their usual activities to ensure the flow of environmental services for others will require appropriate compensation. Improved technologies for resource monitoring and fine-tuned management practices will have to be developed to ensure that all parties comply with their obligations. While the criteria and indicators for this will be developed at the national level, implementation and monitoring will take place at the local level. Substantial improvement may also be required in negotiation skills and conflict management at the local level, and highly transparent systems of resources management will have to be implemented.

6

Genetic Modification in the Forestry Sector

The first regeneration of a genetically modified (GM) forest tree was achieved in 1986 in *Populus.* Since then, the genus has become a model for genetic modification and related tree biotechnology studies. The first attempt to genetically modify a conifer (*Larix*) was reported in 1991. Introducing targeted genes into the genome of a forest tree is a way to obtain GM plants. It is also a basic research tool for a better understanding of gene functioning in woody plants.

TRANSFORMATION METHODS

Transformation methods consist of inserting into the genome of the host plant a mutated gene or a gene from another organism either by microinjection/projection (direct transformation), or through a vector such as *Agrobacterium tumefaciens* (indirect transformation).

Most successful work on genetic modification of forest trees species genomes so far has been obtained by using juvenile material, often from explants produced from early germinations tissues which have much higher regeneration capacities than older material. Successful genome modification reports of adult selected plant material are very rare, except in poplars precisely because of its greater capacity to regenerate.

INSECT RESISTANCE

The first strategy consists of using microprojectiles to insert a gene encoding an endotoxin which binds to the receptors in the intestine of Lepidoptera, Coleoptera and Diptera, lysing the organ and killing the insect. This was done in *Populus alba* × *P. grandidentata, P. tremula* × *P. tremuloides* and *Picea glauca*. GM *Populus* trees and *Pinus radiata* expressing the *Bacillus thuringiensis* endotoxin '*Bt*' have been obtained. A second transformation method is based on the introduction of a gene coding for a protease inhibitor that modifies insect digestion, causing the death of the pest. Studies used potato gene *pin*2, a protease inhibitor introduced into *P. alba* × *P. grandidentata* through *A. tumefaciens*, and the gene of a rice protease inhibitor introduced into *P. tremula* × *P. tremuloides*. A third approach focuses on simultaneous modification of two genes for enhanced resistance to insects. This was achieved in *Liquidambar styraciflua*, combining a peroxidase anionic enzyme gene involved in cell growth and wall development with a '*Bt*' gene.

Herbicide Resistance

A first method has consisted in introducing a mutated version of the gene encoding the enzyme target for various herbicides: glyphosate for *Populus alba* × *P. grandidentata, P. trichocarpa* × *P. deltoides, Eucalyptus grandis, Larix decidua* and *Pinus radiata* (the herbicide blocks the synthesis of tryptophan, tyrosine and phenylalanine), or chlorosulfuron for *Populus tremula* and *Pinus radiata* (the herbicide blocks the synthesis of leucine, isoleucine and valine). A second and more frequently used strategy consists of introducing a microbial gene encoding an enzyme for the detoxification of the herbicide, and has been applied to *Populus alba*, *P. alba* × *P. tremula*, *P. tremula* × *P. alba*, *P. trichocarpa* × *P. deltoides* and *Eucalyptus camaldulensis*.

Flowering Modification

In order to produce sterile trees and prevent possible dispersal of transgenic pollen in the environment, an approach based on genetic ablation has been tested on poplar. This technique, which consists of expressing a cytotoxic gene under the control of a very specific poplar floral promoter, resulted in more than 90 percent of transformed lines lacking floral structures. Other approaches, based on the suppression of key flowering genes, are being tested.

Quantitative and Qualitative Modification of Lignin

Modification of lignin composition or content is being actively pursued because of the expected financial gains from pulp processing improvements. Lignins, which enhance cell wall mechanical properties and hardness, are difficult to process and are a significant limitation in processing wood into paper pulp by chemical treatment. Genetic transformation to modify lignin characteristics is a key research feature on species used in the paper industry. The aim is to regulate the activity of key enzymes involved in the lignin biosynthesis pathway. Active on-going research targets the effects of lignin biosynthesis in *Populus* on soil carbon transformation and storage.

Other Traits

Other traits that have been considered for genetic modification include: resistance to diseases in *Populus*, *Betula pendula*, *Alnus* and *Juglans*.; bioremediation in *Populus*; enhanced growth through higher efficiency of nitrogen assimilation in *Picea* and *Populus*, and modification of gibberellin synthesis in *Populus*.

DATA SET ANALYSIS IN GENETIC MODIFICATION

Genetic modification activities represent 520 entries in the data set, or 19 percent of all biotechnology activities recorded worldwide. Research is reported in 35 countries and on 29 tree genera. North America is the lead region, accounting for 48 percent of genetic modification activities on forest tree species, followed by Europe (32 percent), Asia (14 percent), Oceania (5 percent) and Africa and South America (less than one percent each), while no activity is reported in the Near East. Genetic modification research at the country level is led by the United States (219 entries in the data set, or 42 percent of the world's recorded activities on GM forest trees). Studies carried out by, or in cooperation with, agencies or companies from developed countries have been identified in Brazil, Chile, China, Indonesia, South Africa and Uruguay.

Populus accounts for 47 percent of data set entries, followed by *Pinus*. (19 percent), *Eucalyptus* (7 percent) and *Liquidambar* and *Picea* (5 percent each). The pedigree of a number of hybrid poplars used in China has not been ascertained. Several countries have already established GM tree field trials.

Interestingly, while the first GM forest trees were developed some fifteen years ago, about three-quarters of the total number of species recorded in the dataset have been genetically transformed during the past five years. The main studies underpinning genetic modification target gene expression (21 percent), *in vitro* regeneration tissue culture (18 percent), herbicide resistance (13 percent), biotic resistance (12 percent) and lignin properties (9 percent).

THE COMMODIFICATION OF GENETIC RESOURCES

The administrative body of the CBD has been markedly influenced by the green-developmentalist perspective of Western governments and environmental organizations such as the World Conservation Union (IUCN). Interpretations of the CBD mandate prepared by the Secretariat of the CBD reflect this influence. One such document's proclamation that "modern biotechnology offers the potential to invent sustainable systems of the future, to be accompanied by a new paradigm for industry" reveals how central biotechnology has been to the interpretation of the treaty. "The private sector is the key player in benefit-sharing arrangements," the aforementioned secretariat document states flatly. Another secretariat note identifies a "policy setting" conducive to benefit sharing as one that encourages "access legislation, incentives, partnerships and contracts". This emphasis on market transactions and business partnerships illustrates how CBD administrators have promoted the market based management of genetic resources.

Case studies compiled by the secretariat interpret biodiversity benefit sharing almost exclusively in terms of North-South bio-prospecting deals. In effect, access to "biodiversity benefits" under the CBD is being made contingent on the participation of diversity-rich countries in a global genetic-resource market. But when CBD documents define biodiversity benefits as benefits to be derived from the commercialization of genetic resources by biotechnology industries, they fail to recognize that biodiversity benefits already exist—that is, that the benefits of natural resources are known and valued by people who depend on them directly for sustenance, shelter, aesthetic pleasure, and spiritual significance.

This equation of biodiversity benefits with genetic resources reduces biological variety to its purported essence as a commodity, separable from its complex interrelationships with the rest of nature and society. It is as if the values of biological variety come into existence only when its genetic information is "developed"—codified, counted, and commercialized—by biotechnology. A broader conceptualization would center on the incalculable present and future benefits of healthy ecosystems and diversity-based farming systems both to people locally and—insofar as such eco-social systems may conserve and generate biodiversity—to people in all countries.

Measuring the value of a country's biological resources on the basis of their commercial potential, or in terms of the market costs of replacing them, is compatible with the dominant discourse of environmental managerialism. However, this green-developmentalist approach privileges those aspects of nature that can be removed from their local contexts, transformed by investment, and sold. The approach fosters a view of ecosystems as warehouses of potential commodities to meet the demands of foreign consumers rather than as the bases of local and national life, as the sources of material necessities and meanings, and as the biophysical contexts of cultures. In this way, green developmentalism divorces the problems of biotechnology and genetic-resource management from the development needs of gene-rich but hard-currency-poor countries.

The globalized, market-based management of biodiversity requires clear property rights to natural resources, and the CBD secretariat has devoted what might seem, in a conservation treaty, inordinate attention to intellectual property. Conference decisions and secretariat documents display an almost schizophrenic ambivalence. The predominant assumptions in these documents support Northern intellectual-property models and their international extension, but references to IPRs are counterbalanced in nearly every statement by references to "alternative systems" or to the "concerns of indigenous and local peoples" about the impacts of IPRs.

This controversy has been kept alive by the active influence in the CBD and related international forums of social movements that oppose the privatization of genes and associated knowledge. On the basis of ethical, equity, and ecological concerns, the NGOs and indigenous peoples' and farmers' organizations that constitute this transnational movement have denounced the patenting and commoditization of life.

With the support of Sweden and a number of Southern states, these organizations have energetically resisted Northern proposals that the CBD recognize the WTO as the appropriate body for settling international disputes over property rights to organisms and knowledge about nature. In light of fears among many Southern delegations of losing sovereignty to multilateral trade bodies; growing doubts about the benefits of industrial agriculture and the safety of genetic engineering; and questions about the fairness of intellectual-property regimes, these NGOs have found a growing audience among developing-country delegations.

BIOTECHNOLOGY AND PROPERTY IN LIFE

What became known as the Cartagena Protocol on Biosafety of the CBD was finally negotiated, after significant concessions from the United States, in January 2000. Because the protocol stipulates that countries may decline to accept exports of genetically engineered "living modified organisms", U. S.-based agribusiness interests have seen it as a threat to their export markets. Nevertheless, a united front of European and developing country delegations forced significant concessions from the United States delegation.

The United States-led Miami Group of six grain-exporting states gave up its opposition to protocol rules for the labeling of exports of genetically altered organisms meant for release into the environment, and dropped its proposed language that would have asserted the primacy of the WTO over the CBD. Included over U. S. objections was a provision allowing governments to take account of the socioeconomic impacts of transgenics in deciding whether to permit imports of particular organisms. This principle may yet collide with WTO rules against "unfair" barriers to trade, rules in which criteria of economic efficiency are preeminent.

Although thirty-eight states had ratified the protocol as of November 2002, it will not become international law until 50 governments also ratify it. Substantial disagreements remain over whether states or enterprises will be held liable for environmental or health damages resulting from the use of their genetically engineered products; over specific requirements for labeling, transport, and "contained use" of genetically engineered organisms; and over the interpretation of the precautionary "principle" or precautionary "approach"—both terms appear in the protocol—which contrasts with the guidelines that frame U. S. biotechnology regulatory policy.

In June 1999, a coalition of developing-country CBD members asked the CBD's scientific advisory body to call for a worldwide moratorium on the field testing and commercialization of "terminator technologies." These genetic engineering methods are being developed by the U. S. Department of Agriculture and commercial biotechnology firms to produce crops with seeds that will not germinate. By hardwiring property rights into plant genomes, these technologies would enable companies to control their privately-held crop genetic resources in cases and places where their IPR claims are not recognized. The advent of terminator technology, which confers no agronomic value to plants, has added to widespread skepticism about industry claims that transgenic crops are designed to benefit the hungry and increase the productivity of poor farmers in the developing world. Although the economic purpose of the technology—enlarging seed markets by preventing farmers from saving seed—is perfectly in keeping with the letter and spirit of the WTO, it is arguably at odds with the CBD's commitments to conserving crop biodiversity and the relevant practices of local communities.

In regional forums such as the European Union Parliament and the Organization of African Unity, governments, environmental ministries, and NGOs have called for recognition of the CBD's precedence over the WTO in matters pertaining to biodiversity. The negotiation of the International Undertaking on Plant Genetic Resources for Food and Agriculture was deadlocked for seven years over debates about IPRs, genetic-resource benefit sharing, and biopiracy.

The treaty that finally emerged from this process in November 2001 reflects an uneasy compromise that has not yet resolved issues of access to an valuation of crop genetic resources. Disputes about private property in genetic information and in biotechnology tools have also embroiled the Consultative Group on International Agricultural Research (CGIAR), which is the world's largest multilateral network of seed banks and crop-research centers, the World Intellectual Property Organization, and even the World Bank.

In August 2001, the UN Sub-Commission for the Promotion and Protection of Human Rights passed a stinging resolution against the WTO TRIPS Agreement, noting conflicts between TRIPS and "economic, social and cultural rights" in relation to the need for technology transfer and the consequences for the enjoyment of the right to food of plant variety rights and the patenting of genetically modified organisms, "bio-piracy" and the reduction of communities' (especially indigenous communities') control over their own genetic and natural resources and cultural values, and restrictions on access to patented pharmaceuticals and the implications for the enjoyment of the right to health.

The report on which the resolution was based, which described the WTO as a "nightmare" for poor countries, prompted an official complaint by top WTO officials. The repercussions of these disputes echo in the WTO. A United Statesbacked proposal to establish a WTO body on biotechnology, seen by critics as an attempt to outflank the Cartagena Protocol on Biosafety, was rejected by Europe and by developing countries at the failed 1999 WTO ministerial session in Seattle. Developing countries continue to call for reform of the TRIPS requirement that all countries enforce patents on microorganisms and property rights to animals and plants. The African Group wants to remove references to plants and all life forms from TRIPS, whereas India wants language inserted that would require disclosure of the source of genetic material on patent applications. The Philippines Department of Agriculture has recommended that governments "get life forms (plants and animals) and biodiversity (and indigenous knowledge) out of the jurisdiction of WTO". India, Brazil, and the African negotiating bloc, among others, have asked that the WTO's Council for TRIPS take into account biodiversity, traditional knowledge, benefit sharing, farmers' rights to save and share seeds, socioeconomic welfare, and the ethics of patenting of life forms.

The majority of council delegations have asked the council to "harmonize" TRIPS with the CBD. The U. S. delegation adamantly opposes this and objects to granting the CBD secretariat observer status in the council, insisting that the only issue to be discussed is the progress of TRIPS implementation.

Even if the majority is able to block or slow TRIPS implementation at the international level, the United States may achieve its goal of standardized IPRs by pressuring reluctant countries to include IPR promises in bilateral trade pacts or by threatening to reduce aid to countries that do not comply. Regional trade accords, including the North American Free Trade Agreement and the U. S.-proposed Free Trade Area of the Americas, include requirements that signatories maintain IPR regimes compatible with those of the United States. Bilateral trade deals, such as the recent agreement between the United States and the government of Vietnam, contain provisions requiring the new U. S. trade partner to join the UPOV (International Union for the Protection of New Varieties of Plants) convention, the international IPR agreement which the United States regards as the acceptable regime for plant-variety protection.

Nevertheless, enforcement of IPRs at the local level may be difficult in the face of growing defiance by social movements. Even where governments feel forced to comply with globalized IPR rules, many citizens are refusing to do so. In India and Bangladesh, Thailand and the Philippines, and many Latin American countries, organizations of farmers have linked their demands for land tenure and support of domestic agriculture and rural livelihoods to their opposition to crop-variety IPRs, life patenting, and food-crop biopiracy. International farmers' organizations such as Via Campesina and NGOs such as Genetic Resources Action International, the Malaysia-based Third World Network, and the United States-based Institute for Agriculture and Trade Policy stress the connections between food security (of people), food sovereignty (of nations), and preservation of agro-biodiversity and diversity-based farming systems.

Thus, a high-stakes battle continues over whether food, farming, and biotechnology will be understood and governed as a problem of corporate technoscience, economic efficiency, and universal legal standards, or whether the broader issues of who really benefits and who loses from genetic engineering of crops, privatization of research, and world-scale consolidation of agro-economic power will be addressed by emerging institutions of global governance. In the CBD and in the WTO, these international battles have brought to the surface deep discontent with persistent and widening inequalities in the postcolonial world order.

THE SCIENCE OF MODERN GENETIC ENGINEERING

Genetic engineering is one form of biotechnology and usually refers to copying a gene from one living organism—plant, animal or microbe—and adding it to another organism. In genetic engineering, a small piece of genetic material (DNA) is inserted into another organism to produce a desired effect. This is in contrast to traditional plant breeding in which all the genes desirable and undesirable contained in the male plant—pollen—are combined with all the genes of the female plant.

The progeny resulting from this cross may contain the gene for the desirable character, but it will also contain many of the undesirable genes from both parents. Genetic engineering has the advantage of being able to transfer only the gene of interest and greatly accelerate plant breeding. But genetic engineering also is more powerful than traditional breeding since it can move genes not only between similar plant species but also from distant relatives, including non-plant species. It is possible to move genes between such seemingly unrelated organisms because all living organisms share the same code for DNA and the synthesis of proteins and other basic life functions. What might seem on the surface to be very different organisms are, in fact, very similar, at least at the molecular level. All living things are more alike than different, and this is one of the reasons that genes can be moved so successfully between such seemingly different organisms as plants and bacteria. Genes are not unique to the organisms from which they came, so there really aren't "tomato genes" or "bacterial genes." It's the collection of all genes in a tomato or a bacterium that makes it a tomato or bacterium, not a single gene. As we learn more about the genetic makeup of all organisms, we see that most plant species differ by only a small percentage of their genes and those even such seemingly different organisms as tomatoes and bacteria have many common genes.

These findings suggest that in the long-term evolutionary process even tomatoes and bacteria had some common ancestor. From the discovery 50 years ago of the structure of DNA, scientists soon came to realise they could take segments of DNA that carried information for specific traits—genes—and move them into another organism. In 1972, the collaboration of Hubert Boyer and Stanley Cohen resulted in the first isolation and transfer of a gene from one organism to a single-celled bacterium where it would express the gene and manufacture a protein. Their discoveries led to the first direct use of biotechnology—the production of synthetic insulin to treat people with diabetes—and the start of what is often called modern biotechnology. Plants were first transformed through genetic engineering in the late 1970s. Mary-Dell Chilton and colleagues used a common soil-dwelling bacterium, Agrobacterium tumefaciens, that attaches itself to plants and transfers some of its DNA into the plant. Chilton and her colleagues added a gene to this bacterium, which in turn transferred the gene into a plant where it became part of the plant's DNA. This bacterium is still commonly used in genetic engineering along with another technique that uses a high-velocity mechanism to inject DNA into plant cells. The result from either technique is the same—the plant cells take up the gene and begin to express it as their own.

Benefits and Risks of Biotechnology

Plants developed through genetic engineering were first grown on 1.7 million hectares in 1996 in the United States, but by 2002 they were grown on 58.7 million hectares in 16 countries. By far the major use of the present plants is to manage pests—weeds, insects and diseases. Weed management with genetically engineered plants is accomplished because the plants have a modified enzyme (a protein) that allows them to survive an application of a specific herbicide that normally acts on that enzyme. Growers can plant the herbicide-tolerant seeds, allow the plants to emerge along with any weeds in the field and then treat the field with an herbicide.

The result is that the weeds, but not the crops, die. The advantage to growers is that they spend less time on weed management, have enhanced weed control, use safer herbicides, and in many cases use less herbicides. Additionally, this technology allows growers to use soil conservation practices such as reduced or no-tillage, thus helping to retain soil structure and moisture and reduce erosion. Herbicide tolerant crops (soybean, canola, cotton and maize) were grown on 48.6 million hectares in 2002. Insect-resistant crops developed through genetic engineering utilize the common soil bacterium, Bacillus thuringiensis (Bt), which has been commercially used for more than 50 years, as an insecticide spray.

Although safe to humans and the environment, when a susceptible insect ingests Bt, the Bt protein binds to specific molecular receptors in the gut and creates pores causing the insect to starve to death. Insecticidal products containing Bt were first commercialized in France in the late 1930s, but even in 1999 the total sales of Bt products constituted less than 2 per cent of the total value of all insecticides. Bt, which had limited use as a foliar insecticide, became a major insecticide only when genes that produce Bt toxins were engineered into major crops. The Bt crops available at present are maize and cotton. These were grown on a total of 14.5 million hectares in 2002. Virus-resistant crops were created by inserting a non-infective part of a plant virus into a plant, essentially "vaccinating" the plant to protect it from the virus.

This technique is called "pathogen-derived resistance." Squash and papaya have been engineered to resist infection by some common viruses and are approved for sale in the United States. There are fewer than 1 million hectares of these crops. The bioengineered plants available at present provide growers with better tools to manage pest problems. As with any technology, there are risks and benefits to currently available genetically engineered plants, but the present body of information indicates their use has enhanced pest management, substantially reduced the amounts of pesticides used in some crops, enabled growers to use safer pesticides, and contributed to enhanced safety for humans and the environment.

The regulatory process for managing these plants and their effects on the environment and humans has evolved with the technology and the scientific community's knowledge of these tools. Many of the more controversial issues surrounding genetic engineering of plants—such as pesticide resistance, gene flow and intellectual property issues—are not unique to this new technology but pertain to all types of agriculture. Some species of insects have developed resistance to sprays of Bt, indicating the potential for some species to become resistant to Bt plants. However, despite Bt plants being grown on more than 62 million hectares worldwide from 1996 to 2002, there have been no documented cases of resistance development. The reasons for this lack of resistance appear to involve not only biological factors of the insects and the Bt plant, but also the fact that the regulatory agency (the Environmental Protection Agency) in the United States requires a resistance management plan for growing Bt plants. No other insecticide has such strict regulations. Still, growers, companies and federal regulatory agencies must be vigilant about resistance developing for biotech crops used to manage insects, weeds and viruses as they also must with non-biotech pest management tactics.

It will be important to consider the accrued environmental and health benefits of these biotech crops prior to the development of any resistance and how resistance can be managed if and where it occurs. In addition to pesticide resistance, gene flow from biotech to non-biotech crops may also be a concern. However, the risk of gene flow varies with each crop and each gene. Pollen flow in soybeans is very limited so the risk of a biotech soybean crop crossing with a non-biotech soybean is minimal, but this may be different for another crop. Likewise, if the gene in the biotech crop that provided a pest management trait, such as insect resistance, moved into a non-biotech plant, such as a weed, any selective advantage of the insect-protected weed in the ecosystem should be assessed. These same questions should also be answered with non-biotech crops, but these have not received the same level of attention as biotech crops because of the latter's higher profile.

Horizon of Biotechnology

In the future, the potential uses of plant biotechnology are far more wide-ranging than the pest-management biotech crops of today. Plants are being developed that serve as production "factories" for medically important drugs, sources of alternative energy, tools for cleaning toxic waste sites, and biomaterials including dyes, inks, detergents, adhesives, lubricants, plastics and the like. Consumers may see these products as more directly enhancing their quality of life than the pest-management biotech crops of today. Perhaps an even more dramatic advantage to consumers will be seen when plants are genetically engineered to have enhanced health benefits such as disease-fighting chemicals or increased amounts of essential vitamins and minerals. A healthy and well-informed discussion of the risks and benefits involved in agricultural biotechnology is needed to ensure a proper role for this new technology in our future food and health systems. No one should believe that any technology, including biotechnology, will completely solve the world's agricultural problems. Many people familiar with biotechnology, however, believe it to be an important component of the solution.

GENETIC MODIFICATION IN THE FORESTRY SECTOR

In vitro regeneration of transformed plants is still a technical limitation for many species and genotypes.

Genetically transformed lines have been regenerated for *Populus*, *Eucalyptus*, *Pinus* (including *P. taeda*) and *Picea*. Technical processes involved in regeneration are not always well controlled or even studied; especially in mature tissues where the situation is the most critical.

No GM tree is reported to be commercialized in international databases to date. Insertion of genes controlling traits of interest in a given forest species is limited by insufficient knowledge of the molecular control of these traits. Most of the traits important in wood production, such as growth rate, adaptability, and stem and wood quality, are under polygenic control, and therefore depend on more than a single gene. Biosafety regulations governing GM trees are often a concern due to the long period of time necessary to monitor environmental effects (including gene stability, pollen and seed dispersal, and impact on other elements of the ecosystem).

Most existing regulations are both stringent and involve high development costs compared to annual crops. Collaborative genomics programmes such as those carried out on *Populus, Pinus* and *Eucalyptus*, with the on-going characterization of genes of interest (including those coding for flowering, and lignin and cellulose biosynthesis), and the identification of specific promoters of transgenic expression, are expected to boost advances in genetic modification of forest trees. Studies to assess the stability of GM tree growth under abiotic stress, and their environmental impact, have been initiated.

BIOTECHNOLOGY RESEARCH AND APPLICATIONS IN FORESTRY

Unlike work carried out on crop or farm animal species, the domestication of forest trees is, with few exceptions, very recent. However, within a few decades, knowledge transfer from agriculture to forestry has been so fast that genetic knowledge of a few tree model genera, such as poplar (*Populus*), pine (*Pinus*) and *Eucalyptus*, is about to become as important, in terms of complexity and quality, as the genetic knowledge of the main crop species. Modern biotechnologies have potentially outstanding applications in the forestry sector, mainly because of the potential genetic gains they could confer. Expected gains for tree breeders include new genetic pools and significant reduction in tree selection time. Owing to technical difficulties, high costs, and time needed to evaluate forest species, tree breeding programmes have long been limited to the mass selection of the best individuals.

Between the 1970s and 1990s, development of horticultural techniques such as grafting, propagation by cuttings and controlled pollination, as well as information technology development, contributed to the implementation of sophisticated and efficient tree improvement activities. Long-term conventional tree breeding programmes and large-scale clonal propagation techniques have resulted in, and will continue to provide, significant productivity gains, especially for fast-growing species such as *Eucalyptus* and *Populus.* While the objectives of tree breeding are clearly defined and address end-users' (tree growers') requirements (in terms of yield, wood quality, and biotic and abiotic resistance), the situation in biodiversity conservation is less clear. Several issues (including knowledge of species biology, ecosystem conservation, restoration of endangered species, and decision-making tools for ecosystem or forest management) are common to biological diversity and forestry.

However, in spite of a considerable number of studies carried out on hundreds of forest tree species, there are still very few reported biotechnology applications. Many countries are currently engaged in biotechnology-supported forest biodiversity conservation programmes. Nonetheless, there are only a few reported direct impacts at policy or technical levels. The impact of biotechnology on genetic studies has allowed the use of cutting-edge methodologies with forest tree species in spite of the late domestication process compared with crop species. The sequencing of the *Populus* genome has significant implications for forestry. *Eucalyptus* gene sequencing, which was reported during a IUFRO (International Union of Forest Research Organizations) Tree Biotechnology meeting in 2003, has since been pursued by the Eucalypt Genome Initiative, the Eucalyptus Genome Sequencing Project Consortium and the Brazilian project Genolyptus. The Dendrome Project is a collection of forest tree genome databases and other forest genetic information resources, with a special emphasis on conifers and *Pinus*. It is in genome sequencing that research seems currently most active. The development of large-scale vegetative propagation techniques, essentially based on micropropagation or somatic embryogenesis, makes it theoretically possible to deploy superior planting materials rapidly and effectively. This is already the case for some coniferous and broad-leaved tree species including *Eucalyptus* and teak (*Tectona grandis*). Micropropagation techniques are also necessary to regenerate GM plants. Genetic modification has been applied to forest trees mainly during the last two decades, at the experimental stage in the laboratory, to alter various traits such as herbicide, metal, salt, or insect resistance or cold tolerance. The rationale for producing trees with reduced lignin content for the pulp and paper industry is generally based on the expected environmental and economic benefits. Equally if not more important, genetic modification is often considered as a tool to improve knowledge of tree biology and functioning, with significant potential applications; for example, for studying cell wall properties and wood formation. Since genes regulating secondary tissue and lignin formation in the annual plant *Arabidopsis thaliana* were identified, research on the genetic basis of wood quality in forest trees has significantly increased (mainly in *Eucalyptus, Pinus* and *Populus*).

With the exception of micropropagation, for which some applications are reported, most public domain information relates to research activities. Field applications appear extremely limited, according to the literature. Nevertheless, and in spite of unclear economic rationale, the use of biotechnology in forestry research has increased and provided new knowledge on tree biology and functioning. It also appears that forest biotechnology is increasingly funded by the public sector in academic research, while the private sector seems to focus investments more on specific traits of commercial interest such as wood quality.

MICROARRAY DATA USING RANDOM FOREST

Selection of relevant genes for sample classification (e.g., to differentiate between patients with and without cancer) is a common task in most gene expression studies e.g.. When facing gene selection problems, biomedical researchers often show interest in one of the following objectives:

1. To identify relevant genes for subsequent research; this involves obtaining a (probably large) set of genes that are related to the outcome of interest, and this set should include genes even if they perform similar functions and are highly correlated.
2. To identify small sets of genes that could be used for diagnostic purposes in clinical practice; this involves obtaining the smallest possible set of genes that can still achieve good predictive performance (thus, "redundant" genes should not be selected).

We will focus here on the second objective. Most gene selection approaches in class prediction problems combine ranking genes (e.g., using an *F*-ratio or a Wilcoxon statistic) with a specific classifier (e.g., discriminant analysis, nearest neighbor). Selecting an optimal number of features to use for classification is a complicated task, although some preliminary guidelines, based on simulation studies by [4], are available. Frequently an arbitrary decision as to the number of genes to retain is made e.g., keep the 50 best ranked genes and use them with a linear discriminant analysis as in; keep the best 150 genes as in. This approach, although it can be appropriate when the only objective is to classify samples, is not the most appropriate if the objective is to obtain the smaller possible sets of genes that will allow good predictive performance. Another common approach, with many variants e.g., is to repeatedly apply the same classifier over progressively smaller sets of genes (where we exclude genes based either on the ranking statistic or on the effect of the elimination of a gene on error rate) until a satisfactory solution is achieved (often the smallest error rate over all sets of genes tried).

A potential problem of this second approach, if the elimination is based on univariate rankings, is that the ranking of a gene is computed in isolation from all other genes, or at most in combinations of pairs of genes, and without any direct relation to the classification algorithm that will later be used to obtain the class predictions. Finally, the problem of gene selection is generally regarded as much more problematic in multi-class situations (where there are three or more classes to be differentiated), as evidence by recent papers in this area e.g.. Therefore, classification algorithms that directly provide measures of variable importance (related to the relevance of the variable in the classification) are of great interest for gene selection, specially if the classification algorithm itself presents features that make it well suited for the types of problems frequently faced with microarray data. Random forest is one such algorithm.

Random forest is an algorithm for classification developed by Leo Breiman that uses an ensemble of classification trees [14-16]. Each of the classification trees is built using a bootstrap sample of the data, and at each split the candidate set of variables is a random subset of the variables. Thus, random forest uses both bagging (bootstrap aggregation), a successful approach for combining unstable learners [16,17], and random variable selection for tree building. Each tree is unpruned (grown fully), so as to obtain low-bias trees; at the same time, bagging and random variable selection result in low correlation of the individual trees. The algorithm yields an ensemble that can achieve both low bias and low variance (from averaging over a large ensemble of low-bias, high-variance but low correlation trees). Random forest has excellent performance in classification tasks, comparable to support vector machines. Although random forest is not widely used in the microarray literature, it has several characteristics that make it ideal for these data sets:

- Can be used when there are many more variables than observations.
- Can be used both for two-class and multi-class problems of more than two classes.

- Has good predictive performance even when most predictive variables are noise, and therefore it does not require a pre-selection of genes (i.e., "shows strong robustness with respect to large feature sets",*sensu*).
- Does not overfit.
- Can handle a mixture of categorical and continuous predictors.
- Incorporates interactions among predictor variables.
- The output is invariant to monotone transformations of the predictors.
- There are high quality and free implementations: the original Fortran code from L. Breiman and A. Cutler, and an R package from A. Liaw and M. Wiener.
- Returns measures of variable (gene) importance.
- There is little need to fine-tune parameters to achieve excellent performance. The most important parameter to choose is *mtry*, the number of input variables tried at each split, but it has been reported that the default value is often a good choice. In addition, the user needs to decide how many trees to grow for each forest (*ntree*) as well as the minimum size of the terminal nodes (*nodesize*). These three parameters will be thoroughly examined in this paper.

Given these promising features, it is important to understand the performance of random forest compared to alternative state-of-the-art prediction methods with microarray data, as well as the effects of changes in the parameters of random forest. In this paper we present, as necessary background for the main topic of the paper (gene selection), the first through examination of these issues, including evaluating the effects of *mtry*, *ntree* and *nodesize* on error rate using nine real microarray data sets and simulated data.

The main question addressed in this paper is gene selection using random forest. A few authors have previously used variable selection with random forest and use filtering approaches and, thus, do not take advantage of the measures of variable importance returned by random forest as part of the algorithm. Svetnik, Liaw, Tong and Wang propose a method that is somewhat similar to our approach. The main difference is that first find the "best" dimension (p) of the model, and then choose the p most important variables. This is a sound strategy when the objective is to build accurate predictors, without any regards for model interpretability. But this might not be the most appropriate for our purposes as it shifts the emphasis away from selection of specific genes, and in genomic studies the identity of the selected genes is relevant (e.g., to understand molecular pathways or to find targets for drug development).

The last issue addressed in this paper is the multiplicity (or lack of uniqueness or lack of stability) problem. Variable selection with microarray data can lead to many solutions that are equally good from the point of view of prediction rates, but that share few common genes. This multiplicity problem has been emphasized by and and recent examples are shown in Although multiplicity of results is not a problem when the only objective of our method is prediction, it casts serious doubts on the biological interpretability of the results. Unfortunately most "methods papers" in bioinformatics do not evaluate the stability of the results obtained, leading to a false sense of trust on the biological interpretability of the output obtained. Our paper presents a through and critical evaluation of the stability of the lists of selected genes with the proposed (and two competing) methods.

In this paper we present the first comprehensive evaluation of random forest for classification problems with microarray data, including an assessment of the effects of changes in its parameters and we show it to be an excellent performer even in multi-class problems, and without any need to fine-tune parameters or pre-select relevant genes. We then propose a new method for gene selection in classification problems (for both two-class and multi-class problems) that uses random forest; the main advantage of this method is that it returns very small sets of genes that retain a high predictive accuracy, and is competitive with existing methods of gene selection.

EVALUATION OF PERFORMANCE AND COMPARISONS WITH ALTERNATIVE APPROACHES

We have used both simulated and real microarray data sets to evaluate the variable selection procedure. For the real data sets, original reference paper and main features and further details are provided in the supplementary material. To evaluate if the proposed procedure can recover the signal in the data and can eliminate redundant genes, we need to use simulated data, so that we know exactly which genes are relevant. Details on the simulated data are provided in the methods and in the supplementary material.

Estimation of error rates

To estimate the prediction error rate of all methods we have used the.632+ bootstrap method. The.632+ bootstrap method uses a weighted average of the resubstitution error (the error when a classifier is applied to the training data) and the error on samples not used to train the predictor (the "leave-one-out" bootstrap error); this average is weighted by a quantity that reflects the amount of overfitting. It must be emphasized that the error rate used when performing variable selection is not what we report in as prediction error rate.

To calculate the prediction error rate as reported, for example, the.632+ bootstrap method is applied to the complete procedure, and thus the samples used to compute the leave-one-out bootstrap error used in the.632+ method are samples that are not used when fitting the random forest, or carrying out variable selection. The.632+ bootstrap method was also used when evaluating the competing methods. Stability of variable (gene) selection evaluated using 200 bootstrap samples. "# Genes": number of genes selected on the original data set. "# Genes boot.": median (1st quartile, 3rd quartile) of number of genes selected from on the bootstrap samples....

Effects of parameters of random forest on prediction error rate

Before examining gene selection, we first evaluated the effect of changes in parameters of random forest on its classification performance. Random forest returns a measure of error rate based on the out-of-bag cases for each fitted tree, the OOB error, and this is the measure of error we will use here to assess the effects of parameters. We examined whether the OOB error rate is substantially affected by changes in*mtry*, *ntree*, and *nodesize*.

For both real and simulated data, the relation of OOB error rate with *mtry* is largely independent of *ntree* (for *ntree* between 1000 and 40000) and *nodesize* (nodesizes 1 and 5). In addition, the default setting of *mtry* (*mtryFactor* = 1 in the figures) is often a good choice in terms of OOB error rate. In some cases, increasing *mtry* can lead to small decreases in error rate, and decreases in *mtry* often lead to increases in the error rate. This is specially the case with simulated data with very few relevant genes (with very few relevant genes, small *mtry* results in many trees being built that do not incorporate any of the relevant genes). Since the OOB error and the relation between OOB error and *mtry* do not change whether we use *nodesize* of 1 or 5, and because the increase in computing speed from using *nodesize* of 5 is inconsequential, all further analyses will use only the default *nodesize* = 1. These results show the robustness of random forest to changes in its parameters; nevertheless, to re-examine robustness of gene selection to these parameters, in the rest of the paper we will report results for different settings of *ntree* and *mtry* (and these results will again show the robustness of the gene selection results to changes in *ntree* and *mtry*).

Gene selection using random forest

Random forest returns several measures of variable importance. The most reliable measure is based on the decrease of classification accuracy when values of a variable in a node of a tree are permuted randomly, and this is the measure of variable importance that we will use in the rest of the paper. (In the Supplementary material we show that this measure of variable importance is not the same as a non-parametric statistic of difference between groups, such as could be obtained with a Kruskal-Wallis test). Other measures of variable importance are available, however, and future research should compare the performance of different measures of importance.

To select genes we iteratively fit random forests, at each iteration building a new forest after discarding those variables (genes) with the smallest variable importances; the selected set of genes is the one that yields the smallest OOB error rate. Note that in this section we are using OOB error to choose the final set of genes, not to obtain unbiased estimates of the error rate of this rule. Because of the iterative approach, the OOB error is biased down and cannot be used to asses the overall error rate of the approach, for reasons analogous to those leading to "selection bias". To assess prediction error rates we will use the bootstrap, not OOB error. In our algorithm we examine all forests that result from eliminating, iteratively, a fraction,*fraction.dropped*, of the genes (the least important ones) used in the previous iteration. By default,*fraction.dropped* = 0.2 which allows for relatively fast operation, is coherent with the idea of an "aggressive variable selection" approach, and increases the resolution as the number of genes considered becomes smaller. We do not recalculate variable importances at each step as mention severe overfitting resulting from recalculating variable importances. After fitting all forests, we examine the OOB error rates from all the fitted random forests. We choose the solution with the smallest number of genes whose error rate is within u standard errors of the minimum error rate of all forests. Setting $u = 0$ is the same as selecting the set of genes that leads to the smallest error rate. Setting $u = 1$ is similar to the common "1 s.e. rule", used in the classification trees literature [14,15]; this strategy can lead to solutions with fewer genes than selecting the solution with the smallest error rate, while achieving an error rate that is not different, within sampling error, from the "best solution". In this paper we will examine both the "1 s.e. rule" and the "0 s.e. rule".

On the simulated data sets backwards elimination often leads to very small sets of genes, often much smaller than the set of "true genes". The error rate of the variable selection procedure, estimated using the.632+ bootstrap method, indicates that the variable selection procedure does not lead to overfitting, and can achieve the objective of aggressively reducing the set of selected genes. In contrast, when the simplification procedure is applied to simulated data sets without signal (see Tables Tables11 and and22 in Additional file 1), the number of genes selected is consistently much larger and, as should be the case, the estimated error rate using the bootstrap corresponds to that achieved by always betting on the most probable class.

For additional results using different combinations of*ntree* = {2000, 5000, 20000}, *mtryFactor* = {1, 13}, *se*= {0, 1},*fraction.dropped* = {0.2, 0.5}). Error rates when performing variable selection are in most cases comparable (within sampling error) to those from random forest without variable selection, and comparable also to the error rates from competing state-of-the-art prediction methods. The number of genes selected varies by data set, but generally the variable selection procedure leads to small (< 50) sets of predictor genes, often much smaller than those from competing approaches. There are no relevant differences in error rate related to differences in *mtry*, *ntree* or whether we use the "s.e. 1" or "s.e. 0" rules. The use of the "s.e. 1" rule, however, tends to result in smaller sets of selected genes.

Stability (uniqueness) of results

We have evaluated the stability of the variable selection procedure using the bootstrap. This allows us to asses how often a given gene, selected when running the variable selection procedure in the original sample, is selected when running the procedure on bootstrap samples. The results here will focus on the real microarray data sets. For other combinations of*ntree*, *mtryFactor*, *fraction.dropped*, *se*) shows the variation in the number of genes selected in bootstrap samples, and the frequency with which the genes selected in the original sample appear among the genes selected from the bootstrap samples. In most cases, there is a wide range in the number of genes selected; more importantly, the genes selected in the original samples are rarely selected in more than 50% of the bootstrap samples. These results are not strongly affected by variations in *ntree* or *mtry*; using the "s.e. 1" rule can lead, in some cases, to increased stability of the results.

Discussion

We have first presented an exhaustive evaluation of the performance of random forest for classification problems with microarray data, and shown it to be competitive with alternative methods, without requiring any fine-tuning of parameters or pre-selection of variables. The performance of random forest without variable selection is also equivalent to that of alternative approaches that fine-tune the variable selection process. We have then examined the performance of an approach for gene selection using random forest, and compared it to alternative approaches. Our results, using both simulated and real microarray data sets, show that this method of gene selection accomplishes the proposed objectives. Our method returns very small sets of genes compared to alternative variable selection methods, while retaining predictive performance. Our method of gene selection will not return sets of genes that are highly correlated, because they are redundant. This method will be most useful under two scenarios: a) when considering the design of diagnostic tools, where having a small set of probes is often desirable; b) to help understand the results from other gene selection approaches that return many genes, so as to understand which ones of those genes have the largest signal to noise ratio and could be used as surrogates for complex processes involving many correlated genes. A backwards elimination method, precursor to the one used here, has been already used to predict breast tumor type based on chromosomic alterations.

We have also thoroughly examined the effects of changes in the parameters of random forest (specifically *mtry*, *ntree*, *nodesize*) and the variable selection algorithm (*se,fraction.dropped*). Changes in these parameters have in most cases negligible effects, suggesting that the default values are often good options, but we can make some general recommendations. Time of execution of the code increases H" linearly with *ntree*. Larger *ntree* values lead to slightly more stable values of variable importances, but for the data sets examined, *ntree* = 2000 or *ntree* = 5000 seem quite adequate, with further increases having negligible effects. The change in *nodesize* from 1 to 5 has negligible effects, and thus its default setting of 1 is appropriate. For the backwards elimination algorithm, the parameter *fraction.dropped* can be adjusted to modify the resolution of the number of variable selected; smaller values o*ffraction.dropped* lead to finer resolution in the examination of number of genes, but to slower execution of the code. Finally, the parameter *se* has also minor effects on the results of the backwards variable selection algorithm but a value of *se* = 1 leads to slightly more stable results and smaller sets of selected genes.

In contrast to other procedures our procedure does not require to pre-specify the number of genes to be used, but rather adaptively chooses the number of genes have conducted an evaluation of several gene selection algorithms, including genetic algorithms and various ranking methods; these authors show results for the Leukemia and NCI60 data sets, but the Leukemia results are not directly comparable since focus on a three-class problem. They report the best results with the NCI60 data set estimated with the.632 bootstrap rule. These best error rates are 0.408 for their evolutionary algorithm with 30 genes and 0.318 for 40 top-ranked genes. Using a number of genes slightly larger than us, these error rates are similar to ours; however, these are the best error rates achieved over a range of ranking methods and error rates, and not the result of a complete procedure that automatically determines the best number of genes and ranking scheme. A comparative study of feature selection and multi-class classification. Although they use four-fold cross-validation instead of the bootstrap to assess error rates, their results for three data sets common to both studies (Srbct, Lymphoma, NCI60) are similar to, or worse than, ours. In contrast to our method, their approach pre-selects a set of 150 genes for prediction and their best error rates are those over a set of seven different algorithms and eight different rank selection methods, where no algorithm or gene selection was consistently the best. In contrast, our results with one single algorithm and gene selection method (random forest) match or outperform their results.

Recently, several approaches that adaptively select the best number of genes or features have been reported. For the Leukemia data set our method consistently returns sets of two genes, similar to using an exhaustive search method, and lower than the numbers given by of 3 to 25 have proposed a Bayesian model averaging (BMA) approach for gene selection; comparing the results for the two common data sets between our study and theirs, in one case (Leukemia) our procedure returns a much smaller set of genes (2 vs. 15), whereas in another (Breast, 2 class) their BMA procedure returns 8 fewer genes (14 vs. 6); in contrast to BMA, however, our procedure does not require setting a limit in the maximum number of relevant genes to be selected have developed a method for gene selection and classification, LS Bound, related to least-squares SVMs; their method uses an initial pre-filtering (they choose 1000 initial genes) and is not clear how it could be applied to multi-class problems. The performance of their procedure with the leukemia data set is better than that reported by our method, but they use a total of 72 samples (the original 38 training plus the 34 validation of [44]) thus making these results hard to compare. With the colon data sets, however, their best performing results are not better than ours with a number of features that is similar to ours. Proposed two Bayesian classification algorithms that incorporate gene selection (though it is not clear how their algorithms can be used in multi-class problems).

The results for the Leukemia data set are not comparable to ours (as they use the validation set of 34 samples), but their results for the colon data set show error rates of 0.167 to 0.242, slightly larger than ours (although these authors used random partitions with 50 training and 12 testing samples instead of the.632+ bootstrap to assess error rate), with between 8 and 15 features selected (somewhat larger than those from random forest). Finally, [31], applied both shrunken centroids and a genetic algorithm + KNN technique to the NCI60 and Srcbt data sets; their results with shrunken centroids are similar to ours with that technique, but the genetic algorithm + KNN technique used larger sets of genes (155 and 72 for the NCI60 and Srbct, respectively) than variable selection with random forest using the suggested parameters. In summary, then, our proposed procedure matches or outperforms alternative approaches for gene selection in terms of error rate and number of genes selected, without any need to fine-tune parameters or preselect genes; in addition, this method is equally applicable to two-class and multi-class problems, and has software readily available. Thus, the newly proposed method is an ideal candidate for gene selection in classification problems with microarray data.

A reviewer has alerted us to the paper by Jiang et al., previously unknown to us. In fact, our approach is virtually the same as the one used by Jiang et al., with the exception that these authors recompute variable importances at each step (we do not do this in this paper, although the option is available in our code) and, more importantly, that their gene selection is based both in the OOB error, as well as the prediction error when the forest trained with one data set is applied to a second, independent, data set; thus, this approach for gene selection is not feasible when we only have one data set. Jiang et al. [45] also show the excellent performance of variable selection using random forest when applied to their data sets. The final issue addressed in this paper is instability or multiplicity of the selected sets of genes. From this point of view, the results are slightly disappointing. But so are the results of the competing methods. And so are the results of most examined methods so far with microarray data, as shown in [29] and [30] and discussed thoroughly by [27] for classification and by [28] for the related problem of the effect of threshold choice in gene selection. However, and except for the above cited papers and [6,46] and [5], this is an issue that still seems largely ignored in the microarray literature. As these papers and the statistical literature on variable selection (e.g., [40,47]) discusses, the causes of the problem are small sample sizes and the extremely small ratio of samples to variables (i.e., number of arrays to number of genes). Thus, we might need to learn to live with the problem, and try to assess the stability and robustness of our results by using a variety of gene selection features, and examining whether there is a subset of features that tends to be repeatedly selected. This concern is explicitly taken into account in our results, and facilities for examining this problem are part of our R code.

The multiplicity problem, however, does not need to result in large prediction errors. That the very different classifiers often lead to comparable and successful error rates with a variety of microarray data sets. Thus, although improving prediction rates is important, when trying to address questions of biological mechanism or discover therapeutic targets, probably a more challenging and relevant issue is to identify sets of genes with biological relevance. Two areas of future research are using random forest for the selection of potentially large sets of genes that include correlated genes, and improving the computational efficiency of these approaches; in the present work, we have used parallelization of the "embarrassingly parallelizable" tasks using MPI with the Rmpi and Snow packages [50,51] for R. In a broader context, further work is warranted on the stability properties and biological relevance of this and other gene-selection approaches, because the multiplicity problem casts doubts on the biological interpretability of most results based on a single run of one gene-selection approach.

USEFUL TRAITS BY GENETIC TRANSFORMATION NUTRITIONAL IMPROVEMENT

In addition to its basic calorific value, wheat with its high protein content is an important source of plant protein in the human diet. One of the prominent targets for the application of transgene technology for the nutritional improvement of wheat is targeted at enhancing the grain quality by, i) increasing the protein content, ii) increasing essential amino acids such as lysine, iii) increasing the high molecular weight (HMW) glutenins to improve breadmaking properties of wheat flour, iv) modifying starch composition and v) producing pharma-and neutraceuticals. Amongst the cereals, the flour of bread wheat, Triticum aestivum, has a superior capability of forming leavened bread. This superiority stems from the structure and composition of its seed storage proteins, which upon hydration can interact to form gluten, an insoluble, but highly hydrated, visco-elastic aggregate that endows the wheat dough with its unique properties.

Although the majority of wheat seed storage proteins participate in gluten formation, biochemical and genetic evidence has demonstrated that high molecular weight glutenin subunit (HMW-GS) plays a major role in determining the visco-elastic properties thereby determining bread making qualities. The HMW glutenins are necessary to create a strong dough, which is essential for making high quality, yeast-raised breads. Strong dough traps tiny bubbles of carbon dioxide gas formed naturally by yeast during mixing and subsequent raising thereby enabling the dough to rise forming leavened breads. Dough strength and the ability to contain gas bubbles is known as visco-elasticity and is an important characteristic of wheat with respect to its end product quality. Increasing understanding of the molecular basis of dough visco-elasticity would thus help in developing strategies for minimizing the effect of unfavourable environmental factors on wheat end-use quality.

Genetic transformation of wheat is a key component in a scheme proposing a complete set of approaches to apply biotechnology to improve wheat quality via direct manipulation of HMW-glutenin genes. To alter the amount and composition of these proteins by genetic engineering, a gene encoding a novel, hybrid-subunit of HMW-GS under the control of native HMW-GS regulatory sequences was inserted into wheat using the biolistic approach (Blechl and Anderson, 1996). The HMW-GS 1Ax1 gene which is known to be associated with superior bread making quality, was introduced into the cultivar (Bobwhite) lacking this gene. The introduced 1Ax1 gene under the control of HMW-GS promoter was expressed at high levels and stability was maintained for several generations.

The results demonstrate the feasibility of manipulating the composition of wheat kernels by genetic engineering and have successfully made changes in both the levels and the types of seed storage proteins. This work also demonstrated the usefulness of tissue-specific promoters for the expression of transgenic proteins in the endosperm tissue of wheat. The prospect of achieving the elusive goal of nutritional improvement of wheat brightened further with the improvement in functional properties of wheat dough due to transformation of wheat with high molecular weight subunit genes. Transformation with one or two subunit genes results in stepwise increase in dough elasticity. The expression of a recombinant protein with x-and y-type HMW-GS subunit in transgenic wheat has resulted in altered gluten polymer assembly and composition. This study thus made it feasible to change the glutenin composition by expressing modified HMW-GS in transgenic plants. The quality improvement studies were extended to durum wheat by He *et al.*

1999 with the introduction of one of the two subunits of HMW-GS subunit genes (1Ax1 or 1Dx5). Analysis of the expression of the additional subunits in the T2 generation using a mixograph, indicated an increase of dough strength and stability. This study demonstrated the feasibility of manipulating durum wheat for superior bread and pasta making qualities. Similarly, the overexpression of the HMW subunit 1Dx5, in transgenic wheat (T-aestivum) resulted in a four-fold increase in this proportion of component in the seed protein and also a corresponding increase in the proportions of the total HMW proteins and glutenins.

However, the overexpression of the 1Dx5 gene was found to be associated with a dramatic increase in dough strength by making it too strong, and, therefore, unsuitable for use in conventional breadmaking. Recently, Alvarez *et al.* (2000) introduced the HMW-GS genes 1Ax1 and 1Dx5 into a commercial cultivar of T-aestivum that already expresses five subunits. The overexpression of 1Dx5 gene increases the contribution of the HMW-GS to a level of 22% of the total protein content. In near future, molecular approaches including genetic transformation and marker-assisted selection will provide an opportunity for improving further the wheat processing qualities.

Modification of wheat starch is presently targeted in various laboratories to improve its potential utility. Wheat grain is predominantly composed of starch, which is a mixture of two polymers, the almost linear amylose molecules, and the heavily branched amylopectin molecules. The ratio of amylose to amylopectin in starch determines its physico-chemical characteristics and thereby its end-use. Wheat flour, low in amylose content is desirable for noodlemaking as it improves noodle texture. In wheat, scientists are thus working towards increasing the amylopectin content of starch thereby reducing amylose content leading to the formation of a value added, low amylose flour. The starch branching enzymes catalyse formation of 1, 6-linkages in the glucan polymer and control the amount of amylopectin produced. In recent years, efforts are underway towards detailed characterization of the starch branching enzymes and starch synthases. This increasing information and characterization of the various components of starch biosynthesis will enable researchers to make rational design of novel starches and alteration in starch levels in other crop plants as well.

Wheat is widely used as an animal feed also for non-ruminants in several developed countries of the world. The phytase of Aspergillus niger is used as a supplement in animal feeds to improve the digestibility and also to improve the bioavailabilty of phosphate and minerals. The phyA gene from Aspergillus niger, encoding for the phytase enzyme has been successfully expressed in transgenic wheat lines by the microprojectile bombardment of immature embryos. The constitutively overexpressed phytase was found to accumulate at high levels in the endosperm. This work may thus open newer avenues for a wider applicability of wheat. Wheat is also an ideal system for the production of novel compounds due to its excellent storage properties and the existence of an efficient processing industry. The potential products include the high value, low-volume compounds such as biologically active proteins and peptides, as well as, high volume, low-cost raw materials for packaging and building. The production of recombinant antibodies in rice and wheat was recently reported with the expression of a medically important, single chain Fv recombinant antibody against the carcino-embryonic antigen.. The recombinant antibody was targeted to the plant cell apoplast and endoplasmic reticulum and was detected in the leaves and seeds of wheat. More significantly, the recombinant antibodies remained active after prolonged seed storage at room temperature thus opening avenues for the further exploitation of wheat for producing high-value, novel compounds.

ENGINEERING NUCLEAR MALE STERILITY

The production of hybrids is an essential component of crop breeding programmes but till date, hybrid wheat has remained elusive! The development of a suitable hybridisation system for wheat requires a high degree of male sterility in all parts of the female parent to avoid self-fertilization. De Block and coworkers have developed a nuclear male sterile system in wheat by introducing the barnase gene under the control of a tapetum specific promoter, the expression of which prevents normal pollen development at specific stages of anther development. This system employed the ribonuclease-inhibitor barstar gene to restore the fertility of male sterile plants. To avoid complicated gene integration patterns, the target tissues were incubated on niacinamide containing medium before bombardment. The authors suggest that the enzyme poly (ADP-robose) polymerase (PARP), which plays a key role in the processes of cell division and recombination, is inhibited by niacinamide, thereby resulting in simple integration pattern of the transgene. Expressing the barnase gene at specific stages of anther development destroys the tapetum, thereby preventing normal pollen development and causes pollen sterility.

RESISTANCE TO BIOTIC STRESS

The integration of transgenic approaches with classical breeding techniques offers a potential chemical-free and environment-friendly solution for controlling pests and pathogens. Wheat is attacked by a number of viral, bacterial and fungal pathogens and also by insect and nematode pests. Introgression of genes from the wild relatives exhibiting resistance to pests and pathogens has been successfully utilized over the years for generation of resistant varieties in wheat. With the development of plant transformation techniques, newer avenues for creating disease resistant and insect resistant crops have been created. Coupled with this is the availability of novel transgenes encoding highly potential antimicrobial peptides, defense-related proteins and enzymes for the production of anti-microbial compounds in crop plants, have greatly enhancing the possibility of engineering crop plants for resistance to pests and pathogens. Fungal pathogens of wheat cause severe crop damages by infecting the spikes, leaves and roots. Amongst the various strategies to introduce fungal resistance by transgenic approach, strengthening the host plant defense by genetic manipulation hold tremendous potential. Biochemical and structural responses against fungal attack include reinforcement of plant cell wall, accumulation of phytoalexins with microbial toxicity, ribosome-inactivating proteins (RIP) that inhibit protein synthesis, antimicrobial peptides and synthesis of other PR proteins. Genetic engineering allows the expression of foreign genes from distant unrelated species as well as the modification of the usual pattern of expression of an already present gene.

Most of the works on genetic engineering of wheat for resistance against biotic stress have focussed on developing protection against fungal pathogens. Introduction genes encoding for chitinases from barley resulted in increased resistance against Erysiphe graminis. Bliffeld *et al.* (1999) reported the adverse effect of a ribosome inactivating protein on plant regeneration and development. Nonetheless, moderate protection against Erysiphe graminis was reported with the transformation of wheat with a gene encoding for a ribosome inactivating protein from barley. The genes encoding for thaumatin like protein (TLP) and stilbene synthase in transgenic wheat have been shown to improve resistance of T1, T2 progeny plants against the fungal pathogens. An increase in endogenous resistance against Tilletia tritici was achieved with the introduction of virally encoded antifungal protein.

For the engineering of resistance against different pests and pathogens in wheat, genes encoding for viral coat proteins, antifungal proteins, and proteinase inhibitors have been successfully introduced. Most of the introduced genes confer increased resistance to the corresponding pests and pathogens in the transgenic plants. Introduction of canditate genes (lectin, proteinase inhibitor) for insect resistance into wheat have resulted in growth inhibition of insects on transgenic seeds, thereby decreasing the fecundity of insect population.

RESISTANCE TO ABIOTIC STRESS

Traditional approaches at transferring resistance to crop plants are limited by the complexity of stress tolerance traits, as most of these are quantitatively linked traits (QTLs). Nonetheless, the direct introduction of a small number of genes by genetic engineering offers convenient alternative and a rapid approach for the improvement of stress tolerance. Although, present engineering strategies rely on the transfer of one or several genes that encode either biochemical pathways or endpoints of signalling pathways, these gene products provide some protection, either directly, or indirectly, against environmental stresses. Drought is a major abiotic factor that limits crop productivity, thereby causing enormous loss. The genes encoding the late embryogenesis proteins (LEA) which accumulate during seed desiccation, and in vegetative tissues when plants experience water deficiencies have recently emerged as attractive candidates for engineering of drought tolerance. Transgenic approach has been used for successfully introducing and overexpressing the barley HVA1 gene encoding for a late embryogenesis abundant (LEA) protein by Sivamani *et al.* (2000) into wheat by particle bombardment.

Most of the transgenic lines tested displayed improvement in important agronomic traits, including total dry mass and water use efficiency, shoot dry weight, root fresh and dry weights, when plants are grown under soil water deficit conditions. In general, this investigation showed that the transgenic lines expressing the HVA1 gene had improved growth characteristics including an enhanced biomass yield under water deficit conditions. The discovery of novel genes, determination of their expression pattern in response to abiotic stress and an improved understanding of their roles in stress adaptation (obtained by the use of functional genomics) will provide the basis of effective engineering strategies leading to greater stress tolerance (Cushman and Bohnert, 2000).

TRANSGENE SILENCING IN WHEAT

Success at developing improved wheat cultivars through genetic engineering depends on stable and predictable expression of the inserted gene. However, gene silencing is a common phenomenon in the production of transgenic plants and needs to be effectively controlled for the desired results. This is all the more important because gene silencing is an important phenomenon involved during natural plant defense. Gene silencing is a complicated phenomenon as it includes both transcriptional gene inactivation and post transcriptional gene inactivation. The complex and the large genome size of wheat is expected to be prone to silencing by introduced transgenes.

With the development of transformation methodologies for wheat, more information has started to accumulate regarding the inheritance and expression of the introduced transgenes. Information regarding the long term stability of transgenes is thus of immense significance for the use of genetic manipulation as a tool in wheat crop improvement and is necessary for the introduction and subsequent expression of desirable agronomic trait.

DNA methylation plays a significant role in establishing and maintaining an inactive state of the gene by rendering the chromatin structure inaccessible to the transcription machinery. Methylation of DNA is expected to result in reduced gene expression. Muller *et al.* (1996) studied the variability of transgene expression in clonal cell lines of wheat and found a negative correlation between the degree of methylation and marker gene expression. PEG-mediated approach was used to transform protoplasts isolated from suspension cultures of T-aestivum. The integration and expression of the selectable marker gene nptII fused with different promoters, namely, alcohol dehydrogenase, shrunken from maize, and actin1 from rice, was confirmed by Southern analysis and enzyme activity test. Transgenic cell lines under selection were 'protoplasted' and clonal callus lines were cultivated from genetically identical single cells without selection pressure. A reduction/loss of marker gene expression was observed due to a reduction in the nptII transcript level and was seen to be associated with hypermethylation of the integrated DNA. The silencing effect was reversed by a 4-week culture phase on media supplemented with demethylation agent, 5-azacytidine, thus confirming the role of methylation in transgene silencing Demeke *et al.* (1999) studied the inheritance and stability of an Act1D-uidA: nptII expression cassette in T4 and T5 transgenic plants. Based on the histochemical localization of GUS activity, this study demonstrated the lack of any cytoplasmic effect on the inheritance of the transgene. The transgenes maintained a multiple integration pattern similar to that observed in the T1 generation. The transgenic plants which produced low gus and nptII activity in seeds had an intact expression cassette. Southern blot analysis of genomic DNA performed with the methylation sensitive enzyme, HpaII, showed the transgene in GUS negative plants to be highly methylated relative to the transgene in GUS positive plants. Some of the studies on gene silencing also indicate that multiple integration pattern and copy number is also associated with DNA methylation.

Cannell and coworkers studied the inheritance of gus and bar marker genes over three generations. The integration, inheritance and expression of these marker genes in the population studied were stable and predictable with a few exceptions. The inheritance of integration patterns was stable, and transmission/ inheritance of the transgenes followed Mendelian ratios in a majority of lines. From this study, the authors claim that the transformation procedure, transgene integration, and marker gene expression had little effect on the transmission of transgenes to the subsequent progenies. However, over the three generations studied, a uniform, 'progressive' transgene silencing, specific to the gus gene, was observed. This gradual loss of the gus gene expression was not accompanied by a reduction in bar expression. Since in this case, the gus and the bar genes are located on the same plasmid, observations indicate a post-transcriptional gene silencing.

The silencing of HMW glutenins was also observed in transgenic wheat expressing extra HMW subunits. Characterization of six independent events involving the transformation of wheat HMW-GS genes, 1Ax1 and 1Dx5, in a cultivar expressing five subunits by particle bombardment resulted in partial or complete gene silencing. Silencing of all the HMW glutenin subunits was observed in two different events of transgenic wheat expressing the 1Ax1 subunit transgene and overexpressing the 1Dx5 subunit. Control of gene silencing thus remains a challenge for the immediate future. Meyer, (1995) advocated that the problem of gene silencing in wheat can be minimized by optimising methods for simple integration patterns, use of promoters and gene sequences isolated from cereals, use of matrix associated regions (MARs) or scaffold attachment regions (SARs), which insulate transgenes from surrounding chromatin, might help in reducing the gene silencing problem.

TRANSPOSON TAGGING IN WHEAT

Transposon mutagenesis has been widely exploited in various organisms to isolate genes that encode unidentified products. The maize activator (Ac) and Dissociation (Ds) elements are the best-studied transposable elements in heterologous host plants. Initial studies have reported the introduction and activation of the Ac/Ds elements into cultured wheat cells by using a wheat dwarf virus by particle bombardment. Takumi and coworkers has developed a transposon tagging system in wheat by introducing the Ac transposase gene under the CaMV 35S promoter into cultured wheat embryos by particle bombardment.

For the development of a transposon tagging system, embryos isolated from a stable Ac line were bombarded with a plasmid containing the maize dissociator (Ds) element located between the rice Act1 and the gus genes. The transient expression of the gus gene was observed after the excision of Ds elements. Southern and northern analysis of the T0 and T1 plants has exhibited the stable expression and inheritance of the Ac transposase gene. These results have demonstrated the precise processing of the maize Ac transposase gene and the synthesis of the transposase protein in transgenic Ac lines. The Ac transposase gene causes the transactivation and excision of the Ds in the Ac transgenic lines transformed with the maize Ds elements. Thus in near future we can expect traits of commercial importance to be tagged for a wider utility in important crop plants.

MARKER ASSISTED SELECTION IN WHEAT BREEDING

Conventionally, plant breeding depends upon morphological/phenotypic markers for the identification of agronomic traits. With the development of methodologies for the analysis of plant gene structure and function, molecular markers have been utilized for identification of traits. Molecular markers act as DNA signposts to locate the gene(s) for a trait of interest on a plant chromosome, and are widely used to study the organization of plant genomes and for the construction of genetic linkage maps. Molecular markers are independent from environmental variables and can be scored at any stage in the life cycle of a plant. Over the last several years, there has thus been marked increase in the application of molecular markers in the breeding programmes of various crop plants. Molecular markers not only facilitate the development of new varieties by reducing the time required for the detection of specific traits in progeny plants, but also fasten the identification of resistance genes and their corresponding molecular markers, thus accelerating efficient breeding of resistance traits into wheat cultivars by marker assisted selection (MAS). The availability of back-cross derived near isogenic lines (NILs) have also facilitated the analysis of various lines by using different marker systems. The introduction of alien genetic variation into wheat is a valuable and proven technique for wheat improvement. Wild relatives of wheat provide an enormous resource of new genes for wheat improvement, particularly disease and stress tolerance. These genes can be of use if recombined into lines adapted to the conditions of a particular region.

Initial studies on the application of molecular markers in wheat relied on the hybridisation based restriction fragment length polymorphism (RFLP) system-RFLP maps provided a more direct method for selecting desirable genes via their linkage to easily detectable markers thereby expediting the movement of desirable genes among varieties. The factors that had been instrumental for the use of RFLP in wheat was the limited number of polymorphisms observed among wheat lines and more significantly the availability of aneuploid stocks for the determination of chromosomal location of genes. In wheat, RFLP's have been used to map seed storage protein loci, loci associated with flour colour, cultivar identification, vernalization (Vrn1) and frost resistance gene on chromosome 5A, intrachromosomal mapping of genes for dwarfing (Rht12) and vernalization (Vrn1), resistance to preharvest sprouting, quantitative trait loci (QTL's) controlling tissue culture response (tcr), nematode resistance, milling yield, resistance to chlorosis induction by Pyrenophora tritici-repentis.

RFLP markers are also useful in selection programmes for resistance against pests and pathogens, which is otherwise labour and time consuming, and to detect homozygous individuals and have been used for resistance to barley yellow dwarf virus, resistance to wheat spindle streak mosaic virus, resistance against powdery mildew, resistance against leaf rust, resistance against cereal cyst nematode. The use of RFLP analysis in wheat has, however, been of limited use in the intervarietal analysis due to low level of polymorphism and the high cost for screening in breeding situations.

With the development of polymerase chain reaction (PCR) methodologies, Random Amplified Polymorphic DNA (RAPD) emerged as a convenient and effective technique for tracing alien chromosome segments in translocation lines. RAPD markers provide a useful alternative to RFLP analysis for screening markers linked to a single trait within near isogenic lines and bulked segregants. He *et al.* (1992) reported the development of a DNA polymorphism detection method by combining RAPD with DGGE (denaturing gradient gel elecrophoresis) for pedegree analysis and fingerprinting of wheat cultivars. RAPD markers can be converted to more user-friendly Sequence Characterized

Amplified Region (SCAR) markers, that display a less complex banding pattern. SCAR markers linked to resistance genes against fungal pathogens have been characterized in combination with RAPD and RFLP. In recent years, RAPD and other PCR based markers like Sequence Characterized Amplified Regions (SCAR), Sequence Tagged Sites (STS) and Differential Display Reverse Transcriptase PCR (DDRT-PCR) are increasingly being used for identification of desirable traits in wheat and related genera.

These markers have been used in particular for disease resistance against viral and fungal pathogens and also for insect and nematode pests and have the potential of pyramiding of resistance genes for effective breeding programmes.

PCR based markers have been extensively characterized for genes of resistance against common bunt, Tilletia tritici, powdery mildew, Erysiphe graminis, leaf rust, Puccinia recondita, resistance against Hessian fly, Mayetiola destructor and Russian wheat aphid, Diuraphis noxia.

Simple sequence repeats or microsatellites are more promising molecular markers for the identification and differentiation of genotypes within a species. The high level of polymorphism and easy handling has made microsatellites extremely useful for different applications in wheat breeding.

Microsatellites have also been used to identify resistance genes like Pm6 from Triticum timopheevii and Yr15 from breadwheat. In near future, molecular markers can provide simultaneous and sequential selection of agronomically important genes in wheat breeding programmes allowing screening for several agronomically important traits at early stages and effectively replace time consuming bioassays in early generation screens.

MANAGEMENT OF PLANT GENETIC RESOURCES IN AGROECOSYSTEMS

In 1995, national partners in Burkina Faso, Ethiopia, Hungary, Mexico, Morocco, Nepal, Peru, Turkey and Vietnam together with IPGRI initiated a global project, 'Strengthening the scientific basis of *in situ* conservation of agricultural biodiversity'. The purpose is to strengthen the scientific basis, institutional linkages and policies that support the farmers' role in conservation and use of crop genetic diversity.

Projects are implemented by linking existing national plant genetic resource (PGR) programmes with other partners such as universities, national institutes, agricultural extension workers, NGOs, community-based organizations (CBOs) and farmers. An expected result is the creation of a channel for input from these sources into national agricultural research agendas.

A cross-section of case studies includes 'Creating a national framework' (Nepal), 'Understanding farmer preferences' (Morocco), 'Gendered participation in *in situ* conservation' (Mexico) and 'Participatory methods to add value to PGRs' (Nepal).

APPROACH

On-farm conservation of traditional crops is carried out by farming communities for farming communities. The primary task for those concerned with conservation and with the maintenance of traditional crop diversity *in situ* is to understand when, where and how this will happen, who will maintain the material and how those maintaining the material can benefit. Four areas of investigation were identified that set the necessary scientific agenda needed to support farmers and local communities in *in situ* conservation on farm:

- What is the extent and distribution of the genetic diversity maintained by farmers over space and over time?
- What are the processes used to maintain the genetic diversity on-farm?
- Who maintains genetic diversity within farming communities (men, women, young, old, rich, poor, certain ethnic groups)?
- What factors (market, non-market, social, environmental) influence farmer decisions on maintaining traditional varieties?

Formulating and answering these questions required a participatory approach at all stages of the process. Participatory methods, such as key informant interview, focus group discussions, spatial mapping and matrix ranking served to include farmers' knowledge on local social-cultural, economic and agroecological conditions, their crop and seed management practices and the characteristics and origins of their varieties into project data. Information from participatory research is complemented by household, market and seed system surveys, field trials on-station and on-farm and genetic diversity measurements in the field and in laboratories by economists, sociologists, ethnobotanists, agronomists, ecologists and populations geneticists, often in collaboration with farmers and extension workers. A preliminary 'exploratory approach' (not based on preliminary hypotheses) was used first, because it did not presuppose or assume the different categories or reasons underlying farmers' knowledge and it enables farmers to employ their own values and standards of measurement. Hypothesis testing followed, to answer specific questions that would support the scientific basis of on-farm conservation. The different sources and levels of information include the variety, crop, parcel or plot, household, village or community, landscape or region. Information is collected, disaggregated by gender and in some cases by wealth categories and ethnic groups.

Moreover, information from one aspect may be useful to answer more than one question. Understanding the relationship between what farmers recognize as or name a variety and the genetic distinctiveness of this unit is key to understanding the amount and distribution of crop genetic diversity managed by farmers. The information collected in only one year at the level of the household or farmer's plot may not be the appropriate scale for analysis or for agrobiodiversity conservation. Thus, information is analysed at different spatial and temporal scales.

Implementing the project involved developing multi-institutional, multidisciplinary collaboration at international, national and local levels. This included ensuring that trained national male and female personnel were available to carry out the work at central and local levels, and that the teams promoted equity at all project levels. In addition, because the project is largely community-based, time was devoted to building or creating rapport with the farmers in whose fields much of the work is being undertaken, and whose experiences and knowledge provide a central component of the project. Key has been separate male and female farmer cross-site visits for promoting exchange of information by gender.

Linking research to development is central to the project. A range of activities in the different countries ensures that the project benefits national conservation programmes, partner institutions and the participating farmers. The information collected is used to mainstream the use of local crop genetic resources into the agricultural development arena. In contrast to research and analysis methodologies that have widespread application, actual interventions to include management of crop genetic diversity in agricultural development activities are found to be site-specific. The project is creating a portfolio of options, based on many case studies from participating countries, to which national programmes may refer for ideas for increasing the benefits to farmers from local crop diversity.

Reflections

The method has some disadvantages. Time is needed to formulate proper hypotheses to avoid unnecessary data collection that can result because of the complexity of research and number of issues to be addressed. Integrating quantitative and qualitative data (empirical versus farmer knowledge) is a challenge. Time is also needed to build linkages between disciplines and formal and informal sectors and these linkages are vulnerable to continuing cultural, economic and environmental change.

Advantages of the research are that it focuses on locally important crops and builds on local knowledge. The gender disaggregated approach to data collection, cross-site visits and actively pursuing gender equity in employing local project team members and managers has enabled the project to more clearly address issues of livelihood improvement and empowerment and address gender and equity concerns. Farmers take ownership for their own resources. The method recognizes the achievements of both male and female farmer-breeders as 'keepers of diversity'. They in turn guarantee sustainability in that they perpetuate the process, continuing it when 'intervention' has finished. The process also channels farmers' voices into national agricultural research and extension systems and fosters cooperation between local, national and international levels and GO and NGO sectors.

Innovative approaches were developed for participatory NRM (PNRM), including developing frameworks to support the recognition, conservation and improvement of farmer-developed PGRs *in situ*. An understanding was gained of the appropriate scale for data collection, aggregation and analysis and for different stakeholders' management decisions. The process revealed the importance of taking time to strengthen farmer informal and formal linkages, integrating on-farm conservation into national PGR programmes as part of their regular annual plans and including agricultural extension staff at national and local levels in participatory training and project implementation.

Plant genetic resources are a natural resource fundamental to agricultural production. Most conservation efforts to date have focused on *ex situ* options. However, this method of conservation recognizes crop germplasm as the evolutionary product of the continuing interaction between farmers and their environments. As farmers continue planting, harvesting, selecting and storing seed, PGRs are renewed and developed. Non-use of PGRs leads to their loss from their surrounding environments. Acknowledging the dependence of crop PGRs on human use highlights the necessity of a participatory approach as inherent in the goal of conservation.

NATURAL RESOURCE MANAGEMENT RESEARCH

Since 1994, ICARDA scientists have worked with colleagues in Egyptian research institutions to design and implement a programme of resource management research in key agricultural environments in the country. Researchers carried out literature reviews, rapid appraisals, formal farm surveys and planning before establishing long-term trials at four irrigated sites (one each in the Delta and Middle Egypt, and two in the newly reclaimed desert lands, known as New Lands), and at one rainfed site (near Rafah, North Sinai). Each site has three major research problems. Water (both quality and quantity) is the paramount concern at all sites. Maintaining soil fertility is essential in the old lands of the Delta and Middle Egypt but building up soil fertility is essential for sustained production in the New Lands and rainfed areas. A third issue that the trials address is the choice of sustainable crop sequences for rotational systems. These on-station trials are designed for a minimum of 12 years and are entirely managed by researchers.

APPROACH

At each site, the long-term trial is integrated with participatory research in surrounding villages and on individual farms. Like the on-station trials, the participatory work, called long-term monitoring (LTM), is intended to have an extended life. Its purpose is to establish a continuing dialogue with farmers concerning their farming practices, management decisions and the related conditions of their natural resource base. The dialogue centres on long- and short-term farmer objectives, their perceptions of the qualitative aspects of the resource base and their technical knowledge of resource management. The participatory research also involves a longitudinal study of farmers' management of natural resources in response to changing environmental, economic and social circumstances.

As part of their exchange with farmers, researchers are also monitoring changes in the status of natural resources on representative farms through periodic biophysical measurements. Researchers are combining farmer participation with biophysical measurements to provide information about the interaction between natural resource conditions and farmers' management practices. Once institutionalized, the LTM system will provide a mechanism by which researchers and farmers can exchange knowledge on improved management practices and their effects on natural resource health. A multidisciplinary research team is conducting the monitoring at each location. Each team includes members of local farmer associations, local extension staff, researchers from various institutes and participating farmers.

These farmers were selected according to a carefully prepared list of environmental criteria relevant to each location, including hydrological and soil factors and cropping patterns. Socioeconomic factors such as farm size and type, natural resource endowment, social background, level of education and household composition were given equal weight. Farmers were selected at random from lists prepared for each site.

They received a thorough explanation of the purpose and activities of the LTM system and were asked if they would like to participate. They were also informed about the amount of time and information required and the need for a long-term commitment. The 85 farmers who agreed to participate in developing the system represent the whole range of social, economic and natural resource conditions at each study location. Without altering the integrity of the research design, the programme made provisions for new participants to join. For each participating farmer, information on socioeconomic factors, farm management decisions and perceptions of resource conditions and productivity performance are being collected every six months, after the main winter and summer cropping seasons.

Natural resource conditions are measured on different schedules according to scientific requirements. In addition to basic information about crop sequences and rotations, management practices, input use, productivity and economic returns, data are collected on labour use and sources, household composition, income sources and household investment patterns. This information will explain why farmers make the decisions they do and thus should help develop profitable and sustainable production practices.

A review workshop is held once a year to bring together the research teams, including farmer members, for discussion of results and trends in the information collected.

Reflections

Through this work, Egyptian farmers, researchers and extension workers are building and testing a new holistic approach to studying agricultural production, including socioeconomic and biophysical factors and their effects on the natural resource base over time.

STRATEGIC USE OF MULTIPURPOSE FORAGES

The objective of working with rural communities is mainly to enhance interaction with local stakeholders and to achieve an impact in the initial intervention sites. The partnership with nationally based institutions is the basis for a wider distribution of results and for multiplication of farmer-selected forage seeds– seed availability being a major bottleneck for the adoption of improved forages. Conceptually, this activity relies on farmer–researcher–institutional linkages. The aims are to use forages to:

- Increase the income of smallholders mainly through direct effects on livestock production.
- Increase income and food security through, among others, improved soil fertility, soil conservation, breaking pest and disease cycles and reducing the competition of weeds.
- Ensure conservation of natural resources.

In 1998 the study began in Honduras in the CIAT reference site of Yorito, department of Yoro. Work has since been extended to other countries in Central America.

A team of researchers drawn from forage genetic resources, geographic information systems (GIS), participatory research and soils carries out the project. Together with farmers, local grassroots organizations, NGOs and other relevant institutions in and outside the region, the project develops forage-based technologies and identifies forage germplasm adapted to the needs of different types of farmers.

APPROACH

Farmers are offered a range of grass and legume options for livestock feeding, soil fertility improvement and soil conservation. For those forages immediately attractive to farmers, seed is provided in a limited amount to enhance testing, utilization and, eventually, adoption. At the same time, smallholders are encouraged to produce seed themselves. If seed production is not feasible on-site, alternative sites for production are sought. The described approach cannot be implemented by using a single research method. The research builds on an interaction between several partners and includes the application of diagnostic tools (including both biophysical and socioeconomic aspects) to understand the production systems. To allow farmer selection of the different forage options, a combination of open evaluations and preference ranking is used.

To initiate this process of participatory selection of improved forage options, several training and planning activities with the different partners were carried out in 2000 and are continued in 2001. Several types of trials for the agronomic evaluation and simultaneous participatory selection of improved forages were established in Honduras, Nicaragua and Costa Rica. Multipurpose germplasm including grasses, herbaceous legumes, shrub legumes and cover and green manure legumes have been offered to farmers.

Thirty-nine trials were established in three reference sites in Honduras, Nicaragua and Costa Rica. The sites were selected using participatory approaches and in some cases complemented by interested farmers themselves. Each of the ten trial sites is represented by a group of 10–15 farmers. More than 100 farmers are so far involved in the project. To support this work methodologically, a procedure for the participatory selection of forages is being developed. A participatory diagnosis was done in three communities in the Department of Yoro, Honduras. Similar diagnoses are in process in Nicaragua and Costa Rica. These define the participating communities and help incorporate them in the planning process.

Reflections

Although initial results are highly promising, data have to be interpreted with care. Results obtained in specific locations will be extrapolated using a GIS-based decision-support tool, which will be made available to a range of research and development institutions.

Preliminary results indicate that farmers selected in the first place 'improved' grasses for their pastures. Several farmers were also interested in legume species. High interest is shown in incorporating forages for soil conservation, while demand for cut-and-carry legumes is still low. However, the use of shrub legumes for dry season feeding is gaining the interest of some farmers, confirming similar experiences of CIAT research in the region. A more reserved attitude of farmers towards the legume options can be expected because in most cases the use of legumes means a change of the existing production systems.

The combination of different forage options of varying complexity is believed to be instrumental in building trust among farmers to test more complex and thus higher risk alternatives. The community-based NRM research is seen as complementary to on-station research activities and vice versa. This study is building on this concept: the positive aspects of community-based research counterbalance the negative aspects of on-station research. For example, in on-farm research the lower control of the experimental conditions is balanced by the better insight obtained in the real-life situation. The biggest limitation of the approach at the moment is to match the demand and supply of seed. In order to maintain farmers' confidence it is important to stress that new forage options bear certain risks and that their introduction and succesful application is a long-term process. The development of appropriate approaches is indispensable. The approach goes beyond offering farmers a range of forage options for selection; it initiates rather a dynamic, continued process. Farmer-selection of particular grass and legume ecotypes will allow researchers to better define plant characteristics requested by farmers. Adoption of new technologies is in the first place determined by the client's requirements. These demands should help to define and direct further forage germplasm development. The interactive work with farmers may open possibilities to develop forage technology options which go beyond the immediate scope of farmers and researchers. These technologies could contribute significantly to the development of profitable and sustainable production systems. Of particular attention in this study is the search for forage options to directly or indirectly improve the well-being of less privileged groups in the rural society, such as women and small farmers (even those without livestock).

AGROBIODIVERSITY MANAGEMENT

Diversity in the natural ecological systems of mountains has contributed extensively to maintaining biological diversity in the farming systems of the eastern Himalayan region. Typically, many crop and animal species, varieties and breeds are found on farms. Subsistence farmers of the region, besides producing many crops (mainly landraces), rely extensively on wild plants to meet their needs for fibre, shelter, food, medicines, tools and household implements. For the various ethnic groups residing in the region, survival requires extensive use and management of natural resources. More particularly, it necessitates incorporating natural resources into farming systems. By incorporating biological resources at the genetic, species and agroecosystem levels, mountain communities possess extensive knowledge of their environment. The project is situated in the eastern Himalayas, and covers four sites: east Nepal, Sikkim, Bhutan and Nagaland in north east India. The overall purpose of the project is focused on building the capabilities of indigenous mountain populations to better represent themselves in the development dialogue. The process for achieving these objectives is to:

- Develop, through research, a better understanding of the linkages that exist between the way that gender and ethnicity are constructed and then to examine how such constructions affect the management of mountain agrobiodiversity.

- Build upon the existing experience of a network of researchers, who themselves are members of ethnic mountain communities, and increase their capabilities through skill development in research and analysis and community development concepts and practice.
- Include indigenous knowledge on agrobiodiversity management (particularly that of women) in the policy planning of national governments.

An advisory group consisting of scientists and planners from the region has been assembled to advise on effective strategies through which these can be achieved.

APPROACH

The strategies for capacity building work through research, skill development for network members and advocacy. The aims are to:

- Identify and document the extent of women's knowledge related to plant genetic resources.
- Examine changes in women's knowledge in terms of how changes in the last 200 years have transformed production strategies while introducing new crops and technologies. How have such changes been consistent with changes in ideology and how have they compromised the role of mountain women?
- Identify strategies adopted by women to counteract their increasing marginality, especially in the context of their roles as managers of agrobiodiversity on one hand and the 'gender blindness' inherent in extension and development policies of national governments on the other.

Skills of network members are developed through training and learning workshops that build conceptual clarity and analytical skills in research and help develop writing skills. Training is given on gender analysis and for facilitation and community development skills. Participatory rural appraisal (PRA) skills are included for a programme especially developed for rural planning called Village Initiated Planning.

The process requires community involvement in identifying problems and in research and analysis. It includes problem prioritization, project planning (proposed action, objective, developing M&E criteria, financial management) and designing implementation procedures (methods and approaches, group formation, types of training required, guiding principles, implementation and M&E). The advisory group, comprised of scientists, planners and development professionals in the region, propose strategies that focus on bringing an awareness of gender issues and farmers' rights to the nation's policy-makers and high-level officials via links to biodiversity planning bodies at national level and other means. The development of a participatory video documenting the process of a participatory crop improvement (PCI) initiative that builds on farmer-led approaches in eastern Nepal also helps to diffuse the method.

Reflections

While capacity building is part of an on-going process, a preliminary assessment, particularly in the PCI initiative conducted at one site, demonstrates that certain activities are more easily adopted than are others. For instance, a participatory seed management initiative conducted in eastern Nepal demonstrates the potential for dissemination of methodologies to other sites in the eastern Himalayas. The action research project was based on the following hypotheses:

- The development and enhancement of seed management technologies will occur most effectively through a process of interactive learning between indigenous and formal systems of agricultural development.
- Access to improved technologies can be most effectively sustained through community action. Thus existing technical skills for seed improvement and the organizational capacities of CBOs need to be enhanced to ensure community access to improved technologies.
- The success of community action to manage development processes will depend fundamentally on the ability to control processes of knowledge production, design and implementation of interventions.

In the 18 months since the project's implementation, members of the local CBO effectively adopted the knowledge of seed management for maize, and the community developed a seed bank with their own production of maize seed. Another significant outcome is the dissemination of the seed technology through the activities of the CBO that is presently working with over 50 farmers from adjoining communities. Lessons generated from the farmer-led seed management initiative in east Nepal have helped in developing a similar management initiative in Sikkim. Over the last few years, the spread of disease in ginger has severely hampered production of this important cash crop. The new participatory approach is based on experimenting with farmer-identified 'best' practices for disease management. A central component is through the development of local organizational capacity to more effectively participate in the research process.

Finally, members of the CBO in eastern Nepal have emphasized their interest in developing plant breeding skills, especially for maize varieties (landraces) whose seed quality has deteriorated because of factors such as the introduction of high-producing modern varieties that are available through research stations. It is still too early in the project to assess the extent to which local knowledge of mountain communities can inform national policy planning, especially in terms of agricultural development for marginal mountain communities.

7

Development of Forest Product Wastes Industries

Soon as he did and has continued ever since. The specific uses to which these products are put have changed in many respects with the passing years and wood has been largely replaced by other materials in some uses, as in shipbuilding. Nevertheless, forest products still enter importantly into our daily living. Imagine, for example, what would happen to our economic life, our standard of living and to civilization itself, without today's three major economic products of the forest: fuel, lumber and paper.

Although wood fuel is not of large importance in some countries, there are many where life would be exceedingly difficult and present activities would have to be greatly changed if wood fuel were not available. It may appear at times that we might do without lumber and in some regions very little is used, but in those regions the living standards are low in comparison with those of Europe and North America and, furthermore, certain lumber products are still used there. Without paper from wood, however, the business of the world would halt; communications would be greatly limited; monetary systems would be disturbed; books, records, newspapers, magazines and libraries would shrink; and our civilization would be set back hundreds of years. It is true that paper can be made from other materials, including rags, straw, grasses, bagasse and bamboo, but not in the variety and increasing quantities required to carry on the business of the world today.

World statistics on wood consumption and on the quantities used for various purposes in different countries are not by any means complete and in many instances must be intelligent estimates rather than actual records. Detailed statistics from the U.S.S.R. and its satellites are unavailable, which presents a serious difficulty in comparing the wood industries of the world, for the production and consumption of wood in those countries make up an important percentage of the total for the world. Even from the countries that report their statistics, the data are not uniformly collected or segregated by industries and end uses, although improvement in this respect is being made.

The magnitude of the world's wood consumption is indicated by the estimate that the total wood felled throughout the world in 1953 amounted to about 1.4 billion cubic meters. Possibly half of the total quantity of wood cut was used for fuel.

According to the reports made to FAO, wood for industrial purposes amounted to about 56 per cent of the total in 1953, but it is pointed out that many countries do not include in their reported figures fuel wood cut for farmers' use. Thus the share of industrial wood in the total fellings may be as low as 45 or 50 per cent. The total estimated world consumption of roundwood for different purposes and shows how the relative importance of the major products varies in different parts of the world. The total estimated world fellings of wood for industrial purposes and for fuel.

The per capita consumption of wood for all purposes varies in different countries and so do the proportions used for fuel and for industrial purposes. Finland and Brazil, for example, use much more wood for fuel than for industrial purposes, while in Great Britain and the Netherlands nearly all the wood is used industrially. The forest-products industries may be classified in various ways, according to the purpose to be served. In the present discussion we shall divide them into:

- Natural wood industries, in which the inherent character of the wood is not changed, although some of its properties may be improved;
- Wood-conversion industries, in which the wood is processed to change it into other products;
- Other forest-products industries, which extract a great variety of products, other than wood, from the forest. Each of these groups can be subdivided into primary and secondary industries and into an endless variety of other divisions, according to the nature of the processes or equipment used and the kinds of products manufactured.

STATUS AND LOSS OF NATURAL FORESTS

With growing concerns about the status and loss of natural forests, the rapid expansion of protected areas and large areas of forest unavailable for wood supply, plantations are increasingly expected to provide substitutes for products from natural forests, particularly in Asia and the Pacific.

In Asia and the Pacific it is estimated that 52 per cent of natural forests are not available for wood harvest because they are inaccessible or uneconomic to exploit. Of the unavailable forest in the region, it is estimated that about 38 per cent is legally reserved. In addition, logging bans have been imposed on large areas of natural forest covering about 10 million hectares. The reasons for these bans vary but were related to deforestation and forest degradation causing environmental problems in Thailand, the Philippines and China and to conservation requirements in Sri Lanka and New Zealand.

As a result of the net effect of deforestation and removal of natural forests from wood production, some areas in the Asia and the Pacific region have wood deficits and roundwood harvesting is exceeding sustainable levels of cut. The worst affected areas are South Asia and insular Southeast Asia, with continental Southeast Asia also under strain. In contrast, New Zealand has surplus plantation wood available for export.

Of six examples studied in the Asia and the Pacific region, New Zealand is more than self-sufficient in wood production based on plantations. In China and Viet Nam, the importance of plantations will increase as planted resources mature. There have been serious problems with implementing plantation development programmes in Sri Lanka, the Philippines and Thailand. In Sri Lanka, India and elsewhere in the tropics, trees outside the forest are playing a critical role in roundwood and woodfuel supply. Most countries in the region are becoming importers of wood, with imports expected to rise. Sometimes logging bans have shifted the problem to other countries. Problems with acquiring large areas of land in some countries make it difficult to implement industrial plantations. In the Philippines, Thailand and Viet Nam there have been social conflicts with local indigenous people or between traditional forest use and development, as well as between the rich and the poor. Sometimes incentives and the development of social forestry programmes are being used to help resolve such problems.

While it is clear that plantations will have an increasingly significant role in substituting products from natural forests, the impact will be felt on a case-by-case basis as governments and investors determine where and how plantations can be technically, economically and socially feasible as well as environmentally friendly. In the near term, plantations in Asia and the Pacific can make a contribution but cannot replace harvests from natural forests. It is likely that both in the region and globally the current pace of industrial plantation development will barely keep pace with losses from deforestation and transfer of natural forests to protected status. While it would be theoretically possible, actual plantation development is at present not sufficient to offset both growing consumption and declining harvest from natural forests.

NEW SOURCES OF FIBRE

Since FRA 1990, advances in wood utilization technology have resulted in increasing importance of new sources of fibre - rubber (*Hevea brasiliensis*), coconut palm (*Cocos nucifera*) and African oil palm (*Elaeis guineensis*) - especially in the Southeast Asian subregion. These species account for 9.7, 12.0 and 6.0 million hectares of plantations, respectively. All grow in the humid tropics. In terms of plantation area, Asia has 92 per cent of the world's rubber, 86 per cent of the world's coconut palm and 78 per cent of the world's African oil palm. Indonesia,

Thailand and Malaysia have almost three-quarters of the rubber plantations; Indonesia and the Philippines have about half the coconut resources; and Malaysia has 55 per cent of the oil palm resource. All three species are grown principally for other products rather than wood, so when overmature they are available for fibre-based industries at minimal cost.

Rubberwood is harvested when latex productivity declines and yields 100 m^3 per hectare of roundwood, but recovery for lumber is only 25 to 45 per cent because of poor form and small size. Most of the planted stands in Southeast Asia are owned by smallholders and are geographically dispersed, with poor accessibility and poor-quality stems. Currently the major proportion of industrially utilized rubberwood comes from large-scale plantations. Quality furniture, parquet, panelling, reconstituted panels, general utility timber and woodfuel, including charcoal, are made from rubberwood.

However, the rubberwood must be processed within days of harvesting to minimize sapstain attack. The most developed downstream industries are in Malaysia, where the production of sawn rubberwood timber rose from 88 000 m^3 in 1990 to 137 000 m^3 in 1997 and medium density fibreboard (MDF) production from rubberwood reached 1.16 million cubic metres per annum by 1999. Exports of rubberwood furniture have grown from about US$74 million in 1991 to US$683 million in 1998. Rubberwood has become a substitute for light tropical forest hardwoods. Its acceptance as a sustainable plantation-grown, environmentally friendly timber has given it wide appeal.

Coconut palms are harvested as the copra yields decline (beyond 60 years) and yield 90 m^3 per hectare of coconut wood. Coconut palm has variable properties and is intrinsically difficult for conversion but can yield a relatively low-cost, general-utility timber for construction, panelling, stairs, door jambs, furniture, flooring and power poles.

In 1993 Indonesia had 65 million cubic metres of overmature coconut stems which needed disposal before replanting. There is increasing interest in this raw material in European and North American markets. It is unlikely to replace conventional timber, but likely to find its way into niche markets. It will continue to be used as a low-cost construction timber.

Oil palm plantations are harvested for fibre beyond the 25 to 30 year rotations and yield about 235 m^3 per hectare. It is estimated that over 1.6 billion cubic metres of fibre will be available in the years to come from established resources in Southeast Asia. From 1996 to 1999 the area increased by 18 per cent. In Malaysia the area has increased by 3 million hectares in the past 30 years. Most oil palm plantations (unlike rubber and coconut) in the main growing countries, Malaysia and Indonesia, are managed by plantation companies or cooperatives. Oil palm by-products such as kernel shells, pressed fibres and empty fruit bunches are currently used in heat generation at the extraction plants. Water in the stems can reach five times the weight of dry matter. The high moisture content as well as the high amounts of parenchyma tissue rich in sugar and starches make conversion into quality forest products a challenge. An MDF plant in Malaysia is currently being planned to utilize oil palm stems.

Plantations and Carbon Sequestration

In the past ten years, the development of forest plantations as carbon offsets has evolved towards a market mechanism, although an organized market with carbon prices defined according to supply and demand forces is still a long way off. The adoption of the Kyoto Protocol in 1997 triggered a strong increase in investment in plantations as carbon sinks, although the legal and policy instruments and guidelines for management are still debated. A number of countries have already prepared themselves for the additional funding for the establishment of human-made forests. The 1997 Costa Rica national programme was the first to establish tradeable securities of carbon sinks that could be used to offset emissions and the first to utilize independent certification insurance.

To date, greenhouse gas mitigation funding covers about 4 million hectares of forest plantations worldwide. The recognition of afforestation and reforestation as the only eligible land use, land use change and forestry activities under the Clean Development Mechanism of the Kyoto Protocol, as agreed in Bonn during the second part of the Sixth Conference of the Parties to UNFCCC in July 2001, will lead to a steep increase in forest plantation establishment in developing countries. The sink decision of the Bonn Agreement is expected to funnel additional funds into forest activities in developing countries and thus to strengthen the international efforts in this field. However, it will also require a monitoring and verification system to ensure that these plantations will not be established at the expense of the local population or efforts to conserve biological diversity. Thus the decisions taken in Bonn to make the Kyoto Protocol ratifiable will also bear new challenges for forest plantation development.

Trees Outside the Forest

The significance of trees outside the forest (TOF) can be observed in several contexts. In countries with low forest cover, TOF resources constitute the main source of wood and non-wood "forest" products, even though trees may be so scattered that the maps produced by FRA 2000 indicate that no forests exist. Trees are found on agricultural lands, in densely populated areas, in fruit-tree plantations and in home gardens, which often cover a large proportion of the land. In urban areas trees provide important aesthetic and environmental services in addition to providing shade and greatly increasing the livability of cities. Communities, farmers and herders who do not have access to forests diversify their production and protect their land by maintaining various tree systems on their farms.

Deforestation has been mapped and quantified, but very little is known about the fate of land formerly under forest; forest clearing is often followed by the establishment of production systems of which trees are an integral part. Not much is known about the dynamics of trees on farmlands and their corresponding contribution to the production of wood and other products and services. Similarly, little is known about changes in tree cover in fields and urban systems. Knowledge of trees outside the forest comes mostly from local studies on agroforestry, sylvipastoralism and urban, social, community or rural forestry.

This widespread and multipurpose resource, familiar to farmers but poorly defined by managers and mostly absent from official statistics and development policies, needs to be better assessed and known. Growing populations, shrinking forests and degraded ecosystems all suggest that trees outside the forest are destined to play a larger local and global role in meeting the challenges of resource sustainability, poverty reduction and food security.

Trees outside the forest relieve the pressure on forest resources, conserve farmland, boost agricultural productivity, blunt the harmful impact of urban growth on the environment, increase food supplies, provide income and in general make valuable contributions to food security. FRA 2000 did not undertake a global assessment of trees outside the forest, mainly because of resource limitations; nor has there ever been a comprehensive global assessment of trees outside the forest and their products. However, a number of studies have been carried out for specific sectors or geographic areas, often with an emphasis on their economic contributions. This chapter provides a summary of selected studies and discusses the practical and conceptual difficulties related to a comprehensive global assessment. The chapter responds to the concern expressed by the Expert Consultation on Forest Resources Assessment 2000 regarding the lack of information on TOF.

THE NATURAL WOOD INDUSTRIES

FUEL. More wood is cut for fuel than for any other single purpose. It is estimated by FAO that "Farmers' own consumption accounted for about two-thirds of the fuelwood utilization." Fuel-wood production for industrial use or for sale for home use is locally important and furnishes employment for thousands of people in certain countries where other fuels are not plentiful or are relatively expensive, but, as a rule, it is the work of small producers with limited equipment.Where labour is cheap and fuel scarce, as in many parts of Asia, wood may be carried long distances by cart or animal-back for sale in markets and even twigs and bushes are sought after for domestic fuel. In other countries, of which Japan and India are examples, much wood is converted into charcoal, which is preferred for domestic fuel because it is smokeless, simpler to burn and provides more easily controlled heat. Charcoal is also extensively used in Japan for truck and bus fuel. In well-timbered countries the extent to which the forests are gleaned for fuel is influenced largely by the cost of labour and by the price and availability of fuels other than wood.

Table. Total Utilization of Home-grown and Imported Roundwood

Continent	*Total Utiliza-tion*	*Commercial Disposals*								*Other Disposal*		
		Fuel Wood	*Wood for Char-coal*	*Pulp-wood*	*Saw and Veneer Logs*	*Mine Timbers*	*Poles, Piling, and Posts*	*Others*	*Total*	*Fuel Wood*	*Others*	*Total*
World total[b]	833.1	125.3	55.3	128.5	323.7	17.4	12.1	15.0	677.3	148.2	7.6	155.8
Europe[b]	213.1	31.9	5.6	36.2	68.3	10.4	1.6	3.5	157.5	51.5	4.7	56.2
North and Central America	383.9	54.7	3.2	85.5	206.5	2.7	8.9	7.4	368.9	14.0	1.0	15.0
South America	60.8	9.6	25.7	0.3	9.5	0.5	0.1	2.3	48.0	12.5	0.3	12.8
Africa	73.6	7.6	1.6	-	3.2	0.1	0.3	-	12.8	59.7	1.1	60.8
Asia[b]	80.6	13.0	19.2	5.9	25.9	3.5	0.8	1.3	69.6	10.5	0.5	11.0
Pacific Area	20.5	8.5	-	0.6	10.3	0.2	0.4	0.5	20.5	...	...	...

This is why in Canada, the United States, some of the Baltic countries and other parts of the world enormous quantities of standing and down wood are left in the forests, which in central and southern Europe and much of Asia would be greatly prized and find a ready market for fuel.

In the production of lumber and other wood products, the sawdust, planer shavings, trimmings, edgings and other residues for which no other profitable use is found, are often used as boiler fuel in the producing plant or sold to other nearby plants for that purpose. To a limited degree in the United States and Europe, sawdust and planer shavings are being compressed under high pressure into briquettes to be used as domestic fuel.

LUMBER

Lumber is the chief primary product of the natural wood industries of the world. Lumber is made in establishments ranging from the most primitive, in which boards are laboriously sawn out one at a time by hand, to the highly mechanized, mass-production sawmills of the west coast of North America and the Scandinavian countries.

The simplicity with which logs can be sawed into lumber makes possible a great variety of power sawmills of low cost, limited only by the ingenuity of the owner and the economic conditions under which he operates. Wind, water, steam, electricity and internal-combustion engines — as may be available locally — furnish the power for these small mills. As a result, there are thousands of small sawmills throughout the world, but a much smaller number of medium and large mills. Thus, in Austria, with a forested area of about 34,000 square miles, there are more than 6,000 sawmills, operated by about 24,000 workers. A few are large, but most are very small. In the United States, approximately

90 per cent of the 53,109 sawmills reported in 1947 could be classified as small, producing in each case not more than one million board feet per year. Nearly 80 per cent of the total number produced less than 500,000 board feet per year. The 10 per cent of mills producing one million board feet or more per year, however, manufactured nearly 75 per cent of the total U.S. lumber output in 1947.

The Food and Agriculture Organization of the United Nations, in its 1954 *Yearbook of Forest Products Statistics*, gives the 1953 production of lumber by the countries It will be seen that, of the total world production of lumber (including sawn crossties and timbers), estimated at 246 million cubic meters of solid wood, the United States produced about 38 per cent and Canada about 7 per cent. The United States, however, consumes more lumber than it produces, for despite its great production it imports more lumber (chiefly from Canada) than it exports. Canada, on the other hand, although its per capita consumption of lumber is the highest reported, is the greatest lumber exporter, accounting for more than one third of the reported world exports of coniferous lumber.

Ously influenced by the habits and preferences of the local population, as well as by the amount of lumber the country produces. Each important lumber-producing country has both large and small sawmills, but the character of the mills and the method of sawing are greatly influenced by the size and quality of the logs available and by the market served. For example, production sawmills on the west coast of Canada and the United States, in parts of Australia, New Zealand and in some other countries must be equipped to handle and to saw large logs. Logs from virgin timber in these countries often are from 2 to 6 feet in diameter and sometimes considerably larger. In such large logs and many of the smaller logs from virgin timber, the outer portion usually contains a high proportion of clear wood that is free from knots and other defects, while the central portion is more knotty and defective. It is customary, therefore, in sawing logs of this character, to turn them frequently so as to obtain the highest proportion of clear boards, which are the most valuable. This custom of "sawing for grade" is the common practice in many countries, in both large and small sawmills, where the log quality and lumber markets make it profitable.

On the other hand, logs from plantation forests or "second growth" timber stands are not usually allowed to grow large enough to produce much clear lumber. The knots they contain commonly extend to the bark or near it, so that little clear material can be obtained by sawing. This is also generally true of logs from virgin timber where the species or the climate is such that they do not attain large size. With such logs there is no incentive to "saw for grade," and the common practice is to saw them into boards or timbers as quickly as possible, with the least amount of turning. Types of Saws. The simplest sawmill, requiring the least investment, consists essentially of a circular saw, a saw carriage to move the log back and forth past the saw and a source of power. Such mills can cut accurately, if properly constructed and well maintained, but a large percentage of them are so poorly maintained and carelessly operated that they produce much badly cut lumber and are thus wasteful of material. Although a top saw, in addition to the main saw, is needed for sawing large logs, most of these simple mills have only one saw, but they permit sawing for grade, if desired. Where logs are too small to saw for grade and where fast production of standard-sized boards or timbers is desired, circular saws may be set in gangs so as to make two or more cuts simultaneously.

Band saws allow cutting for grade and size as readily as do circular saws and may be used on large or small logs. They are more costly to install and maintain than circular saws, however, so they are not generally suitable for small, low-cost sawmills. Their use for lumber manufacturing is generally confined to larger mills cutting medium to large logs. They are widely used in North American mills and to some extent in other countries.

The sash-gang saw, or frame saw, is prevalent in European mills cutting small logs and to a gradually increasing extent in North America and other countries is being used for the same purpose. This type of saw has numerous short blades set in a frame that operates with an up-and-down motion and cuts an entire log at once.

The distances between the saws are adjustable to produce the lumber thicknesses desired. Since substantial time is required to change the adjustments, the logs are usually sorted for size, so that the mill can operate on logs of the same size for long periods. In the larger mills in Europe, frame saws are commonly operated in pairs; the first saw slabs two sides and may take an additional board from each side, leaving a standard-sized timber or "cant" in the centre. This cant is then transferred to a second saw, which reduces it to boards. Sash-gang saws are generally heavier and more costly than circular-saw installations, but cut a thinner "kerf," and thus waste less wood in the form of sawdust than do circular saws. Their cost, weight and immobility limit their use for small sawmills in North America, although they may be found in many small mills in Europe.

However, some portable sash-gang saws are coming into use along the Pacific coast. Sash-gang saws are commonly used as secondary saws in large sawmills in the western part of North America. In such mills, the circular or band "head saw," in breaking down the log, produces many large cants of standard thicknesses, which then move on to the sash-gang saw for reduction to boards. There are, of course, innumerable combinations of saws of the several types, according to the size and character of logs being cut, the size of the sawmill, the markets served and the ideas of the owners or the designers of the sawmills. Lumber for the world markets and lumber for domestic use in most lumber-producing countries is square edged and cut to standard thicknesses, widths and lengths. In some European countries, however, particularly with hardwoods, logs cut into lumber for domestic consumption are often sawed into boards the full length and width of the log and then reassembled into the same relative position they had in the original log, without "edging" or squaring the edges. They are seasoned and sold in this form. The buyer or user then cuts up the boards to best advantage for his particular purpose.

Secondary Lumber Industries. Although lumber is a primary product and its manufacture constitutes one of the important economic activities of the world, the secondary industries that use lumber as their principal raw material account for a larger share of the world's business and provide employment for more people than does lumber production.

The average percentages of lumber used in the United States for different major use classifications during the five-year period 1948 to 1952, inclusive, were estimated from which it is apparent that nearly three fourths of the lumber consumed in the United States is used for structural purposes of one kind or another. To a large extent it goes into residential and farm buildings, but there are numerous other structural uses. In other countries the percentages of lumber going into the various categories would probably differ widely.

Table. United States: Lumber used for Various Purposes, 1948-1952

	Per Cent
Building and construction	71.4
Boxes and crating	10.6
Industrial	12.5
Railroad	3.6
Export	1.9
TOTAL	100.0

In its industrial application lumber goes into hundreds of products, which vary in character and percentage in different countries according to differences in national customs, preferences and economics. In most countries large quantities of lumber are used for furniture and millwork. In Japan, for example, the wooden shoe or "geta" so generally worn accounts for possibly 7 per cent of the country's lumber consumption, while in most other countries people depend on shoes or sandals made of other materials.

COOPERAGE

The production of wood kegs, barrels and similar containers for liquids and solids dates from ancient times and the cooperage industry still employs thousands of workers all over the world. Dependable statistics on the quantity of such containers produced or the amount of wood used in their production, however, are very limited and not up to date. Cooperage-type containers are not included in the estimates. Estimates by the U.S. Forest Service indicated that about 746 million board feet (log scale) of timber were required in 1939 for cooperage in the United States.

Of this amount, approximately half was for tight cooperage and half for slack cooperage. The use of both slack and tight cooperage was apparently on the decline, however, because of the substitution of metal, cloth and paper containers for many of the products formerly shipped in kegs, barrels and pails and the present consumption of wood for cooperage is probably much less than in 1939.

VENEER AND PLYWOOD

The manufacture of veneer and veneer products, although consuming much less timber than does the manufacture of lumber, is nevertheless one of the world's important wood industries. Accurate comparisons between the world production of lumber and of veneer are not available and comparisons within any single country do not correctly imply the world picture. In the United States estimates indicate that in 1948 more than ten times as much timber was used in lumber manufacture as in veneer manufacture.

The principal use of veneer is in the manufacture of commercial plywood, although large quantities go into the production of baskets, fruit boxes, shipping boxes, drums, trunks and other products. Much veneer is also glued into doors, furniture, radio cabinets and the like. Plants that manufacture plywood, baskets, boxes and other containers commonly produce their own veneer, while the veneers used in furniture, radio cabinets and generally for decorative purposes are more often produced in plants that manufacture the veneer only.

Veneer is manufactured by sawing, slicing and rotary cutting, but mostly by the latter method, in which the veneer log is revolved on its axis while a heavy knife, parallel to the log axis, is pressed against it. The veneer peels from the log in a manner similar to the unwinding of a roll of paper. Rotary-cut veneer can be made as thin as 1/100 inch but is commonly produced in thicknesses up to 3/16 inch.

Thicknesses of 1/4 inch and sometimes up to 1/2 inch can be produced, depending on the character and quality of the wood and the equipment. Rotary cutting is normally used for species and for purposes where beauty is not of major importance. Beautiful figure, however, is obtained in this way from some species, of which birch and birdseye maple are good examples. It is also obtained in the manufacture of figured veneer from walnut stumps, which are fastened eccentrically in the lathe.

In veneer slicing, the usual practice is to fasten a long block or flitch to a heavy frame that moves diagonally past a veneer knife, thus taking off slices of the desired thickness from the flitch. This method is particularly suitable for the rapid production of highly figured wood, for the flitch can be fastened in the frame in such a way as to produce. As the veneer falls from the knife, it is usually reassembled in its original relative position in the flitch and through subsequent drying, packaging and storage is so handled that the veneer from each flitch is finally sold and shipped as a unit.

The slicing method is also used in the production of veneer for some kinds of boxes and for battery separators, in which case the knife usually moves instead of the wood. This permits faster cutting, because the wood blocks are fed continuously against the knife and it is not necessary to stop the cutting to fasten each new block into the frame, as is the case when the knife remains stationary. When veneer is cut with a moving knife, the blocks used are considerably smaller than are those common with the stationary knife, which may be as much as 18 feet long. While veneers of similar maximum thicknesses can be made by slicing or by rotary cutting, the former method may be used to produce somewhat thinner veneers.

In both rotary cutting and slicing, the veneer is bent sharply away from the log as it passes over the thick knife and this may cause checks in the convex or outer face of the curve. The severity of the checks depends on the character and condition of the wood, the thickness of the veneer and the skill of the machine operator. Checks cannot be entirely prevented except in very thin veneer, but they can be kept within usable limits by skilful operators working with good material.

The sawing method of cutting veneer avoids the knife checking and is particularly suitable for species that do not cut well by knife. It is accomplished by moving a log or flitch on a carriage past a thin circular saw, all working very accurately. Sawed veneer can be produced commercially as thin as 1/20 inch and, of course, also in considerably thicker dimensions. Veneer sawing is slow in comparison with the rotary or slicing methods and is seldom used with species that may be cut well by these. Sawed veneer can be produced from either green or seasoned logs or flitches, whereas green material only can be cut well with a knife.

Plywood, the principal product made from veneer, is produced by gluing sheets of veneer together. There are usually three or five sheets, with the grain of each sheet, or "ply," at right angles to the adjacent ones. For special purposes, other numbers of plies and arrangements of grain direction are used to a limited extent.

The centre sheet or "core" may be lumber, veneer, or even some form of fiberboard or particle board. Alternating the grain directions of adjacent plies makes plywood resistant to splitting in both directions and reduces its tendency to shrink and swell in length and width. These properties and its availability in thin sheets of large size make plywood attractive for innumerable uses. Although plywood or veneered wood seems to have been used to some extent since the days of the early Egyptians, the glues originally employed in its manufacture were easily weakened by moisture and its use, for that reason, was necessarily confined to dry places. About the beginning of the twentieth century moisture-resistant glues made from casein or animal blood became available and means were found for increasing the moisture resistance of animal glue with formaldehyde.

It was not until after World War I, however, that the invention of modern waterproof synthetic-resin glues made it possible for plywood and other glued wood to retain their strength even when thoroughly wet. With the advent of these improved glues, the field of usefulness of plywood widened enormously and its production expanded rapidly.

Plywood is produced to some extent in all industrialized countries that have suitable timber and also in some countries where the veneer logs are mostly imported. The total world production of plywood in 1953 was estimated at 7,960,000 cubic meters, of which more than half was produced in the United States. Plywood production, imports and exports in 1953 for the principal producing countries that reported. Although the United States produced more than half of the world's plywood, it used more than it produced and exported very little. The production of plywood from Douglas fir on the west coast of the United States and Canada has grown phenomenally.

The first Douglas-fir plywood is said to have been made commercially near Portland, Oregon, in 1905. Statistics on the production of Douglas-fir plywood in the United States begin with 1925, when 153 million square feet (calculated to 3/8 inch thickness) were reported. By 1952 the production had expanded to 3,066 million square feet, or approximately twenty times the amount in 1925. Before 1937 the U.S. production of hardwood plywood was greater than that of Douglas-fir and other softwood plywood, but by 1952 it was estimated that more than twice as much softwood as hardwood plywood was produced and most of this was of Douglas fir.

RAILROAD TIES

The production of wood railroad ties is a separate industry only to a limited degree. Railroad ties include bridge ties (special-sized ties used on bridges), switch ties used in sets of increasing length at the points where turnouts or sidings depart or switch off from the running track and crossties used for the running track. In the United States, switch ties and bridge ties are sawmill products, usually made in connection with other items of lumber and timber. The majority of crossties are also produced at sawmills. In large sawmills ties are usually only one of the numerous items produced, but in small sawmills they may be the principal item manufactured, with lumber as a by-product. Many crossties are still made by hand hewing. Although this practice is gradually disappearing, at least one sixth of the nearly 42 million crossties treated in the

United States in 1952 were hewed. Much of the tie hewing is done by farmers during slack seasons or by other part-time workers and their ties are sold to the railroads directly or through tie-producing companies. Some species of wood are so straight-grained and split so readily that ties can be made from fairly large logs by splitting with hand tools. This method finds some use in the redwood forests of California, with some of the eucalypts in Australia and possibly elsewhere. In India, Pakistan and some other countries, many crossties are produced by hand sawing.

In the United States and Canada and to a considerable degree in other countries, the railroad-tie industry is as much a part of the wood-preserving industry as it is of the lumber industry, for, in most countries, relatively few wood ties are used without preservative treatment.

Statistics on the world production of crossties are too incomplete and inadequate to be of much value, but they indicate large-scale production in Australia, Canada, Finland, France, West Germany, Japan and Sweden. India is probably also a large producer. Tie production statistics in the United States are less complete than are statistics on the quantities used annually by the railroads.

POLES, PILES AND FENCE POSTS

These products, although of great importance and produced in large volume, require relatively little processing. The pole, pile and post industry, therefore, is largely a collection and merchandising industry, with little processing involved other than that done at wood-preserving plants. The estimated consumption of poles, piles and posts.

Poles. In Europe and North America few species have sufficient natural durability to be economical for use as poles without preservative treatand others about 1830 to 1840. On all continents pressure impregnation of wood is now practiced, using a variety of preservatives that have been found effective for different purposes.

Published statistics on the quantity and forms of material treated commercially each year are available only for the United States, where they have been published annually, beginning with 1909. It is known, however, that large amounts of poles, piles, railroad ties, building lumber, construction timber and other wood products are treated in many countries of Europe and in Canada, Mexico, India, South Africa, Japan, New Zealand and other countries. In addition to the wood that receives pressure treatment, much wood is treated by nonpressure methods such as hot-and-cold-bath, soaking and immersion treatments, for which statistics are not available.

Coal-tar creosote, made by distillation from coal tar, is the principal wood preservative in the world and thus far the most dependable for general use against all wood-destroying organisms and all conditions of service. It is often diluted with petroleum oil for the treatment of timber to be used in land or fresh-water structures.

Where petroleum is cheap, the dilution reduces the cost of the preservative materially without seriously reducing its effectiveness against fungi and insects, but the addition of petroleum to creosote destroys its effectiveness against marine borers. Coal tar is also used as a diluent for creosote and the mixture is suitable for salt water as well as land use. Numerous water-borne preservatives are also used throughout the world to give protection against decay and insects without the discolouration and disagreeable oiliness of wood that has been treated with creosote and its mixtures. Most of the present-day water-borne preservatives have appreciable resistance to leaching when used under wet conditions, but on the whole they are less effective for such uses than are the creosote preservatives. The water-borne preservatives find their greatest use in buildings and other structures where both cleanness and protection are desired.

The resistance of wood to flaming and burning may also be greatly increased by heavily impregnating it with suitable mixtures of water-borne, fire-retarding salts. The treatment is effective but too costly for general use and, therefore, not so widely practiced as is preservation against fungi and insects. For example, less than one million of the total of 315 million cubic feet of material treated in the United States in 1952 was treated with fire-retarding chemicals.

In the pressure-treating process the wood is run on tram cars into a large steel cylinder built to stand high internal pressures. The preservative is then admitted to the cylinder and pressure, usually between 100 and 200 pounds per square inch, is applied by pumps until the desired quantity of preservative has been forced into the wood.

The wood may be treated either in the green or seasoned condition. There are many variations in the details of the preparing or conditioning of the wood prior to the application of preservative pressure, in the amounts of pressure used and in the methods of applying it, depending upon the preservative used, the species and condition of the wood under treatment and the preservative absorption and penetration desired.

The principal nonpressure treatment in commercial use is the hot-andcold-bath method, in which the wood is first heated in preservative or some other heating medium and then covered with a cool preservative and allowed to cool further while submerged. There are a number of variations in details depending upon the product treated, the preservative used and the equipment available.

In the treatment of window sash, doors and other items of millwork, a growing practice in the United States is to immerse the product for three minutes in a solution of pentachlorophenol in a volatile petroleum solvent, or to accomplish the same result by applying a vacuum and then flooding the wood momentarily with preservative and releasing the vacuum.

TYPES AND REGIONS OF FOREST

FOREST TYPES

The forest flora of India is rich and varied both in composition and value. The number of species is estimated at 30,000, although the vast majority have no economic value. The distribution of forests and forest types as elsewhere, depends on climatic, edaphic (including relief) and biotic factors acting in combination. Hence, no one of these factors considered alone can give a satisfactory clue to the nature of the forest in any particular locality. Indeed, a forest itself is often a better index of climate, soil and certain biotic conditions than vice versa.

THE CLIMATIC FACTOR

Marked climatic contrasts reflect pronounced differences in the relief of India. The mercury rises above 120°F. at Jacobabad in Sind and falls as low as — 49°F. at Dras in Kashmir. A climb of only a few miles brings one from hot, moist marshes (tarai) along the outer foothills of the Himalayas to the temperate climate of the former British hill stations (ca. 6,500 feet) and thence to arctic cold above the snow line (ca. 12,000 feet). The contrasts in rainfall are even greater and have a profound effect on the vegetation, the extremes ranging from an annual mean of only 3 inches in Upper Sind to nearly 40 feet in the Khasi Hills.

Three seasons may be distinguished: winter (November-February), summer (March-June) and the rains (July-October). During the winter, winds of continental origin blow from the land toward the sea, in general from the northeast and northwest. This "northeast monsoon" becomes fully established by January, bringing cool, dry weather, although local disturbances cause occasional showers in northern India and in the eastern Deccan. The winter rainfall is heaviest in the northwest. As summer advances, the wind direction is reversed.

The winds now blow from sea to land, culminating in the "southwest monsoon," which sets in with the rainy season. The summer is characterized by progressively rising temperature and falling atmospheric pressure. Temperature maxima range from 100°F. in March in the Deccan, to 110°F. in April across central India and 115°F. in May and June throughout the IndoGangetic plains, reaching extremes of 120°F. in the northwestern desert.

Coinciding with the area of lowest pressure, this blazing furnace in the northwest induces indrafts of the monsoon from the Arabian Sea and Bay of Bengal.The bulk of the rainfall is brought by the southwest monsoon during the four months July-October. One great current of the monsoon bursts against the Western Ghats and flows over the Deccan, another sweeps up the Bay of Bengal and, combined, they flood the eastern Himalayas and outlying hills to the south. July and August are the rainiest months here. Northwestern India, away from the main paths of the southwest monsoon, gets the lowest rainfall in the country. For the purposes of the following description of its forest types, India will be divided into six major physical regions designated somewhat arbitrarily:

- The Deccan,
- The Indus Basin,
- The Gangetic Basin,
- The Himalayas,
- The Seacoasts,
- The Andaman Islands.

While these provide a convenient means of classification, it should be borne in mind that they are based on criteria of surface relief and not of phytogeography. Thus, more than one region may include representatives of a single forest type and divers forest types are found within each region. The estuarine vegetation in the delta of the Ganges (or Ganga), for example, is like that of the deltas of the eastern coast of the Deccan; the evergreens of the Malabar Coast have far more in common with those of the northeastern Himalayas than with the vegetation of the rest of the Deccan.

MAIN FOREST TYPES

In describing the forests of these regions we shall follow Champion's classification in broad outline, omitting details. Five main forest types and several transitional, intermediate and subtypes will be distinguished. The main types are:

- *Rain Forest:* Annual rainfall in general 80 to 120 inches; dense, tall evergreens.
- *Monsoon Forest:* Rainfall 40 to 80 inches; dominants deciduous; evergreens form the second story.
- *Dry Forest:* Rainfall 30 to 40 inches; deciduous; no evergreens.,
- *Xerophytic Forest:* Rainfall 10 to 30 inches; low, stunted, thorny, deciduous species and fleshy evergreens.
- *Mangrove Forest: R*ainfall varies; both evergreen and deciduous species.

Changes in the floristic composition of a given type brought about by edaphic and biotic factors and by rainfall and other climatic factors will be pointed out. For example, coarse, friable, porous soils require a much higher rainfall to support a rain forest than do soils with higher moisture retentivity. The presence of alkalis and salts may retard and even eliminate tree growth altogether, despite high moisture content. Human exploitation, fires, grazing and lopping often act as limiting factors, affecting the character of the vegetation independently of the rainfall. The range of temperature and, more particularly, minimum temperatures, sharp frosts and snowfall have a profound influence. In the Himalayas the vegetation of the northern latitudes of the temperate zone is reproduced. The vegetation in the Deccan plateau, with its equable climate, differs from that of the Indo-Gangetic plains, which exhibit a continental range of temperatures, both seasonal and diurnal. The Nilgiri Hills, in the south, have a vegetation unlike that of the Himalayas at the same range of elevation. A distinction must often be made between tropical and Himalayan varieties of the same main forest type. The similarities between a given type of forest occurring in the Deccan and its counterpart in the north, however, may be fully as striking as the differences. The vegetation of the rainless Thar Desert is reminiscent of that of alpine wastes both in form and appearance.

THE DECCAN

The term "Deccan" should be here understood as embracing the whole of peninsular India and of the Central Highlands, with the exception of the northern and northwestern slopes of the latter, which will be considered as parts of the Gangetic and Indus basins, respectively.

The Deccan plateau rises above the Indo-Gangetic plains in a series of irregular terraces. Its principal rocks are of archean origin, comprising granites, gneisses and schists, with basaltic lavas of more recent age capping them, particularly in the northwest. Decomposition of these lavas has given rise to the rich "black cotton soils." The plateau is buttressed on the north by the Aravalli, Vindhya, Satpura and Kaimur hills, with characteristically flat tops. The Aravallis (ca. 5,500 feet), an extension of the plateau to the northwest, form an isolated, ancient and much worndown range flanking the desert of Rajputana on the east. Farther south, the eastern and western flanks of the plateau are fringed by the Ghats, which converge to an apex in the south. With an average elevation of about 1,500 feet, the plateau surface slopes down eastward from an altitude of about 3,000 feet in Mysore. The Western Ghats begin near the Tapti River and increase in elevation southward, reaching nearly 9,000 feet in the Nilgiri Hills. The Eastern Ghats average about 1,500 feet in altitude.

In considering the forest types of the Deccan, we may conveniently divide the region as a whole, exclusive of the seacoast, into three lesser regions: the West Coast, the Southeast and the Central Plateau.

THE WEST COAST

The region comprising the Malabar and Konkan coasts, the Western Ghats and the Nilgiri and Cardamom hills has the most luxuriant vegetation in the Deccan, owing largely to the heavy rainfall and to a soil rendered rich by the decomposition of a vast mass of leaf litter. The southwest monsoon floods the entire region. The mean annual rainfall increases with elevation on the Western

Ghats; thus, while Bombay at sea level receives only 74 inches, Matheran at an altitude of nearly 2,000 feet some 25 miles to the east receives 200 inches. The higher altitudes of the southern portion of the Western Ghats also helps explain why the rainfall on the Malabar Coast (some 130 inches) is nearly twice as heavy as that on the Konkan Coast to the north (75 inches).

Of the main forest types, Dry Forest and Xerophytic Forest are not found in this region. The Mangrove Forest will be mentioned under another heading. Besides Rain Forest in the strict sense, there are two intermediate types: Shola and Semi-Evergreen Forest.

Rain Forest

Forests of this type are generally met with along the Malabar Coast and throughout the Western Ghats up to altitudes of 1,000 to 1,500 feet, wherever the mean annual rainfall is above 100 inches (with 120 inches as the optimum). In these areas the mean annual temperature varies from 80° F. to 100° F., the January minimum seldom goes below 65° F. and the mean annual humidity is about 80 per cent (e.g., at Cochin). Under such humid tropical conditions a bewildering variety of species flourish, their only common characteristic being their evergreen nature.

These forests present several ill-defined stories, or tiers, the highest containing lofty trees, often buttressed, which reach a height of 150 feet and have girths of 15 feet or more. The lower tiers consist of an impenetrable mass of evergreens covered with a luxuriant growth of epiphytes (mosses, ferns, orchids) and climbers. The most typical genera are Dipterocarps and Hopeas, although the former are the less frequent and may even be locally absent.

Species commonly met with are: Dipterocarpus indicus, Hopea parviflora,Hopea wightiana,Calophyllum tomentosum, Cullenia excelsa,Palaquium elliptica,Dysoxylum malabaricum, Cedrela toona,Vateria indica,Canarium strictum,Tetrameles nudiflora, Mesua ferrea,Mangifera indica,Sterculia alata, Artocarpus spp., Acrocarpus fraxinifolius and many other Myrtaceae and Lauraceae. The undergrowth, conspicuously free from grasses, is usually a tangled mass of cane, congested and creeping bamboo and palms.

Sholas. Above the rain forests, related forests of a special intermediate type, locally known as sholas, are found up to about 4,500 feet in the valleys and on the slopes of the Nilgiri, Anaimalai, Palni and other hills of southernmost India. At relatively lower elevations, these sholas usually consist of Mesua ferrea,Vitex altissima,Artocarpus lakoocha and Dysoxylum malabaricum, often with Strobilanthes and Selaginella as a ground cover and Arundinaria spp. and Oxtenanthera spp. as the commonest bamboos.

At higher altitudes, the floristic composition changes to include such subtropical and temperate species as Michelia nilagirica, Ternstroemia japonica,Eurya japonica,Gordonia obtusa,Eugenia spp., Ilex spp., Meliosma wightii,Photinia notoniana,Cinnamomum wightii, Euonymus crenulatus. Here, owing to comparatively rigorous weather conditions, the tree growth suffers in height and the crowns are more rounded. Mention must also be made of artificial plantations of Australian eucalyptus and of South African wattles. Exotic pines and cypress have also been introduced in the Nilgiri and Anaimalai hills. Farther north, on the Western Ghat slopes — as for example, at Mahableshwar — sholas are found in which the representative species are Eugenia jambolana, Actinodaphne hookeri,Canthium didymum,Terminalia chebula,Olea dioica,Strobilanthes callosus and Sideroxylon tomentosum.

Semi-Evergreen Forest

Along gentle slopes, where the mean annual rainfall is between about 80 and 100 inches, the evergreen nature of tile rain forest is further modified by the increasing association of deciduous species, typified by Terminalia paniculata,Xylia dolabriformis,Stereospermum chelonoides,Trewia polycarpa, Eugenia spp., Diospyros spp., Lagerstroemia lanceolata, Schleichera trijuga,Vateria indica, with Bambusa arundinacea as the representative bamboo in the lower story. The middle canopy is usually composed of evergreens, Myrtaceae and Lauraceae and the ground cover includes various species of Rubiaceae and Acanthaceae. Canes and climbers persist.

Monsoon Forest

The semi-evergreens merge imperceptibly into a moist, deciduous "monsoon forest," comprising comparatively few species but some of considerable economic importance, such as teak, rosewood, Xylia spp., Terminalia spp., Lagerstroemia lanceolata, Pterocarpus marsupium,Adina cordifolia, Stephegyne parvifolia and Grewia leptopetala, which occur in more or less gregarious associations. Schleichera and Careya are usually present in the second story. Bambusa arundinacea,Dendrocalamus strictus and Oxytenanthera monostigma are the representative bamboos, although often locally absent. The undergrowth may consist of near-evergreen shrubs but more frequently of grass. This monsoon forest occurs throughout the western Deccan in regions of 40 to 80 inches of rainfall, more particularly above the Palghat Gap. The mean annual temperature varies from 75° F. to 80° F., with the January mean often below 60° F. The trees, which reach a height of 100 to 120 feet, are generally leafless during March and April, putting on new foliage somewhat later. Along the Konkan CoastTerminalia spp. lend their name to the forest wherein they abound, in which teak may or may not be present locally.

THE SOUTHEAST

The southeastern subregion of the Deccan comprises the Carnatic, the Eastern Ghats and the bulk of the plateau south of the Godavari River. During the period of the southwest monsoon, the region lies in the rain shadow of the Western Ghats. The rainfall, generally associated with the northeast monsoon, when the rest of India is dry, is relatively low and extremely variable. The annual precipitation seldom exceeds 40 inches, the average varying between 20 and 30 inches. Of the main forest types, only the two drier ones are represented here.

Xerophytic Forest

In the Carnatic and farther south in the lee of the higher Western Ghats the vegetation is of a dry, thorny, stunted type. The most representative genus is *Acacia,* with low (20 to 30 feet) crowns. The ground is usually bare, with thorny shrubs dotted here and there and light grass appearing during periods of rain. The mean annual rainfall varies between 20 and 30 inches (with about eight months with less than 2 inches) and the mean annual humidity between 50 and 60 per cent. The mean annual temperature is about 78° F., the mean maximum about 105v F. and the hottest month is May (85° F. to 90° F.).

Dry Forest

The xerophytic thorn forest of the Carnatic gradually merges into a dry deciduous forest, where the mean rainfall is between about 35 and 40 inches, the mean temperature is close to 75° F. and the mean annual humidity is close to 60 per cent. This forest forms an intermediate stage between the monsoon and xerophytic forests, having characteristics of both. The trees are seldom more than 75 feet high. More species thrive under these conditions than in the xerophytic forest and the dry forest includes the most valuable tree of southern India, sandalwood (*Santalum album*). Gregariousness is not common, unless artificially brought about. This type of forest usually occurs on shallow, sandy soils and laterites and in places where the rainfall is relatively heavy but rapidly drains off.

In the northern part of the region the dry forest is fairly dense. The representative species are teak (Tectona grandis) and associated Terminalia spp. and Anogeissus latifolia. Other species include Dillenia pentagyna,Kydia calycina, Lagerstroemia lanceolata, Adina cordifolia, Bassia latifolia, Buchanania latifolia,Grewia asiatica, Odina wodier, Pterocarpus marsupium, Bombax malabaricum, Santalum album, Butea frondosa, Bridelia retusa, Diospyros melanoxylon and Phyllanthus emblica. The chief bamboo is the well-known commercial variety, Dendrocalamus strictus. Grasses are plentiful, including Ischoemum angustifolium, Flemingia spp., Millettia auriculata,Lantana,Carissa spinarum, etc.

Farther south, teak is less common and in many places absent, owing, perhaps, to past human exploitation. Terminalia spp. and Anogeissus spp. are associated with Acacia catechu, Chloroxylon swietenia,Cleistanthus collinus,Hardwickia binata, Boswellia serrata and Soymida febrifuga. The ground cover is composed of grasses, usually Andropogon spp., Panicum spp. and Aristida spp.

THE CENTRAL PLATEAU

For present purposes the Central Plateau may be regarded as comprising the entire region between the Godavari River and the Bay of Bengal on the south and southeast, the Western Ghats and Aravalli Hills on the west and northwest and the northern slopes of the Central Highlands on the north and northeast.

North of the source of the Godavari, the Western Ghats become progressively lower and finally disappear and, hence, no longer constitute an effective barrier to the southwest monsoon. The hill ranges of the Central Highlands run in an east-west direction, which allows a more even distribution of rainfall over this part of India than farther south. The climate becomes more continental, with marked extremes, as one goes northward. Of the major forest types, xerophytic, dry and monsoon forests are represented on the Central Plateau. At higher elevations the shola type, which is intermediate between the monsoon and rain forests, is also met with.

Xerophytic Forest (rainfall, less than 25 inches). Here, the general pattern of xerophytic vegetation found in the Carnatic is closely repeated. Acacia leucophloea, Balanites roxburghii,Cordia myxa, Zizyphus jujuba and Calotropis procera are characteristic species. Dry Forest (rainfall, 35 to 40 inches). This is met with throughout the region, with low-quality teak much more frequent than in the corresponding dry forests farther south. Here, also, teak may be locally absent, suggesting human exploitation in the past. Any one of such species as Hardwickia binata,Boswellia serrata,Butea frondosa and Anogeissus pendula may be locally preponderant. Babul (Acacia arabica) forests grow more particularly on the rich black cotton soils, notably in Berar and in the vicinity of Poona.

Monsoon Forest (rainfall, 40 to 80 inches). From the Maikal Hills eastward and especially on the Chota Nagpur plateau, there is a copious rainfall due to the convergence of the two great currents of the southwest monsoon. Hence, the vegetation is rich and clearly assumes the form of monsoon forest. Where teak is more or less gregarious, the forest is known as "moist teak forest," and is not unlike the Terminalia forest of the West Coast region of the Deccan. Teak in this region of the Central Plateau is associated with Terminalia tomentosa,Adina cordifolia,Pterocarpus marsupium,Dalbergia latifolia,Odina wodier,Lagerstroemia parviflora,Schleichera trijuga,Ougeinia dalbergioides and Phyllanthus emblica, with Dendrocalamus strictus as the common bamboo. Helicteres isora covers the ground, along with Millettia auriculata and Bauhinia vahlii. Where sal (Shorea robusta) predominates, in the so-called "moist sal" or "peninsular sal" forests, the associated species are Pterocarpus marsupium,Terminalia chebula (and T. tomentosa), Anogeissus latifolia,Bassia latifolia,Eugenia jambolana,Phyllanthus emblica and Indigofera pulchella, with Phoenix acaulis,Flemingia chappar and Anthistiria ciliata L. as the ground cover. Sholas (rainfall, about 100 inches). On the upper levels of the Maikal Hills (e.g., near Pachmarhi) and of the Chota Nagpur plateau (e.g., near Parasnath), the sholas of the Western Ghats are repeated, with slight modifications due to soils derived from basaltic lavas. Pterospermum acerifolium,Phoenix robusta and Clematis nutans are representative species.

THE INDUS BASIN

The Indus basin forms a vast area of low rainfall and high summer temperatures. The rainfall increases in a generally northeastward direction, for the clouds borne by the southwest monsoon current from the Arabian Sea tend to pass over the Thar Desert without dropping their moisture. In the Punjab the rain brought by this current is to some extent supplemented by rain from the Bay of Bengal. In the North-West Frontier region, however, the force of both monsoon currents is feeble and, consequently, cold-weather rains account for more than half of the annual rainfall.

Recent excavations at Mohenjo-Daro in Sind and the discovery of elephant and rhinoceros remains along the right bank of the Indus point to a time when there must have been moist forests in this part of India and the chroniclers of Alexander's invasion described the Larkana area in Sind as a "garden." Conditions are very different today. Two main forest types may be distinguished: dry forest and xerophytic forest.

Dry Forest

This type occurs in the northern Punjab where the mean annual rainfall is between 35 and 40 inches, the mean annual temperature between 75° and 80° F., the mean annual minimum and maximum are about 35° and 100° F., respectively and both the diurnal and the seasonal ranges are wide. Owing to these severe conditions the trees seldom exceed 50 feet in height.

In summer the lack of leaves gives a desolate look to these forests, but they turn bright green at the break of the monsoon. Among the species met with, Anogeissus latifolia is typical, its usual associates being Buchanania latifolia,Sterculia spp., Bauhinia spp., Acacia catechu,Aegle marmelos,Limonia acidissima,Ehretia laevis, Kydia calycina, Ougeinia dalbergioides and Terminalia tomentosa. Mallotus spp. and Nyctanthes occur in the lower story. Dendrocalamus strictus is the usual bamboo. The undergrowth is composed of Woodfordia floribunda,Indigofera spp., Carissa spinarum, Adhatoda vasica, Dodonaea viscosa,Bauhinia vahlii and Acacia caesia. Grasses are common. But for certain differences in emphasis, the dry deciduous type of the South is closely reproduced. Owing to the more pronounced climatic extremes, however, the teak, *Chloroxylon,Soyrnida* and *Cleistanthus* met with in the dry forests of the Deccan are conspicuously absent and the *Dalbergia latifolia* is replaced by its hardier cousin, *Dalbergia sissoo.*

Xerophytic Forest

Around the margins of the Indus Plain,' wherever the mean annual rainfall is between 10 to 25 inches — as, notably, in Cutch and Kathiawar, along the base of the Kirthar and Sulaiman ranges, on the Salt Range, in the Sirhind Gap, on the Delhi Ridge and along the lower western slopes of the Aravalli Hills— a scrub, seldom more than 3 feet high, is found, corresponding to the xerophytic forest of the Deccan. Here temperature conditions are comparable to those of the dry forest area of the Punjab.

Various species of acacia abound, the chief being *Acacia arabica* and *A. leucophloea* and are associated with *Prosopis spicigera* and the exotic *Prosopis juliflora,* which has come to occupy the Delhi Ridge and parts of Rajputana. *Acacia catechu* is found in coniderable quantity in Cutch. In the southern Aravallis *Anogeissus pendula* occurs in fairly homogeneous patches.

Farther north, typical species are *Salvadora oleoides,Tamarix articulata,Zizyphus jujuba, Grewia populifolia* and *Ehretia laevis.* The undergrowth consists of *Capparis aphylla* (and *C. spinosa*), *Asclepias* and *Calotropis,* with the fleshy *Suaeda* and *Salsola* growing on saline soils and *Calligonum polygonoides* on sandy soils.

Light grass flourishes only during the rains. On the floodplain of the Indus, which is periodically inundated and in the regions with slightly higher rainfall, *Acacia arabica,* a species that adapts itself admirably to xerophytic conditions, occupies the ground, usually in association with *Populus euphratica.Dalbergia sissoo* and *Acacia catechu* are the first colonizers of the alluvial deposits along the river banks in their Himalayan reaches.

Lower down (in southern Punjab and Sind), fresh alluvial deposits are first occupied by pure associations of *Tamarix dioica* and *Populus euphratica,* which give place in the course of time to *Acacia arabica* and *Dalbergia sissoo,* though the first two often persist in the understory. In the Thar Desert the xerophytic vegetation takes a more retrograde form, with fleshy evergreen *Euphorbias* as the most characteristic species.

8

Genetic Resources of Conserving Forestry

A wide range of methods, from protected reserves to intensive management of breeding populations for production systems, can be used to conserve FGR. The choice of methods depends on available genetic material, selected time scale and specified aims. The method selected and the subsequent implementation of a conservation strategy also depend on the availability of both human and financial resources. The two most commonly considered methods for conserving FGR are *in situ* and *ex situ* conservation. The term *in situ* refers to the continued maintenance of tree populations at their natural sites, in the environment to which they have adapted. *Ex situ* conservation takes place outside the natural habitat of a tree species, and may consist of activities such as establishing live collections or *ex situ* conservation stands, or storing seeds, pollen or tissue.

FGR can also be conserved and maintained by using tree species in forestry or other land-use systems such as agroforestry. It is likely, however, that forest management interventions reduce genetic variation in tree populations. Not even carefully planned and implemented forest management activities, therefore, can replace more active conservation measures with clearly specified objectives. The same caveat also applies to tree domestication and the use of trees in agroforestry systems.

IN SITU CONSERVATION

There is a general consensus among scientists and practitioners that no single conservation method is adequate, and that different methods should be applied in a complementary manner. *In situ* conservation, however, has a number of benefits and so often forms the basis of conservation programmes. It allows evolutionary processes to be maintained, including the adaptation of tree populations to changing environmental conditions.

This is particularly important for breeding programmes, since future human needs and environmental conditions are difficult to predict. The reproductive biology and overall survival of many tropical forest species are dependent on complex ecological interactions. Thus long-term genetic conservation of such species is difficult, if not impossible, without *in situ* conservation. In addition, recalcitrant seed behaviour (*i.e.* inability to tolerate desiccation) also complicates long-term *ex situ* conservation efforts for many tropical tree species. *In situ*conservation can also contribute to the conservation of biological diversity at higher levels, *i.e.* species and ecosystems.

Protected areas have often been established on the basis of ecosystem or species conservation, rather than gene conservation. Thus the design of *in situ* conservation programmes has been considered primitive. In tropical forests in particular, the complexity of interactions and lack of scientific information have hampered the development of *in situ* conservation strategies.

Uncertainty surrounds the adequate size of *in situ* conservation areas, the number of individuals to be included, how to select locations, and how genetic variation is distributed within selected areas. Undisturbed tropical forest ecosystems are often taken as a starting point when planning *in situ* conservation programmes.

Today, however, rural landscapes are a mosaic of disturbed and less disturbed patches of forest, ranging from seemingly natural and secondary forests to seriously degraded forests and other wooded fragments. The most obvious genetic effects of fragmentation are a loss of genetic diversity at both population and species levels, changes in the genetic structure of a population and increased inbreeding. Clearly, *in situ* efforts cannot always rely on intact natural forests, and so it is essential to understand how the processes of genetic drift, gene flow, selection and mating affect genetic diversity in fragmented forests. IPGRI's partners at the Universities of Costa Rica, Alberta and Massachusetts have been studying the effects of fragmentation on the genetic diversity of *Enterolobium cyclocarpum* in Costa Rica. Studies of the reproductive biology of *E. cyclocarpum*, a tropical dry forest species, in continuous forests and forest fragments in pastures indicate that pasture trees are less likely to receive pollen and set fruit, and that their fruits bear less seed, than trees located in continuous forests.

Also, outcrossing rates within the two groups of trees have been found to be similar, but progeny vigour among seedlings in continuous forest is higher than in pastures. These results have immediate implications for the conservation and management of *E. cyclocarpum* and other species in similar habitats. Trees in pastures can aid the movement of pollinators between intact forest fragments and so contribute to gene flow and maintenance of genetic diversity among forest fragments. However, because seedlings in pastures have less vigour, seeds from these areas cannot be recommended for establishing plantations or rehabilitating degraded natural forests.

In southern India, sandal (*Santalum album*) forests have long been exposed to selective logging, poaching and changes in land use. The effects of disturbance on the genetic diversity of sandal have been investigated by the University of

Boston, together with the Ashoka Trust for Research in Ecology and the University of Agricultural Sciences in Bangalore. Results from two protected areas indicate that genetic diversity of sandal is highest in the core undisturbed park zone, whereas allelic diversity is reduced in the outer buffer and disturbed zones. These findings suggest that present protection measures are adequate and that excessive logging will cause genetic deterioration in natural sandal populations. Allelic diversity, however, is similar in the two disturbed zones, which suggests that, contrary to general assumptions, no significant genetic segregation has resulted from different degrees of disturbance. This finding can be attributed to the fact that sandal is insect-pollinated and animal-dispersed, *i.e.* substantial gene flow still occurs despite varying degrees of disturbance.

EX SITU CONSERVATION

In situ and *ex situ* conservation should be used in a complementary fashion to conserve FGR. Both are an integral part of the conservation process, and both can only be effective after genetic diversity has been located and conservation priorities have been set. The main purpose of *ex situ* conservation is to capture and maintain a representative sample of the existing genetic diversity of a species. For highly endangered tree species, *ex situ* conservation may be the only approach in the short to medium term. The main pitfalls of collecting germplasm samples for *ex situ* conservation are: i) limited coverage of genetic variation; ii) biases in the collected plant material; and iii) samples that are too large to deal with. Because *ex situ* conservation is more costly than *in situ*conservation, it is particularly important that sampling of populations and germplasm within populations is given special attention to maximize the use of limited financial and human resources.

The traditional approach to *ex situ* conservation of FGR is to establish conservation stands of a species outside its native habitat to facilitate gene management. However, long-term maintenance of the collected genetic variation in *ex situ* stands tends to be complicated by genetic drift and potential contamination by external gene flow. Another commonly used method of *ex situ* conservation is to store seeds collected from a range of natural populations. In the case of many tropical tree species, however, this method is limited by recalcitrant seed behaviour.

More than 70 per cent of commercially valuable tropical tree species are estimated to have recalcitrant or intermediate seeds. In recent years, therefore, considerable effort has been expended in developing *in vitro* techniques for *ex situ* conservation of recalcitrant tree species. A notable example is cryopreservation. Cryopreservation, the storage of cells or tissue at ultra-low temperatures, offers great opportunities for longer-term *ex situ* gene preservation. Its applicability, however, has been limited by difficulties in identifying protocols to reduce the water content of tissue before freezing.

Tissue water content must be reduced to avoid lethal ice formation in cells. Different results from different laboratories engaged in drying the same species of seed are an additional obstacle. Although it is likely that further research will ease these problems, the main application of cryopreservation seems to be in supporting tree improvement programmes and conserving biotechnically derived germplasm, rather than as a widely applied *ex situ* conservation method to support *in situ* conservation of FGR.

Micropropagation, the use of tissue and organ cultures for organogenesis and somatic embryogenesis, can also be applied to maintain genetic variation in germplasm collected for *ex situ* conservation. Practical propagation applications are already available for several tropical trees, based on shoot tissue culture in broadleaved tree species and on somatic embryogenesis in conifers. Micropropagation does not necessarily require expensive laboratory facilities, and can therefore be used in less-developed conditions. Maintaining juvenile material, however, can be problematic. At present, the technology used in somatic embryogenesis is in most cases too expensive for cost-effective *ex situ* conservation. Since 1997, the Danida Forest Seed Centre (DFSC) and IPGRI have cooperated in enhancing *ex situ* conservation methods for recalcitrant tropical forest trees. Project activities have focused on determining whether seeds are recalcitrant, intermediate or orthodox, and have included seed development research to assess optimal conditions for seed collecting, germination and storage.

An international Forest Tree Seed Research Network has been established under the project to facilitate information exchange among scientists developing protocols for collecting, handling, testing and screening of tolerance to desiccation and optimal storage conditions. At present more than 20 countries worldwide are involved in the project. More than 50 tropical forest species have been screened so far. Activities also include publications and training workshops to increase research capacity, particularly in developing countries.

CONVERSION OF FORESTS TO OTHER LAND USES

Forests can be lost either because forest resources and trees are not regarded as being of economic importance, or because of a policy framework that makes it possible to replace forests with other land uses. Often this is based on short-term maximization of economic returns and lack of supportive forest policies based on good understanding of the potential of forests as sources of income and products for local and regional markets and their associated services for other sectors of the economy.

CONSERVE: ECOSYSTEMS, SPECIES OR GENES

The first step in genetic conservation is to specify the objectives of the conservation programme. This is of the utmost importance, since it is possible to conserve ecosystem properties and still lose species entities. It is also possible to conserve a species and still lose genetically distinct populations, and therefore, genes that may be of value in disease and pest resistance, and in future adaptation. They could also be important in species deployment through breeding programmes, if that becomes a need or necessity. Many forest genera and species around the world provide goods and services, such as timber, wood, food, fodder, environmental stabilization, shade, shelter, and cultural and spiritual values.

However, fewer than 1 000 tree species have been systematically tested for their present-day utility, and less than 100 are the subject of intensive genetic research programmes. Evidently, therefore, many different forest species are being used in situ to provide important goods and services, without any active genetic management. Worldwide, the conservation of forest genetic resources has as its overall objective the maintenance of genetic diversity in the thousands of tree species of known or potential socioeconomic and environmental importance. Moreover, the levels and distribution of genetic variation in any given species are expected to be in a process of constant natural change resulting from the main forces of evolution. Therefore, the central concern of conservation should be the evolutionary processes which promote and maintain genetic diversity, and not the endeavour to preserve the present distribution of variation as an end in itself.

ASSESSING SPECIES' PRIORITIES FOR CONSERVATION ACTION

Within any given country or local area there may be divergent opinions on priorities among tree species. Forestry departments are likely to have a somewhat different emphasis and priorities from those of local forest dwellers and users, which can be different again from those of farmers and various other users of trees. It is apparent that in situ conservation programmes will be more successful if they target species of direct interest, use or concern to the land management authority and/or landowner(s): this will have major implications for the planning of in situ conservation programmes. A participatory rural appraisal approach can be useful for helping local communities whose land is communally owned to better identify their priority tree genetic resources and to develop appropriate in situ conservation responses.

It is also important to consider the case for species and provenances which are of major economic importance when planted as exotics, but currently of much less significance in their native range and habitats; an example is Pinus radiata from the south-western USA and Mexico. In such cases it is not unreasonable to expect that the likely beneficiaries of in situ conservation should contribute, financially or otherwise, to conservation. Given that there will be limited financial resources available for specific conservation programmes for forest genetic resources, it is necessary to consider which of the priority species are also in most need of, or warrant, conservation interventions and actions. This can be conveniently undertaken for different species by comparing the extent of the resource (level of genetic diversity or intraspecific variation) with the vulnerability or threats to the populations and/or ecosystems of which they are a part.

In Situ Conservation Atrategies

In the case of non-domesticated species, in situ conservation is probably the most important strategy and sometimes the only viable approach. In the tropics, where extinction rates of species are high because of land-use changes, setting conservation priorities is critical. This is particularly evident in developing countries, where resources allocated for conservation are scarce and baseline information on species distribution and richness data are lacking. In a world of scarce resources, one approach to priority-setting is through networking activities, with initiatives involving multiple countries and stakeholders.

In situ conservation is usually the preferred conservation strategy for most wild plant species, including some of the wild relatives of crop species, because, it allows the populations of interest to continue to be exposed to evolutionary processes. Alternatively, for many domesticated species (crop and livestock), on-farm conservation of traditional varieties is now widely supported as an important practice for conservation of genetic diversity.

Molecular genetic studies, carried out on many forest tree species around the world, are contributing to a better understanding of patterns of variation to support the development of improved management practices, and to monitor changes of species turnover in time and in space. In some situations, these priorities could be refined by the use of new tools, such as molecular markers and modelling simulations.

Integrating GIS tools with molecular research will improve our knowledge of landscape patterns of genetic diversity of species distribution, and help develop resource management plans. For example, in the Western Ghats, India, these two approaches are being used in combination to detect areas with high diversity (intra- and interspecific) and to set priorities for conservation. Molecular markers can thus assist in the conservation of tree species and may allow national programmes to reflect biodiversity patterns in their own management plans.

Molecular methods can also help to identify differences between local and non-local provenances and genotypes, identify where diversity is being lost, and facilitate the introduction of new diversity for integrated conservation and breeding programmes. This can also provide for a better management of intraspecific genetic variation.

Evolutionary Conservation

Evolutionary conservation activities are characterized by programmes where the trees produce progeny in successive generations: genes are generally 'conserved', but genotypes are not. Natural selection takes place among trees with new allelic combinations that either favour or disfavour different genotypes. This process ensures that gene frequencies will change in the population: alleles with positive influence on fitness will increase, and alleles associated with low fitness will decrease. If the population size is sufficient, neutral genes should, in general, be maintained, but some genes will inevitably be lost by genetic drift; new genetic variation will arise by mutation after several generations. Human interventions (if any) are designed to facilitate moderate genetic processes rather than to avoid them. Genetic variation between populations is generally maintained when they are growing in different environments, and is even expected to increase over time.

A typical example of a conservation population capable of evolutionary processes is a protected area in a natural forest. In a protected area, the species occupies its natural habitat (it is said to be in situ conserved), typically with a wide range of other species. Natural selection for general fitness is therefore largely related to competition among species, as well as to adaptation within species to current and future environmental conditions. However, evolutionary conservation can also take place in a planted stand, if natural selection is allowed to work, and if the planted trees are regenerated from seed, rather than by vegetative techniques, for the next generation. In such programmes, plantations will preferably be established and managed in ways that mimic the natural processes that will support natural selection. Of course, in most situations the mixture of species (if any) is largely artificial, and the selective forces may therefore favour different genes than would be the case in true in situ conservation. However, this reflects the fact that selection and fitness always depend on the degree of human influence in any ecosystem. Directional selection in favour of commercial traits—including characters such as good stem form or ease of establishment in plantations—is typically avoided in strict evolutionary conservation programmes, but of course this again depends upon the local objectives of the programme.

PLANT SPECIES, GENETIC VARIATION COMMONLY EXISTS

THE WEALTH OF POPULATIONS

Within plant species, genetic variation commonly exists at a high proportion of loci and tree species commonly exhibit the highest levels of all plants. For the species surveyed, which are mostly boreal and temperate pioneer species, the levels of genetic variation within populations often seem to exceed those between populations. The lack of strong differentiation may be due to the wide dispersal abilities of pollen and seed, which maintain a degree of connectedness between mating neighbourhoods among trees over long distances. A few studies of tropical tree species that are not pioneers indicate considerable variation among populations of trees, which may reflect a finer subdivision of mating neighbourhoods.

Another measure of genetic variation is the heritable variation that exists in traits controlled by many gene effects simultaneously but in which individual gene loci may have small effects.

These are the so-called quantitative traits that can involve many genetic loci which act indirectly on the observed trait and hence have small direct effect. Most of these loci would not change much in frequency from one generation to the next.

Cumulatively, however, they could cause large changes in the trait expression. Thus, many of the traits selected for in breeding populations can shift mean expression by as much as one standard deviation in one generation even if there is little shift in gene frequencies at loci sampled by molecular means. Furthermore, since differences among populations in the frequency of molecular markers are more difficult to detect than differences measured in quantitative traits, traits of adaptive significance may be more obviously segregated among populations than molecular genetic measures may indicate.

Some traits and the gene loci affecting them may be under strong selection pressure that is divergent between populations, while other traits may be under mild or virtually no selection pressure. Some traits may be indirectly related to strong selection events, such as caused by fire, drought, or forest practices that leave few resistant individuals, but only weakly related to resistance mechanisms.

Other genes may not be presently involved in any recent selection events. Thus the genes associated with selection response may be under strong or weak selection according to the strength of the selection event and the degree of relatedness of the trait to the selection agent. There may also be multiple genes that affect a trait expression, and the combined effects of the genes may not be simply assigned to any simple action of any one of them. This kind of multigene interaction is called epistasis and it may cause multiple gene locus differences to be more pronounced at the level of trait observations than in single-locus allele frequency differences. These kinds of interactions may mask any obvious trends in gene frequency changes associated with selection trends.

Since the phenotypic traits themselves can be expected to be under a wide array of selection intensities and the genes associated with them can also have a wide range of influence on those traits, we can expect that there would be a wide range of selection pressures among the genes of any one population. Thus, if the relative divergence of populations was tracked by weakly selected traits or loci, population similarities would predominantly reflect migration effects.

In the absence of selection, even very low migration rates would tend to maintain similar allele frequencies among populations. However, if divergence was tracked using traits or loci that respond strongly to divergent habitat selection, then similarities would tend to reflect habitat differences. In any one population, strong directional selection would force disfavoured alleles towards the point of extinction unless maintained by an input from migration or mutation.

Among populations exposed to divergent directional selection, different alleles would be at high and low frequencies among them and hence their frequencies would be highly dissimilar.

If a large number of traits or loci are used and some loci reflect similarities where others reflect multiple patterns of divergence, then contrasting patterns of allelic distribution must coexist. Similarly, traits subject to different intensities of divergent selection would also display different degrees of divergence.

Clearly, if phylogenetic relationships were estimated using selectively significant loci along an environmental gradient, then genetic similarities or distances would be correlated with that environmental variable. Alternatively, if selectively neutral traits were used, the genetic distance measures would be correlated with migration distances, and the derived phylo-genies would be different. The existence of differing strengths of divergence has been observed in variations in the patterns of geographically related population traits in *Pinus armandii* (Ma 1989). Differences in the amount and geographical pattern of isozyme allele frequencies and in growth traits have also been found for *Pinus lam-bertiana* and red alder.

In addition to those forces of evolution, the mating system can have heterogeneous effects on the distribution of alleles. If pollen and seed have different migration distances or different direction of flow due to their migration season or the influence of different wind pollinator effect on the sexes, then the balancing effects of migration and selection would be different (Namkoong & Gregorius 1985). Even for selectively neutral alleles, differences between the migration paths of the sexes causes heterogeneity of allele distribution.

Genetic variation also exists at several levels of expression and the amounts and patterns of variation can depend on which measures of variation are used to estimate that differentiation. With currently available tools of molecular measurement, it is possible to measure very fine levels of difference so that individual trees (or humans for that matter) can be discriminated by unique changes in the fine structure of single genes. It is also possible to discriminate differences among local populations if they share maternal or paternal lineages, by using particular molecular genetic markers that distinguish among populations even though the populations are not obviously different phenotypically.

Furthermore, some genes are expected to be more influenced by the different forces of evolution and therefore the distribution of variation among genes may vary according to their function (Ma 1989). Therefore, there is no one preferred measure of genetic variation but several options, and the choice of which measure to use depends on the purpose of the measure and the type of gene being measured. For forest trees it is satisfying to note that the levels of genetic variation measured are usually among the highest in the plant kingdom, although their distribution and the differences among loci have not been well studied. If the frequency of heterogeneous associations is high, the evolution of forest species may be unusually complex.

Some forest trees have relatively high levels of genetic variation between populations in several traits and at several molecular loci, while others may have low general levels. Since the beginnings of interest in genetics, foresters have known about economically significant levels of heritable genetic variation among populations in traits relating to growth and soil and climatic adaptability.

One of the earliest tests to indicate the utility of population differences is recorded for *Cryptome-ria* in Japan in 1883. However, the existence of graded changes in traits is widely recognized in forest trees as clines and with high levels of variation among populations relative to genetic variation within populations. The variation patterns often differ depending on which traits are measured but growth rates and climatic adaptabilities are often considered to be optimized within local populations. While there is debate as to the generality of the concept that local populations are optimal, regular trends in performances are often observed.

The geographical patterns of variation among populations or provenances may vary from a single, large, contiguous and intermating population with gradations across mating and environmental neighbourhoods to one of discretely divergent populations with little relationship among subpopulations. Some species may be composed of a main, large, central population with satellites connected only to the central one and with only one-way germplasm flow, or may be a series of more or less equivalent-sized populations with equal and multidirectional flows. Some of them may exist along streamside corridors as single units with connections between drainage basins being more tenuous, while others may form island populations on the tops of separate mountain peaks or ranges.

Since the stability of the interpopulation structure depends on the stability of ecological dynamics, the level of population or provenance differences may fluctuate over time and geographical area. In turn, the response of populations to environmental divergence is contingent on the genetic dynamics that can occur within an intermating neighbourhood. For the most part, the potential for selective divergence among the populations is moderated by migration among populations. Only traits with strong divergent selection would maintain high levels of interpopulation difference.

THE WEALTH OF INTRAPOPULATION GENETIC VARIATION

Within populations, four forces (selection, migration, mutation and population size) determine the structure of genetic variation. The interaction of all four factors can be highly complex and their effects are determined by the mating system and the kinds of gene effects that exist. A problem in analysing even this level of genetic evolution is that the kinds of gene effects that exist depend on the effects that other forces of evolution have had on gene actions themselves. To consider how the forces can affect evolution, we can first start with each of them and then consider how their interactions can change predictions.

For selection to have an effect on speciation and the structure of forests, there must be a difference among genotypes in the traits affected by survival or reproduction events and the individuals must accordingly leave more or fewer progeny than other genotypes. The genotype is defined as the individuals with a distinctive ensemble of alleles at its gene loci that distinguishes it from other genotypes. This is a multiple gene locus definition, which is often simplified by considering every one of the loci to be independent in the occurrence of its alleles in both frequency and effect.

In this case, we can consider the genotype to be the summation of all independent loci and hence can examine gene effects and dynamics one locus at a time. This is, of course, an oversimplified view since genes are in fact linked on chromosomes and therefore have linkages that correlate their occurrences, and most genes affect processes that are combined in their net effect on traits; furthermore, most genes affect enzymatic pathways that ultimately affect several traits when measured at the level of whole organism performance.

These effects are called linkage, epistasis and pleiotropy, respectively, and tie the whole organism together into an integrated set of actions and functions that can survive and reproduce. While undoubtedly true in general, there is little evidence that individuals are so bound by such constrained genetic constitutions and physiological limitations that substantial independence among loci exists. The evidence for the existence of strongly selected, tightly linked 'coadapted gene complexes' is not strong and hence the model of independence can be usefully employed, even though it is surely wrong at least in detail.

Within a gene locus, many alleles may exist and the pairwise interactions among them are called the dominance relations. For a diploid organism, if both alleles at a locus are the same, the individual is said to be a homozygote; if the two alleles are different, the individual is said to be a heterozygote. If the heterozygote has a phenotype intermediate between the two homozygotes, the dominance effect is considered to be zero and the gene action is said to be additive. If the heterozygote is other than intermediate, it may be less than the lesser of the two homozygotes. If its phenotype lies between the homozygotes but is not exactly intermediate and if it appears to be greater than the greater of the homozygotes, the gene action is said to be underdominant, partially dominant or overdominant, respectively.

The effect of selection would depend on the kind of gene action, since directional selection for greater or lesser phenotypes would favour different homozygote and heterozygote states. Selection may also be stabilizing if it favours an intermediate phenotype or disruptive if it disfavours the intermediate and its effect on the frequency of the alleles would depend on the gene actions.

Directional selection experiments seem to indicate that most alleles have partial dominance actions, though many may be involved in multiple locus interactions. Since selection would be expected to favour survival and reproduction of certain genotypes, we expect that it would increase the population level of fitness.

Selection is therefore usually considered to have a positive effect on the evolutionary potential of a population or species in small increments every generation but may not generate an optimal condition at any one time. For alleles that have partial dominance effects, the dynamics of the rates of displacement of one allele by another are such that the frequency of the allele itself affects its own rate of displacement. When alleles are at either low or high frequency, their rates of change are slower than when they are at an intermediate frequency. Except in small populations where accidental loss can occur, selection would not generally lose alleles that are favoured for adaptation.

Another force of evolution is mutation, which is usually considered to be a random change in gene structure and function that occurs indiscriminately throughout the genome. Since most species have evolved a genetic and physiological level that is close to a sufficient capacity for continued evolution, novel changes are not expected to improve performance or fitness but rather to generally decrease adaptability. However, there are always going to be mutations and while most may be disfavoured by selection, there is likely to be a balance between the rate at which new mutants are introduced and the rate of their elimination.

Most mutations occur at low rates for any one locus in any one generation and would not be expected to affect the adaptability of a population unless they are accumulated in small populations. For most conifer species where the frequency of deleterious mutants has been estimated, there are a larger number of lethal alleles carried at one or another locus in most individual trees than in most other plant or animal species. This might indicate that there are many effective loci in trees that can mutate to a harmful state, that the mutation rate is high or that the mutants are favoured in some conditions or are of some advantage in heterozygous conditions. However, independent mutations and selection can generate differences among populations, which can increase fitness in each. Hybridization among populations can then generate high levels of genetic variance in traits that may then be useful in new environments. A third force of evolution is migration, which can occur among many different kinds of population structures.

The input of pollen or seed from other populations that have different allele frequencies may bring in alleles that might otherwise be lost, and if the migration rate is high enough among all populations may induce all populations to have approximately the same frequencies. If the immigrant alleles are favoured, then further immigration would advance the rate of its increase; if disfavoured, immigration would maintain alleles at higher frequencies than expected in a manner similar to mutation; and if neither, would tend to make frequencies homogeneous among all populations. In the presence of divergent selection among populations that exist in different environments, selection would act to eliminate migrant alleles and would maintain population divergence.

The fourth force is sampling error induced by small population sizes, which tends to increase the probability that low-frequency alleles are lost by chance. If populations remain at low sizes for several generations, the chances of random loss are accumulated; even populations of more than 20-50 adults can lose alleles that may have started with moderate frequencies. Since most forest trees seem to carry many deleterious mutants, even if at low frequency at any one locus, another problem is that small populations will suffer from inbreeding depression and may eventually suffer such debilitation from several mildly depressive mutants that the population goes extinct. Individual populations would have to be large enough to sustain viability, and to avoid environmental accidents and catastrophes if no migrants were recruited.

There are two problems faced in persistently small populations, the loss of alleles and inbreeding depression; hence, larger population sizes are generally better for avoiding extinction. However, the effects of selection, migration and mutation are confounded with the sampling errors incurred in real populations of finite size. Thus, if populations are separated and selected for different environmental optima, migration can introduce alleles in high frequency that depress fitness. On the other hand, migration can maintain alleles in populations where they might be useful, but would be lost due to sampling accidents in small populations.

It is reasonable to expect that all species are under the influence of several or all of the forces of evolution simultaneously and that the balance between the several forces is not the same for all genes in all populations. Furthermore, since the physical and biotic environments of forests are rarely at an equilibrium, the genetic and ecological dynamics of most species would have to be considered to be in other than a stable state.

The actual distribution of the multiple functional forms of alleles at each of the several tens of thousands of gene loci in each of the individuals of a population must therefore be the resultant of the mixture of forces felt at each locus, subject to historical events. The vast possibilities of mixed distributions are overwhelming, although it is also obvious that for most of the species we can study, species are not random mixtures of an infinite number of possible combinations.

MANAGEMENT PRACTICES IN FORESTRY CONSERVATION

Forestry Conservation Management Practices (CMPs) are specific, science-based guidelines for conservation of rare species during forest harvesting. CMPs are somewhat analogous to Forestry Best Management Practices (BMPs), except whereas BMPs focus mainly on protection of water resources, CMPs specialize in protection of rare wildlife. The primary objective of CMPs is to guide harvesting activities such that rare species listed under the Massachusetts Endangered Species Act (MESA) are not impacted in a way that jeopardizes long-term viability of local populations. CMPs first identify and describe potential impacts of forest harvesting to state-listed species, whether impacts may be direct (*e.g.*, physical injury or death of individual animals) or indirect (*e.g.*, alteration of habitat in a way that reduces overall reproductive success of a local population).

Then, CMPs provide specific guidelines to avoid or minimize impacts that would be considered negative or potentially detrimental to a local population; the guidelines are based on scientific knowledge of the habitat requirements, reproductive strategy, dispersal ability, survivorship, and other ecological factors that influence population dynamics of the species. CMPs are based on the principle that protection and enhancement of the long-term viability of local populations is prerequisite to statewide recovery of imperiled wildlife.

Simultaneous with the objective of protecting local populations of state-listed species, CMPs aim to maintain adequate opportunity for sustainable management of timber products in Massachusetts. To this end, CMPs tend to focus forest harvesting restrictions on the critical areas within known habitat of state-listed species, thereby allowing timber management to proceed with fewer restrictions over as large an area as possible. This strategy is based, in part, on recognition that forest harvesting typically results in temporary habitat change or sometimes even habitat improvement rather than permanent habitat loss. Thus, the CMP strategy is designed to help maximize the protection of state-listed species and the ability of Massachusetts landowners to manage their forests for timber and other wood products.

MANAGEMENT OF FORESTLAND IN MASSACHUSETTS

Management of forestland in Massachusetts involves multiple stakeholders with multiple objectives, so communication among stakeholders and development of management standards are essential to guiding compatible and sustainable uses of the land. State-owned forestland in Massachusetts received Forest Stewardship Council certification in May 2004, and this certification requires that there be coordination of forest management among the three state forest management agencies (Division of Water Supply Protection [DWSP], Bureau of Forestry [BOF], and Division of Fisheries and Wildlife [DFW]).

In addition, there is a need for state forestland managers to work closely with the Natural Heritage and Endangered Species Programme (NHESP) to develop and implement a formal and consistent process for determining appropriate mitigation of potentially adverse impacts of forest harvesting to state-listed species on both state and privately-owned forestlands. In light of these goals, CMPs are needed to ensure adequate protection of state-listed species during forest-harvesting operations. Without CMPs, a landowner and/ or forester must develop a Forest Cutting Plan without having prior knowledge of possible harvest restrictions required by the NHESP to mitigate potential impacts of forest harvesting to state-listed species. Under these circumstances, activities proposed in the Forest Cutting Plan must often be modified after the plan is reviewed by the NHESP, which tends to result in delay of the harvest, changes in the amount and distribution of timber to be harvested, or changes in the number, type, or location of stream and wetland crossings. However, CMPs shall make potential harvest restrictions available to the landowner and forester before a Forest Cutting Plan is developed. This gives the landowner and forester an opportunity to develop a plan that is more likely to be approved by the NHESP without need for modification, thereby saving the landowner time and money. With proper planning and attention to detail, use of CMPs shall maximize the efficiency of the cutting-plan filing and review process and enable landowners to better understand how forestry and conservation of rare wildlife can be integrated as compatible uses of Massachusetts forestland.

Benefits of CMPs

CMPs benefit both state-listed species and forest management by (1) being based on scientific knowledge of the ecological factors influencing population dynamics of state-listed species, thereby enabling focused protection of key habitats and other resources, (2) improving the predictability and consistency of review of Forest Cutting Plans by the NHESP, and (3) providing detailed guidance to landowners and foresters for developing Forest Cutting Plans that have high probability of being approved by the NHESP without modification. An additional benefit, which is a byproduct of the first listed above, is that time-of-year harvesting restrictions mandated by CMPs tend to apply to smaller areas than those to which restrictions applied in the past.

In this sense, intensive conservation measures are concentrated, which should be more effective in achieving conservation objectives than a strategy by which lighter measures are applied over a broader area. Furthermore, this strategy affords the landowner greater liberty in managing a larger proportion of a given tract of forest.

Process for Developing CMPs

The NHESP is the primary partner responsible for development of CMPs. However, staff of the Natural Resources and Environmental Conservation (NREC) Extension Programme at the University of Massachusetts Amherst and foresters of the DWSP, BOF, and DFW collaborate with the NHESP in this effort. This group of scientists and land managers meets regularly to work together on every aspect of the CMP development process.

The first step in developing CMPs is to determine which state-listed species should be addressed. This decision is based largely upon the frequency with which forest harvests are planned in habitat of each state-listed species, whereupon species that most frequently coincide with harvests are considered top candidates for development of CMPs. A second factor is the degree to which protection of each state-listed species is expected to impact the logistics of forest harvesting, whereupon species that are likely to require highly protective measures are considered top candidates.

A third factor, which often is related to the second, is the sensitivity of a state-listed species or its habitat to forest harvesting, whereupon species that are highly sensitive are considered top candidates. When multiple species have similar ecological requirements and conservation strategies, groups of species may be lumped into a single CMP document (*e.g.*, Blue-spotted Salamander, Jefferson Salamander, and Marbled Salamander are lumped together in a "Mole Salamander" CMP document).

When there is agreement between the NHESP and its collaborators about which state-listed species to address in a CMP document, the NHESP conducts a thorough search of scientific literature pertaining to the ecology of the species of interest.

The information is then consolidated into a species account that includes a physical description of the species and a discussion of its range, distribution, habitat requirements, seasonal movements, reproductive strategy, survivorship, and other concepts associated with its life history. Additionally, the account includes a discussion of known and potential threats to population viability, including those posed by forest harvesting.

The NHESP and its collaborators then develop the actual management practices that scientific evidence suggests is necessary for successful conservation of the species. These management practices are specific to forestry-related activities and, like Forestry BMPs, are split into two categories: required management practices and recommended management practices. Required management practices (denoted with "R" in the CMP document) are those which should be considered mandatory during most harvests, barring special circumstances. Recommended management practices (denoted with "G" in the CMP document) are optional guidelines that are expected to provide additional benefits to the species of interest.

THE IMPACT OF FOREST DEGRADATION ON FOREST RESOURCE USE

In some regions formerly plentiful supplies of wildlife, medicines and supplementary foods have practically disappeared. This is not to suggest that substitutes are not adopted or that cleared forest lands reap no harvests. The degradation of the forest quite naturally leads to a decline in resource use. However, in the West African region there are some forest "products" which cannot be readily replaced; those that serve cultural or symbolic functions are especially important (*e.g.* bushmeat, medicines, and housing material). For example, in Korang's (1986) study of the impact of forest conversion on surrounding residents, a diminished supply of locally valued forest resources was noted. These losses had not been replaced with substitute products from the plantations. Thus people had lost a source of income (sale of canes and chewing sticks), food (bushmeat), and medicine.

In some cases, deforestation has reduced the extent to which forest resources are used. Visser (1975) notes that among the Ando in Côte d'Ivoire a decline in knowledge regarding medicinal plants and their uses has resulted from a decline in actual forest resources. In the case of plant medicines this loss of knowledge is particularly troubling, as traditional medicine is the only health care available to the majority of those living in rural areas of this West African region.

In Western Cameroon, Koagne (1986) notes that there are many forest foods which are decreasingly exploited. He adds that almost all of the region's natural vegetation has disappeared. Traditional supplemental foods are not available in exploitable quantities and new foods that are imported from other regions have taken their place. Forest foods with particular social value are still consumed on ceremonial occasions though these products are often specially protected or produced on farms for such occasions.

The greatest amount of information on the impact of forest degradation comes from studies of wildlife decline. Several studies have shown that bushmeat consumption is limited by the supply of wildlife. In southern Cameroon Laburthe-Tolra (1981) asserts that the decline in bushmeat consumption, caused by reduced supply of wild animals, has resulted in a deterioration in the quality of people's diets. The risks associated with a continued decline in supply become particularity great when it is realised that, as discussed earlier, it is difficult to substitute for wild animals; livestock production is difficult in this region.

Fuelwood scarcity has led to a range of different problems. In the southern regions of Ghana, Ardayfio (1986) has found that fuelwood scarcity has forced households to purchase fuel at the expense of food (in one village household, fuelwood expenditures rose from 1 per cent to 16 per cent of total expenditures within a few months), cooking time is being reduced and, in some cases, different foods are being used. The additional time that must be spent collecting fuelwood has increased the negative health effects of fuelwood gathering, and women's income earning activities which require high fuelwood energy inputs, have, in some cases, been curtailed because the fuel costs have become prohibitive.

Use of Forest Tree Species on Farm Lands

In some areas the combination of declining forest resources and increasing market demand for certain products has resulted in increased use of forest trees on farm lands. This is occurring, to different degrees, throughout humid West Africa. Its predominance depends upon the amount of land pressure that exists, the quality of nearby forest resources, traditional tenure regulations, the extent of commercialisation, and the markets for forest products. In Nigeria, on-farm trees are valued for cash income, shade, fruit, fuelwood, wood for building material and agricultural implements, soil conservation, and palm wine. As nearby forests disappear it is not only trees with marketable products which are increasingly left on farms, trees with food crops or particular household uses such as those use for medicine production, are also being preserved and supplemented.

Generally it is believed that many of the West Africans from the region do not plant trees. Trees on farm lands are most commonly those that have been left and protected during forest fallow and cultivation periods, or transplanted wildlings. But, tree planting on farm lands is increasing. In southern Nigeria, where land pressure is at perhaps its highest in the West African region, the number of trees being planted on farm and fallow lands is increasing. Currently approximately 37 per cent of the farm and fallow land trees were planted.

In southeastern Nigeria, there is a considerable amount of research evaluating trees on farm lands. One hundred and seventy-one (171) common farm and fallow land forest tree species with edible products have been identified; the density of these species is often far greater on farm lands than in natural forests. On-farm trees serve a multitude of household functions while providing support to agricultural production (*e.g.* yam stakes and mulch) and a source of cash income.

Okafor (1987) found that the tree species with the greatest number of household uses were often incorporated into compound farms while less frequently used species were found in outlying fields. In other parts of West Africa, tree planting also appears to be of growing importance. In Western Cameroon, Depommier (1983) notes that natural vegetation has almost disappeared. He finds that many useful forest species are protected and planted in hedgerows, especially those with market value. In Sierra Leone, Engel et al (1984) has also found that most farmers are interested in planting trees on their farms. The most valued species for planting are oil palm and other food trees.

Fruit trees are of growing importance. In an examination of agroforestry systems in southeastern Nigeria, Ijalana (1983) found that trees (especially fruit trees) were being incorporated into farms with increasing frequency. Trees were generally protected as wildlings though they were planted as boundary demarcations; generally those with long gestation periods were selected for this purpose, (*e.g.* Mansonia altissimaand Nauclea diderrichii).

In the Ho district (Ghana), approximately 75 per cent of the fallow field tree species have medicinal uses. In Nigeria medicinally valued trees are also predominant on farm and fallow lands. For example, the leaves of Jatropha curcas are used for treating ringworm (the seeds are used as a soup ingredient).

Okafor found that at least half of the produce gathered from on-farm and fallow land trees was consumed by the household rather than being marketed, though strong markets for these tree products did exist. Another study in southeastern Nigeria echoes Okafor's findings. It reveals that a great portion of the tree product harvest is consumed by the household rather than being sold: for example, 78 per cent of the Irvingia gabonensis fruit and seeds harvested from compounds are consumed by the household.

Changes in Usuary Rights to Forest Products

In addition to planting trees, changes in the commercial value of forest products has, in some cases, led to changes in traditional regulations over tree products. For example, in a rural community near Accra, in Ghana, Ocansey (1985) found that the increasing market value of fuelwood led to severe exploitation of the woody vegetation, and changes in the rules associated with resource use. Fuelwood became a privatised commodity, and could now only be collected from people's own farm lands. Thus landless people in the community were forced to purchase their fuel.

In some instances changes in tree tenure systems may reflect increasing market demand for certain forest products and a rapid decline in resources themselves. This may be the case in Casamance, Senegal, where Pelissier (1966) writes that the traditional rights associated with forest products have changed with the decline of forest resources. Formerly, access to forests was "open", then rights to forest areas surrounding farm fields and especially the vast array of palm products from them, were restricted.

Increased tree planting can also be a sign of increased privatisation as it is traditionally a sign of ownership in some parts of humid West Africa. The introduction of a new technology for palm alcohol production in southern Benin catalysed changes in the management and value of raphia swamp resources. In village areas with considerable raphia resources a system of communal management has developed. However, in villages with good access to the Cotonou urban market, excessive overexploitation has sometimes led to privatisation of raphia groves.

In many cases, raphia groves have been divided among villagers who have been given permanent rights to parcels of grove. By tradition, raphia is never planted, it is seen as a gift of God. However, since the privatisation of these areas people have begun to plant young seedlings. Good seed trees are now protected and their seeds are distributed among the villagers who raise and transport them into the groves. In some cases the privatisation has led to clearing of the raphia groves for vegetable cultivation which, because it can be undertaken in the dry season, leads to higher profits. In both cases the increased marketability of a forest product has led to changes in the management of a formerly "free" resource.

As forest resources decline, their utility and value to local people may diminish. Some authors suggest that loss of knowledge about traditional foods, medicines and uses of surrounding resources are indications of cultural disintegration and impoverishment (Okigbo). There is, however, little concrete information on either the changing uses of forest resources or the differential uses of different forest types. The examples presented in this chapter highlight some apparent trends. They suggest that it may be possible to manage both forest and fallow land in such a way that the production of increasingly valued (both commercially and domestically) forest products is supplemented.

FOREST AREAS, REFORESTATION

Reliable data relating to forest areas in India are available only for the government forests in regions that were formerly part of the British provinces. The figures for the private forests in these regions and for all types of forest in the former Indian States (which neither maintained forest departments nor carried out detailed forest surveys) are of doubtful value.

For the country as a whole the ratio of forest area to total area amounts to about 22 per cent, as compared with the following percentages for other countries: Argentina, 25; Belgian Congo, 45; Brazil, 57; Burma, 58; China, 9; France, 21; the United States, 33.5.

For India, where we have to contend with an oppressive tropical sun, desiccating hot winds, the periodic monsoons, a low forest potential and a predominantly agricultural economy, an ideal forest ratio might be about one third of the total land area. In the mountains as much as 60 per cent should probably be kept under forests, because of their protective function, whereas on the plains the ratio might be as low as 20 per cent.

NEED FOR REFORESTATION

The lack of wood over large parts of India compels the farmers to burn their cow dung as fuel-dung that should go to fertilize their impoverished fields. The consequent diminishing harvests cause a further extension of cultivation, which tends to engulf not only the remaining village woodlands but grasing grounds, habitations and even lines of communication and condemn the average cultivator to a precarious existence in a congested mud hovel on the barest level of subsistence.

The destruction of the forests, moreover, strips the land of its natural defences against hot desiccating winds, dust storms and erosion by wind and water; and this deterioration is the more serious because it is insidious. Hence, the campaign to grow more food must be inextricably tied to a "grow-more-fuel" campaign.

Measures must be put into effect for enlarging the forest area in the Indian states where it is deficient -- measures involving not only the prevention of further encroachment of cultivation upon forest lands, but also the reservation for the production of timber and firewood of wastelands unfit for cultivation and of such state lands as obsolete military camping grounds, canal banks and railway and roadside lands. Success will be achieved partly by arousing mass tree consciousness and by enlisting the active cooperation of villagers in raising trees wherever the opportunity presents itself. In the Uttar Pradesh useful work has already been done in this direction and many cultivators have been persuaded to raise a tree or two even on their own fields.

The "Van Mahotsaya," or annual Festival of the Trees, recently inaugurated by Shri K. M. Munshi, former Union Minister of Food and Agriculture, has done much to stimulate enthusiasm for tree planting throughout the country.

Furthermore, wherever there is now a relatively high proportion of forest land, it should be utilised to the best advantage so as to make good the deficiency where climatic and edaphic factors militate against tree growth.

Production and Imports of Forest Products

In the former British provinces and in certain Indian states for which data are available, the annual production of timber and fuel wood during the three prewar years averaged about 4,379,000 tons. By 1945-46 it had increased by about 60 per cent to 7,115,000 tons, but during the four ensuing years it fell off to about 40 per cent above the prewar level, averaging some 6,151,000 tons each year. The increase, which would seem to have come to stay, may be accounted for by the following factors:

- The opening up of inaccessible forests during World War II';
- A general rise in the standard of living, with a corresponding increase in the demand for forest products;
- The introduction of more speedy traction;
- A closer conversion of logs to meet the increased demand;
- the utilisation of species unacceptable before the war;
- Increased extraction from the Andamans.

The brunt of the fellings during World War II fell on such timbers as sal, deodar, chir, teak and sissoo, which were in considerable demand to meet defence requirements. The excessive wartime cutting of these species, however, was largely balanced by diminished felling of them during the period 1945-51, when the increased general level of production was maintained by the utilisation of hitherto little-used species and by the exploitation of timber-especially teak -- from forests that had been unavailable before the war. Forestry, however, is a long-range enterprise.

The current effort to introduce more scientific methods of management into the state-owned and private forests of India is not expected to yield speedy results. Except for improvement in the firewood situation, the influence of this effort on timber yields will hardly be felt to any great extent for the next 25 years.

The major imports are teak, paper, boards and wood pulp. Teak comes largely from Burma and Siam, but steps are now being taken to replace it by timbers from the Andamans and by treated conifers. A 25-year lease was signed by the Government of India in August, 1951, for operations in the virgin forests of the North Andamans. While it is difficult to grow two cubic feet of timber where one grew before, it is easy to make a cubic foot last the life of two by treatment designed to preserve it against fungus and insect attack. Spread of the practice of timber seasoning and timber treatment is bound to be an important factor in the conservation of the forest resources of the country. The bulk of the imports of paper, boards and wood pulp comes from Sweden and Finland. India is now entirely dependent on imported newsprint, although efforts are being made, with the aid of technical assistance from the Food and Agriculture Organisation of the United Nations, to bring about a domestic development of this industry within the next ten years. Recent surveys of the Himalayan fir, which has hitherto remained unexploited because of its inaccessibility, bid fair for the establishment of an indigenous wood-pulp industry on a modest scale.

THE HIMALAYAS FOREST

The Himalayas extend over a distance of some 1,500 miles between the gorges of the Indus and the Brahmaputra. The main range, with Everest, Kanchenjunga, Dhaulagiri and many other giant peaks, forms the core of the system, to which the Trans-Himalayas on the north, the Outer Himalayas on the south and lesser subsidiary ranges give width and body. The currents of the southeast monsoon from the Arabian Sea and the Bay of Bengal both strike the southern slopes of the Outer Himalayas, which receive the bulk of the rainfall.

In general the rainfall increases eastward. Along the south slopes of the mountains, it increases with elevation to a zone of maximum precipitation between 3,000 and 5,000 feet above sea level. Few clouds pass the main range, north of which the slopes are dry and windswept.

There is relatively little precipitation above 12,000 feet and most of it is in the form of snow. The Himalayas roll up the northern latitudes, as it were; within a relatively short distance they present tropical, temperate and arctic types of flora. The character of the vegetation changes both with elevation and with the change to moister conditions eastward. Conifers like those of the north temperate zone of America and Europe constitute a noteworthy feature of the Himalayan forests, although the proportion of conifers progressively decreases eastward.

We shall deal with the Himalayas in terms of three regions:

- The Western Himalayas, between the Indus and the Sutlej;
- The Central Himalayas, between the Sutlej and the Gandak;
- The Eastern Himalayas, between the Gandak and the Brahmaputra. While the names applied to the forest types of the Deccan and Indo-Gangetic plains will be retained, the dry, monsoon and rain forests of the Himalayas are distinct entities. Although they correspond in a general way to their counterparts elsewhere, they do not resemble the latter in points of detail.

THE WESTERN HIMALAYAS

In the western Himalayas the rainfall increases from the Indus valley (20 inches) eastward. The vegetation most characteristic of this region is best classed as Dry Himalayan Forest (rainfall, 20 to 40 inches). In the arid interior valleys in the lee of the main range we have such forests at altitudes of between about 5,000 and 11,000 feet.

Here the greater part of the precipitation falls in winter as snow. Most of this forest is composed of pure associations of conifers, the broadleaf species being confined to sheltered and comparatively moist valleys. Juniperus macropoda and Pinus gerardiana are found throughout Kashmir in the dry interior valleys, with the junipers persisting farther eastward. Groups of deodars (Cedrus deodara) occur on the Outer Himalayas, more particularly in the relatively moister valleys of the Beas and Sutlej, associated with silver fir (Abies pindrow) at elevations of about 7,000 to 8,000 feet. Oaks are absent in Kashmir, the broadleaf species of the Western Himalayas being Corylus colurna,Viburnum cotinifolium, Desmodium tiliaefolium, Lonicera quinqu-elocularis, Smilax vaginata Dcne., Rubus niveus, Berberis spp., Plectranthus rugosus, Fragaria vesca L., Viola canescens Wall., Thalictrum minus,Impatiens spp., Artemisia maritima L., Ribes grossularia,Rosa webbiana and Abelia triflora.

THE CENTRAL HIMALAYAS

A moister type of vegetation, the Himalayan Monsoon Forest, characterises this region, which has an annual rainfall of between 40 and 80 inches. This forest occurs at elevations of between 5,000 and 11,000 feet.

The rainfall is relatively less in the interior, towards the west and above 10,000 feet, but increases eastward and more particularly in the Outer Himalayas.

The conifers are typically represented by fir, spruce (which is notably absent in the Western Himalayas), cedar, hemlock and pine, with occasional cypress and yew and the broadleaf trees in the main by several species of oak, associated with rhododendron.

Other genera met with commonly are Acer, Carpinus, Aesculus, Prunus, Ulmus and Betula and frequently Litsaea,Lindera, Machilus,Euonymus and Ilex. Bamboo is represented in Arundinaria and the ground vegetation is composed of Indigofera, Desmodium, Rubus., Viburnum, Impatiens, Plectranthus, Senecio, Dipsacus and Heracleum. Within the monsoon forest, various species of oaks give clues to the following subtypes at different elevations.

Low-Level Oak Forest

Ban oak (*Quercus incana*) occurs gregariously in the Central Himalayas at elevations ranging from 6,000 to 8,000 feet, although relatively lower on northern slopes and in moist valleys and relatively higher along the warmer southern slopes. The ban oak prefers cool, shady, moist, sheltered slopes, decreases with the diminishing rainfall westward and is conspicuously absent in Kashmir.

In this zone chit (Pinus longifolia) occupies the hotter southern slopes, occurring more or less gregariously down to about 2,500 feet. Sought for leaf fodder and fuel in the hills, the ban-oak forests are gradually disappearing, yielding place to Rhododendron arboreum and Pieris ovalifolia, which yield neither fuel nor fodder. Other broadleaf species of the zone are: Carpinus viminea, Cedrela serrata,Euonymus pendulus, Ilex dipyrena, Betula alnoides and Litsaea umbrosa.

The ground cover is composed of Viburnum cotinifolium, Desmodium tiliaefolium, Berberis lycium, Lonicera quinquel-ocularis, Rubus ellipticus, Indigofera gerardiana, Myrsine africana,Deutzia staminea,Vitis spp., Hedera and Smilax parviflora.

Mid-Level Oak Forest

In the next higher zone (roughly, 7,000 to 9,000 feet), which is somewhat cooler and drier, though snow is common, Quercus dilatata progressively replaces the ban oak, although the latter persists as an unhappy straggler. The representative coniferous associates, often occurring gregariously, are Abies pindrow,Picea morinda and Buxus sempervirens, while Pinus excelsa replaces Pinus longifolia. Deodar is also found in this zone, but only on the inner ranges, where there is less rainfall and more snow and often descending to the ban-oak level on cooler slopes. Occasionally, associations of cypress (Cupressus torulosa) colonise limestone outcrops that provide relatively dry conditions. The other broadleaf species are Euonymus pendulus,Ilex dipyrena,Rhamnus purpureus, Lindera pulch-errima, Eurya acuminata, Rhododendron arboreum, Meliosma dilleniaefolia, Cedrela serrata, Pyrus lanata, Pieris ovalifolia, Machilus duthiei,Fraxinus micrantha,Betula alnoides,Carpinus viminea and Acer caesium.

The representative bamboo is Arundinaria and the ground cover most commonly is composed of Rosa macrophylla, Rubus niveus,Viburnum spp., Berberis aristata,Strobilanthes wallichii, Deutzia corymbosa and Hedera helix.

High-Level Oak Forest

This zone (8,000 to 11, feet) is characterised by even more rigourous climatic conditions — less rain, heavier snows and hail storms in spring — and hence has fewer species.

The most typical is *Quercus semecarpifolia,* though other broadleaf species characteristic of the mid-level oak forest are also found here.

Their coniferous associates are silver fir (Abies pindrow), deodar and spruce (Picea morinda). Other leading species are Quercus dilatata at relatively lower levels, Betula alnoides (and B. utilis), Pyrus lanata (and P. foliolosa), Prunus padus,Acer caesium,Meliosma dilleniaefolia,Rhododendron arboreum (and R. barbatum), Taxus baccata,Euonymus tingens,Rosa sericea (and R. macrophylla), Viburnum foetens,Cotoneaster acuminata, Strobilanthes wallichii and Salix elegans.

The ground cover is largely composed of Clematis montana,Vitis semicordata,Ribes glaciale, etc. and the representative bamboo is Arundinaria.

THE EASTERN HIMALAYAS

In this region the rainfall progressively increases eastward from about 80 to about 120 inches annually. Snow seldom falls on the outer ranges. The vegetation comprises semi-evergreen and evergreen types.

Semi-Evergreen Forest (Rainfall about 80 Inches)

This occurs in the 7,000-to 9,000-foot zone in Nepal, more particularly its eastern part and in Sikkim. The oaks of the Central Himalayas are replaced by *Quercus pachyphylla* and *Quercus Uneata* in association with groups of hemlock (*Tsuga brunoniana*) on warmer slopes and of *Abies densa* at high elevations.

The chief among other species are: Acer campbellii, Magnolia campbellii,Betula utilis,Taxus baccata, Rhodode-ndron arboretum, Brassaiopsis alpina, Acer pectinatum, Daphne cannabina, Rubus niveus and Berberis aristata. Ferns are abundant and mosses cover the stems of the trees. In the understory Rhododendron arboreum is found. Oaks and other broadleaf species become less frequent with the diminishing rainfall in the interior. In the zone from 9,000 to 11,000 feet in the inner hills of northeastern Sikkim the conifers met with are Picea spinulosa Beissn. Pinus excelsa,Larix griflithii, Juniperus wallichiana and hemlock (Tsuga brunoniana). Himalayan Evergreen Forest (rainfall, about 100 inches).

Sikkim, Bhutan, the Balipara frontier tract and the outlying Patkai, Barail, Khasi and Garo ranges support an evergreen type of vegetation, of which three subzones may be distinguished:

- Laurel Forest (approximately 6,000 to 7,000 feet), consisting characteristically of Machilus edulis, Beilschmiedia spp., Cinnamomum obtusifolium, Litsaea spp., Machilus spp., Michelia cathcartli, Magnolia campbellii, Engelhardtia spicata,Schima wallichii, Castanopsis tribuloides, Quercus spicata, Prunus nepalensis,Mallotus nepalensis,Betula alnoides, Acer laevigatum, Eurya acuminata, Symplocos theaefolia and species of Araliaceae.
- Buk-Oak Forest (approximately 7,000 to 8,000 feet). Buk oak (*Quercus lamellosa*), with *Castanopsis tribuloides,Acer campbellii,Michelia* spp. and many of the characteristic species of the Laurel Forest.
- High-Level Oak Forest (approximately 8,000 to 9,000 feet). Quercus, pachyphylla intermingled with many of the buk-oak associates.

In the northeastern corner of Assam (rainfall about 80 inches) the flora above 5,000 feet includes oaks, maples, pears, magnolias, Manglietia, Michelia,Buchlandia populnea,Alnus nepalensis,Betula alnoides and Carpinus viminea. On the outlying Khasi Hills and Shillong Plateau the characteristic species between 2,500 and 6,400 feet elevation are Pinus khasia and Quercus griffithii.

THE SEACOASTS

Trees are found on beaches and sand dunes along the coasts of India, but the only coastal forests of any considerable extent and economic importance are the mangrove tidal forests of the Ganges, Mahanadi and Godavari deltas, on the Bay of Bengal. Here differences in the floristic composition are caused largely by the varying degrees of the brackishness of the water and by the varying heights of the banks of estuary channels and consequent differences in their liability to flooding. In the Sundarbans area of the Ganges delta, a region named from sundri which grows there, the species generally met with in areas of brackish and salt water are as follows:

- *Low mangroves:* Ceriops roxburghiana,Avicennia alba, Aegialitis rotundifolia,Excoecaria agallocha,Acanthus ilicifolius,Brownlowia lanceolata, Bruguiera caryop-hylloides and Derris uliginosa, with Rhizophora spp. conspicuously absent.
- *Tree mangroves:* Rhizophora spp., Bruguiera,Ceriops and Sonneratia.
- *Salt-water Heritiera Forest in the lee of (a) and (b):* Heritiera, Excoecaria, Ceriops, Bruguiera, etc.

Farther back, where the tidal flats are washed for only a short time at high tides by slightly brackish water, recently deposited river silts support what is known as a Fresh-Water Heritiera Forest, comprising Heritiera and Bruguiera, with Ceriops in the second story, the two former species reaching heights of about 100 feet. On high banks Pandanus, canes and ferns cover the ground. Under brackish conditions near the sea Phoenix paludosa is common, while Oryza grass colonises newly formed high-and-dry mud banks washed by fresh water only. Rhizophores, with their stilted stems, occupy the sea face and are often associated with knee-rooted Heritiera. These mangroves are duplicated with minor modifications on the other east-coast deltas.

THE ANDAMAN ISLANDS

The position of the Andaman and Nicobar Islands along an arc across the Bay of Bengal suggests a former land connection between Burma and Sumatra. The hilltops that constitute the Andamans comprise 204 islands, the Nicobars being a separate group with 19 islands. In the present study the Andamans only are considered, for little is known about the vegetation of the Nicobars.

The Andamans extend over a distance of some 220 miles from north to south, with an average width of 15 miles. The land area is about 2,500 square miles. The surface is generally rough, cut by ranges of hills enclosing narrow valleys. In the North Andamans, Saddle Peak, the highest point of the islands, reaches 2,400 feet. An equable, oceanic climate with a heavy rainfall is reflected in generally moist types of forest vegetation, of which three may be distinguished: Rain, Monsoon and Shoreline Forests.

Rain Forest. Of the type found in Assam and the western Ghats, the rain forest is at its best in the parts of these islands where the mean annual rainfall is approximately 100 inches and over, as, notably, along the west coast, which receives the full force of the southwest monsoon.

Port Blair, on the east coast, however, registers a rainfall of 116.7 inches, with 177 rainy days and a mean annual humidity of 83 per cent. The mean annual temperature is 80° F., the mean.

April maximum 92° F. and the mean February minimum 740° F. A vast variety of species flourishes under these optimum conditions of growth, with trees reaching 125 feet in height and 15 feet and more in girth. Dipterocarps predominate, as is not the case in the rain forests of the Western Ghats.

The species most frequently met with are: Dipterocarpus tuberculatus, Artocarpus chaplasha, Calophyllum spectabile, Planchonia andamanica, Hopea odorata, Endospermum malaccense Bth. ex Muell.

Arg, Sideroxylon longipetiolatum King k Prain,Garcinia andamanicum, Myristica andamanica, Pterospermum aceroides, Caryota mitis, Euphorbia epiphylloides, Mitrephora. prainii, Anaxagorea luzoniensis. The ground cover is comp-osed of Dinochloa adamanica,Calamus palustris and Gnetum scandens.

In valleys protected from heavy rain, deciduous associates appear. In the south at lower elevations the evergreens consist of Dipterocarpus costatus, Mesua ferrea, Canarium manii, Hopea andamanica King ex C. E. Parkins, Cratoxylon form-osum, Euphorbia trigona and Cryptocarya ferrarsi.

Monsoon Forest. The moist deciduous forests of the Andamans, otherwise known as "Padauk Forests," like the moist sal and moist teak forests of the mainland, occupy low-level (below 300 feet) localities, which either have less rainfall, or coarse, porous soils derived from sandstone. The transition from the evergreens is imperceptible. The forest species characteristic of the region are: Pterocarpus dalbergioides, Terminalia bialata, Canarium euphyllum,Sterculia campanulata, Bombax insigne, Lagerst-roemia hypoleuca, Tetrameles nudiflora,Chukrasia tabularis, Albizzia lebbek, Odina wodier, Adenanthera pavonina, Dillenia pentagyna, Oxytenanthera nigrociliata and Bambusa spp. The ground cover is composed of Alsodeia bengalensis, Mallotus acuminatus, Ventilago madraspatana, Delima sarmentosa and Acacia pennata.

Shoreline Forest. The shoreline is fringed throughout with beach forest and tidal mangrove forest. The former occurs on exposed seashores and is composed of Mimusops littoralis, Pongamia glabra, Morinda citrifolia, Erythrina indica, Calophyllum inophyllum, Terminalia catappa, Barringtonia speciosa,Cordia subcordata, I'hespesia populnea, Vigna retusa Walp., Mucuna gigantea and Colubrina asiatica. The ribbon-like tidal mangrove forest consists chiefly of Rhizophora spp., Sonneratia,Heritiera, Ceriops and Bruguiera.

RIGHTS OVER FOREST RESOURCES

The tribal people are facing serious problems with regard to utilisation and rights over forest resources. Due to the increasing pressure on forests by various interest groups, there is a corresponding pressure on the tribals to reduce their dependency on forests. This is creating serious situations of conflict, as tribal life is symbiotic with land and forests and their livelihood and culture are based on their relationship with the natural wealth around them. The tribals are being harassed for using forestlands and being evicted in many places. Such reports have come from places like Khammam, Visakhapatnam, Vizianagaram, Adilabad, Srisailam and other places. In Khammam in one particular village, the forest and police departments allegedly branded the tribals on their shoulders as an indication that they were destroying the forests. There are ambiguities in forest-revenue land demarcation. In some places like Nellore district, a lot of land on Velugonda hills is indicated as 'poramboku' in revenue records and as RF as per the forest department. These lands do not have any forest growth. The FD is taking up palm oil plantation in these RF lands (which is not a forest species). However, landless tribals are booked in criminal cases or prohibited from using these lands for agricultural cultivation. These lands should be given to tribals with pattas for cultivation.

In some areas like in Visakhapatnam district, the lack of clear forest boundaries is making tribals vulnerable to the exploitation of both the forest and revenue departments. A joint survey and demarcation of boundaries by both departments should be immediately taken up to arrest these conflicts. Such Joint surveys need to involve the villagers at various levels. The Forest-Revenue Boundary dispute is a perpetual problem in Adilabad and Warangal districts, leading to booking of cases by the Forest department and tension in these tribal villages.

In Visakhapatnam Agency and Nallamala areas, there is the unique problem of Enclosure Villages. There were many tribal villages that were not enumerated in the forest surveys. Due to such sheer negligence, the villages were not given revenue status and to this day, they do not have pattas for their lands. They face constant harassment from local forest officials, as they do not possess land records. Recognising these enclosure villages and issuing pattas to tribals should immediately resolve this problem. In Buttapur (Adilabad district) and in Nellore Dt. (Yanadis), the tribals were given lands decades ago under the social forestry scheme and are cultivating there. But due to lack of pattas, they are being harassed by the police and forest departments and also do not have access to bank loans as they cannot prove their ownership. These tribals have to be given pattas as promised.

Attacks on tribals, their properties and livestock by wildlife are not compensated by the Forest Department. Several cases are pending where tribals have been either killed or disabled and yet have not received any monetary compensation as due to them under the Wildlife Act. The tribals should not be prohibited from entering the forests to collect NTFP for their domestic requirements, like firewood, medicinal herbs, food, or agricultural and housing material.

The tribals should not be defined as 'encroachers', as is being projected by the forest department and also as indicated in the Circular of IG Forests, MoEF dated 3rd May 2002. In districts like Visakhapatnam, East Godavari and Vizianagaram, where there is high prevalence of podu cultivation, the tribals are facing threats of eviction from the forest department. There was a notification issued by the government of A.P in 1987 ordering that pre-1980 settlements will not be evicted until further orders. This should be implemented. The JFM programme now renamed as the Community Forest Management programme of the A.P Forest Department has caused grievous violations with regard to tribal rights. One major violation is the displacement of tribals from their podu lands by reclaiming them back into the forests through the JFM programme. The official reports of the forest department and the World Bank (which has funded the project) reveal that 37,000 hectares of forestland has been reclaimed back from the people§. The APFD further states its intention of displacing tribals in G.O. where it calls for voluntary surrender of lands by involving NGOs to motivate people under the CFM programme. The APFD has come up with a proposal for rehabilitation of tribals displaced from their podu lands through a monetary and schematic approach. This should be condemned as it bypasses the issue of eviction and rights of people over their podu lands. Their act of reclaiming lands under JFM programme and the APFD's forestry project under World Bank assistance has to be scrutinised closely by the Commission.

Samata and a few other NGOs raised this issue by writing to the World Bank on its violations of its own Operational Directives 4.2 and 4.3 in its financial support to the APFD. This has led to the Bank's insistence on an R&R policy to be proposed and implemented by the APFD in its CFM programme

In Srisailam area, the Rajiv Gandhi Tiger Sanctuary has led to eviction of tribals from their original homes. They have not been properly settled so far. The concept of EDC (Eco Development Committees) that was introduced is working to the detriment of the tribals, as the objective of this programme was to reduce the forest dependency of the tribals. The tribals are given income-generating programmes and alternate sources of firewood by the forest department. This is an artificial mechanism that is not sustainable in the long term.

In the Srisailam Tiger Sanctuary area, the Chenchus, who are traditional hunter-gatherers, go into the forest everyday for all their needs. They are being harassed by the forest department for trespassing into the sanctuary. The groups who came from the chenchu area felt that the Chenchus should atleast be given identity cards to prevent harassment from forest department and the police department who mistake them for naxalites.

Another serious problem with regard to Sanctuaries is the settlement of people's rights. The Settlement Officer is the Conservator of Forests whereas the forest department is one of the interested parties in the land acquisition for the sanctuary and hence it is not appropriate to make the forest officer the settlement officer**. It should be the revenue department (District Collector) who should be authorised to settle the people's rights. G.O.No.112 which calls for involvement of private industries like ITC, Reliance and others for taking up commercial plantations through the VSS should be withdrawn immediately, as this is a backdoor method of allowing private industries to enter forest and tribal lands. Such tripartite agreements between the government, the industries and the tribals can never provide a level playing ground for the tribal people and make them more vulnerable to the exploitation of private industries.

The historical injustices to the tribal people in this region and that of the tribals in southern Orissa due to construction of several 'development' projects and industries like Nalco, HAL, Sileru, Machkund and others which have displaced tribals in large numbers without any rehabilitation, and has forced them to migrate in search of livelihood. They are now being treated as encroachers and criminals in forestlands. This definition should be condemned and Commission should recommend proper mechanisms for resolution of conflicts. The Commission should strongly recommend the orders passed by the Ministry of Environment and Forests in 1990, on the recommendations of the Commissioner of Scheduled Castes and Scheduled Tribes, to ensure that such conflicts are resolved. Infact, the first two orders alone show a way forward to settle disputes concerning forestlands.

All the groups consulted strongly felt that the pressure on forests and the destruction of forests is taking place more due to increase in industries and private commercial activities like Mining, paper-mills, timber smuggling and the growing non-tribal population settling down in Agency areas, than due to the podu cultivation practiced by tribals. Eco-Tourism is also causing destruction of forests. For example, the A.P Tourism Department built huge infrastructure close to Farhanbad, near Mannanur (Srisailam) for Tourism purposes in the middle of the protected area and in Araku, Borra and Anantagiri (Visakhapatnam). While tribals are prohibited from entering these forests for their survival needs, it is totally unjust to allow tourism, which is causing a lot of degradation. Hence, the constitutional provisions of the Fifth Schedule, the Samatha Judgement and the PESA Act should be strictly implemented following the right spirit with regard to industries and non-tribal settlers, which includes the Tourism Department.

The Tourism Department also conducts tours through tribal villages, especially in Chenchu settlements, and exhibits them like museum pieces. The groups felt that this degrading exercise further underscored the lack of empathy among the various government departments towards the tribals.

Because of a shift in the policy of the government, profits to G.C.C. are more important than benefits to tribals. Although there are 32 items in the procurement list of GCC Ltd, it procures a limited number of commodities and at very low prices compared to the market rates. The tribals face several constraints in collection of NTFP and marketing these items because of the monopoly rights enjoyed by GCC Ltd. They do not have the right to sell their produce to private traders even when their rates are higher. Hence, a long-standing demand has been for removal of the monopoly restriction of GCC Ltd and instead a minimum support price should be provided by GCC to protect the tribals from exploitation of traders. The monopoly of forest department over Beedi leaf collection and sale should be removed. The tribals or VSS should be given 100 per cent rights over income from beedi leaf and also the right of sale, with the forest department providing the support price.

INDIAN FOREST LAW AND POLICY

After Independence, however, and according to the newly enacted Indian Constitution, forests were placed under the so-called 'state list' according to which state legislatures have a primary right to make laws. In 1976, the Indian Forest Act was added to the concurrent list of the Constitution of India, giving the Central Government and states shared responsibility and control over forest matters. Thereby the Government of India does have the power to legislate on forestry issues but only after consulting the states. The responsibility for administering the forests lies primarily with the state government. The relative position of the Government of India depends on several factors, such as its contribution to the state budget and state polities. The balance of power between central and state governments has remained a key issue in forest management ever since.

After 1947, in a post-independent India, commercial exploitation and degradation of India's forests increased dramatically. Indeed, the 1952 National Forest Policy set out guidelines which were, for the most part, directed towards the supply of cheap timber and non-timber forest products for state-sponsored industrialisation and modernisation. In 1980, with the passage of the Forest Conservation Act, the central government reasserted some of its control over forest-based resources. The Act restricts the state government's power to de-reserve a forest, and it restricts the use of forestland for non-forestry purposes without the prior approval of the central government. Unilateral decisions from the centre have prevailed from then onwards. Only six centres were set up to monitor forests and conservation (as per the Act), which is obviously insufficient for effective regional implementation in a country the size of India. Thus the Forest Conservation Act of 1980 has been problematic for a number of reasons, and has achieved little improvement in the 'conservation' of India's forests.

The National Forest Policy of 1988 envisaged people's involvement in the development and protection of forests *for the first time*. It stipulated that the requirements of people living in and near forests for fuel wood, fodder and small timber should be treated as the top priority, and forest communities should be motivated to identify themselves with the development and protection of forests from which they derive benefits. A primary task of all agencies responsible for forest management, including the forest development corporations, should be to engage tribal peoples closely in the protection, regeneration and development of forests, as well as to provide gainful employment to people living in and around the forests. However the 1988 National Forest Policy has never been translated into law. It remains essentially a broad statement of government intent and does little in the way of specifying any legal rights or duties owed to forest communities-especially the tribal/indigenous peoples.

NATIONAL FOREST POLICY OF JOINT FOREST MANAGEMENT

One outcome of the National Forest Policy was the launching of Joint Forest Management (JFM) initiatives by all the States and Union Territories. However, there have been no accompanying changes in the legal framework for JFM, and in almost all states it has been implemented through administrative orders, which do not take the existing infrastructure into consideration. Forests continue to be looked upon as the property of the Government and the delegation of management responsibilities to local communities is merely an administrative arrangement, falling short of any real devolution of power to local communities. As an illustration. Prakash Kashwan has looked into JFM from various points of view and highlights some other problems. One of the most important findings is that JFM has been allowed only in degraded forest, but in order to bring more communities under the ambit of such programme, communities have had to actively pursue degradation of natural forest. Furthermore, JFM requires the demarcation of boundaries between villages, often for the first time, but poor implementation has resulted in many boundary conflicts.

Joint Forest Management (JFM) in the state of Jharkand;

- JFM in the state of Jharkand is based on the premise that people are to blame for the destruction of the forest. The objective, therefore, is gradually to diminish people's dependence on forests.
- The Resolution to introduce JFM has been made unilaterally by the Forest Department. There is no scope for consultation with the people, no effort at consensus building, and no mechanism for dissemination of information.
- The resources required for villagers' needs are to be extracted from the protected forest according to the micro/annual-working plan. There is no mention of the existing rights of the villagers to use the produce from these forests. This is a gross violation of customary rights.
- A 33 per cent share in net profits was promised. Now the promise is to share 90 per cent of net profits with the forest dwellers. The profits however, are from commercial productions, which are not based on local needs, thus indicating the focus on commercialisation of forest. The JFM Resolution in Jharkhand is simply an extension of the undivided Bihar Forest Policy (Jharkhand used to be part of Bihar before it became a separate state), which adopted Joint Forest Management in 1990. Jharkhand Forest Department adopted the JFM Resolution on 27th September 2001
- The Forest Department claims to have formed 2034 Village Forest Management and Conservation Committees and Eco-Development Committees managing an area of 5,589 sq. km. in a remarkably short time span. However, local people emphatically rejected JFM. Apparently, the figure is manipulated and incorrect.

Eviction of communities as Forest Encroachers

Indigenous peoples feel that measures that were put in place in the name of protecting the forest have only deprived them of their resources. The JFM initiatives have failed from the point of view of indigenous peoples, and there has also been widespread eviction of communities from forests. The forest department uses any means available to alienate the forest communities from the forest, reduce their dependence on the forest in the name of forest protection and conservation, and invite private partners and multinational companies to negotiate directly with village committees to access forests for carbon trading, Clean Development Mechanisms and trade in eco-services. The direction of forest management is thus very far from anything that takes traditional forest related knowledge into account or that recognises the rights of indigenous peoples to maintain their traditional forest practices; rather, it aims to separate them from the forests, and sets up an externally imposed process of commodification and marketing of forest goods.

Indigenous Peoples' Rights and Forest Management

The Government of India is not a signatory to the ILO Convention 169 and as such does not recognise indigenous peoples. It does recognise "Scheduled Tribes", but not all indigenous groups are scheduled. Some of the indigenous groups are recognised by the Government of India as Scheduled Tribes and enjoy a degree of special status. Some of the areas where the Scheduled Tribes live have been demarcated as Scheduled areas and under the Indian Constitution Sixth Scheduled provisions, Autonomous District Councils have been set up for local administration of the Scheduled Tribes.

In theory there are some measures to protect the interests of Scheduled Tribes. The Constitution empowers the President of India and state governors to withhold any law considered detrimental to tribal/indigenous peoples' interests in Scheduled Areas. Items V and VI of the Indian Constitution give special privileges to the Scheduled Tribes and the Panchayats Act (Extension to the Scheduled Areas), 1996 (PESA) is also designed to offer some protection. However, in practice most of the laws restrict the rights of and control of forest communities. Specifically, the Indian Forest Act (IFA), Forest Conservation Act (FCA) and Wildlife Protection Act (WLPA) continue to be used to hound forest dwellers.

JFM and CFM are present in most Indian states, but are simply plantation programmes related to timber and other economic products, which have nothing in essence to do with forests. They should include Village Forest Committees directly representing the community living in or using the forest. These programmes are operational to varying degrees in different places with varying degrees of participation of indigenous peoples depending upon the local situation. The involvement of indigenous peoples is focused on providing them with a share in the economic benefits and *de facto* also acts to wean them away from their dependence on forests. In other words, integral to these programmes is a subtle (in some cases blatant) redefinition of indigenous peoples' relationship to the forests. The degree of incorporation of TFRK is varied, if there is any at all; the expertise is provided by the forest department, sometimes in partnership with NGOs.

Indigenous peoples' rights to the forests are progressively minimized. In some cases, indigenous peoples do receive some economic benefits (though in others non-indigenous peoples dominate the Village Protection Committees), and in the context of the takeover of resources necessary for basic survival and consequent deprivation, these economic benefits at times are seen as a steady source of survival income. This in certain cases has superficially reduced tensions and conflicts between the forest department and the indigenous peoples as a sense of co-existence is manufactured. However, this also means that the forest department is able to stifle protests and resistance. In many places, where the indigenous peoples' political consciousness is well developed, they keep away from or oppose these programmes leading to conflicts, as has happened in parts of Madhya Pradesh.

India's report to UNFF3 (2002) describes measures taken to make use of indigenous capacity and local knowledge in participatory forest management. Clearly, the picture painted by this report is not matched by the situation on the ground.

GLOBAL SYSTEM ON CONSERVATION AND SUSTAINABLE USE OF PLANT GENETIC RESOURCES

The salient points related to the progress of the global system on conservation and sustainable use of plant genetic resources for food and agriculture which emerged in the seventh regular session of the FAO Commission on Genetic Resources for Food and Agriculture (CGRFA) at Rome during 15-23 May, 1997 may be reviewed as follows:

- A multi-year work programme on agricultural biodiversity established as per decision III/11 of COP shall be pursued by the FAO Commission and CBD Secretariat though cooperation. Similarly, FAO's monitoring/reporting on agricultural biodiversity shall be consistent with CBD, other relevant inter-governmental bodies and world food summit' Plan of Action.
- CBD through its decision II/5 has requested GEF to give priority support to efforts for conservation and sustainable use of agro-biodiversity. Many countries stressed the need to have separate funding! FAO commission requested FAO to provide assistance and guidance to countries on request in implementation/monitoring of the Global Plan of Action (GPA). Further, the monitoring of GPA and updating/reporting on the state of world's PGR were held complementary activities which do not need separate reports (*i.e.*, need only one report each) by countries.

- No changes in the CCGCT (Code of Conduct for Germplasm Collecting and Transfer) or any further work on Code of Conduct for Biotechnology were recommended, till the revision of IUPGR. However, a decision guide on regeneration of accessions in seed collections, prepared by IPGRI, has been made available to the FAO commission for comments by members.
- The existing arrangements between FAO and the 12 CGIAR centres shall be extended, pending the revision of IUPGR.
- Statutes of the inter-governmental technical working group on AGR (Animal Genetic Resources) and PGR were presented. Both working groups have 5 members each for Asia. The AGR were specifically addressed in view of the broadened mandate of the Commission and to facilitate integrated approach to agro-biodiversity.
- The summary statement presented by IPGRI informed the Commission that 0.5 million accession were now covered under the FAO-CGIAR agreements with the designation of 50,000 new accession under this arrangement.
- The summary statement presented by the CBD Secretariat showed that at least three decisions of the COP-III, Buenos Aires, Nov., 1996, were of direct relevance to the ongoing negotiation for revision of IUPGR, namely, Decision III/11-Conservation and Sustainable use of agro-biodiversity, Decision III/15-Access to Genetic Resources, and Decision III/17-Intellectual Property Rights. The COP-III noted the narrow options for legal status of revised International Undertaking, as being, i) a voluntary agreement, ii) a binding instrument, and iii) protocol to CBD. It encouraged parties to implement GPA at national levels. Pollinators and soil-micrograms in agriculture needed special attention (as bio-indicators) and comprehensive reporting by countries to Subsidiary Body on Scientific, Technical and Technological Advice (SBSTTA)/Conference of Parties (COP) of the CBD.
- An identified group of competent people needs to resolve the PGR issues and help prioritise various PGR activities. There is need to identify crops/regions on priority.
- The dialogue between the public and private sector should continue and there is a need to create awareness about PGR.
- An information bank should be established to enhance literacy, management and utilisation of genetic wealth. Emphasis should also be given to regional languages in developing the information base.

World Information and Early Warning System

The FAO's initiative to develop a systematic information base on PGR and related technologies under the World Information and Early Warning

System on Plant Genetic Resources (WIEWS) is aimed to collect, collate, maintain and disseminate facts and figures on country profiles, national PGR programmes, *ex situ* collections, database on PGR, existing crop varieties, seed production and supplying agencies, *in situ* on farm conservation, bioprospecting and exchange of seed. The nature of such an activity requires a comprehensive information input and a grid approach for analysis and interpretation of data. The system is open and responsive to queries from all stakeholders and interest groups. In practical terms, it would be relatively simpler to attend to details of the resource availability and the custodian(s) but, considering the importance of implementing CBD and WTO provisions and the imminent revision of the International Undertaking on Plant Genetic Resources, the access to resources for the purpose of replenishment could pose certain difficulties. The access would have to be required from any one of the following sources:

- The *ex situ* seed collections;
- The seed multiplication and supplying agencies; and
- The local communities/community gene banks.

The implications, in each case, could be several and manifold. The access from *ex situ* CG system collections is being debated wherein two clear-cut options have been put forth. The pre-CBD collections are still available freely which means their replenishment would be only a matter of time actually required for the multiplication of their seed.

However, in case of requirement of post CBD collections, several access-related issues shall have to be sorted prior to the multiplication of seed for replenishment. The requirement of seed from commercial/public seed multiplication and supplying agencies could be easily met but such a replenishment would not serve the purpose effectively. This being so, the agencies could provide seed of few dominant varieties, which they are currently handling, in contrast with the requirement of a wider genetic base. The replenishment of seed from farmers' varieties/land races faces two different situations:

- The issue of consent/access; and
- The issue of compensation or sharing of benefit accrued from the use of their heritage seed.

The setting up of community seed banks and the Global Biodiversity Fund have definitely been provided as suitable institutions which could help in build up at an appropriate system and come to the rescue in cases of crisis. The setting up of a separate Global Biodiversity Fund is still being debated as also the Farmers' Rights.

The development of the *sui generis* systems to meet the countries' own (local) needs, has to be accelerated. Through this mechanism, the whole idea is to encourage local level replenishment through procurement of required seed from nearly/similar climate logical areas.

Some Queries Seeking Answers

Increased investments on *ex situ* conservation are unavoidable with each increment in the quantum of work and management skills. The cost/benefit ratio does not work in such cases because the conservation is for posterity. In this endeavour, the obvious questions seeking answers would be:

- What should be the estimate of funds for genetic resources conservation at the CG Centres, at the National genebanks and at community seed banks?
- What kind of activities we need to take up in the beginning which ensure fast replenishment in cases of crisis?
- What shall be the mechanism of funding-whether the Global Biodiversity Fund or private sector contribution shall be forthcoming or not?
- Which sites for *in situ*/on-farm conservation to begin with?
- How to identify genuine NGOs for their involvement in PGR Conservation?
- What could be a standard format of collecting and collating information for WIEWS on PGR *ex situ* conservation, country status, on-farm conservation and wild relatives? Do the existing proforma suffice?
- What would be a suitable indicator or an appropriate method for monitoring of genetic erosion for use under Early Warning Systems?
- What should be the suitable proforma for multicrop passport data?
- How to generate compatible country reports providing information for the State of the World's PGR and the action on GPA?
- How to minimise diversity and redundancy in use of computer software for various Database Management Activities?

9

Beneficial Economic Production in Commercial Forest Management

In order to get the most beneficial economic production out of our local forests, the practices and techniques of forest management need to be applied. As of 1924, Wicomico County was roughly 46% wooded with the other 54% covered by agricultural, commercial, and residential land. Most of the 46% wooded land was and has been privately owned. In Wicomico County, the majority of wooded areas are privately owned and are mostly idle lands on a landowner's property. As a landowner, especially a farmer, the more economic profits you can get for your land the better. During the late 19th century and early to mid 20th century farmers and landowners were managing their idle lots into pine plantations. The February 17th, 1948 edition of the *Salisbury Times* even had an article entitled "Farmers Discover Timber is a Crop, Shore Benefits Too" and in this article it explains how modern day forestry practices are encouraging the re-growth of loblolly pine plantations in active timber lots. The dominant forestry management practices during the 20th Century were commercially based, meaning most of our forests were used to support timber and a very small portion of Wicomico's forests were being conserved to support diversity. Through different forestry management practices to promote commercial forests there has been a major loss of ecological diversity in Wicomico County.

PRESCRIBED FIRES IN FORESTRY MANAGEMENT

One forestry management practice was borrowed from the local Native Americans and used to protect the commercial loblolly dominated forests. During the 1920's through the 1930's the United States and especially Maryland went through a dry era. This was also around the Great Depression and the famous Dust Bowl in the Mid-West. The pine industry in Wicomico County was not only an important industry it was an economic resource that affected most of the citizens in the county. Devastating forest fires became a threat to dry pine plantations and the fires would consume thousands of dollars in timber.

The reason why these fires were so devastating is because dead vegetative matter including layers of dead pine needles and cones as well as other dead vegetation was allowed to accumulate on the forest floor. This built up matter that would naturally dissipate due to natural forest fires was not allowed to because fire was viewed as a costly event and forests were highly protected.

In the 1950's foresters figured out that fire could be the best tool to prevent more fires and to help promote the loblolly. Like thousands of years of fire forest management practices used by Native Americans, prescribed burns were used in Wicomico to prevent large costly fires from happening. Prescribed burns are man-made controlled fires applied to pine plantations used to burn up extra forest floor material and to eliminate unwanted plants, trees and future fires. Another positive of a prescribed burn is that the burnt material gives the soil back vital nutrients that help promote the growth of more trees.

Environmentally the prescribed fires of the 20th Century took its toll on diversity. The Natives used fire to control the forests to be diverse to support hunting. The fires during the 20th Century were being used to not only burn excess forest floor matter but to also eliminate competing vegetation of the loblolly pine. These fires eliminated young hardwood saplings from maturing as well as other shrubs that many species depend on. Also the smoke from prescribed fires was not too pleasing to the local communities. With the right winds the smoke would drift to people's homes, along highways and into towns making many people unhappy. One of the last things a person would like to smell on a nice spring day is smoke.

Clear-Cut/Regeneration Harvests

Clear-cutting, also know as the regeneration cut, has become the most economically beneficial harvesting management technique for the loblolly pine. Before forestry management existed, clear-cutting was used to open up the land for agricultural uses and to make settlements for the early settlers. Many years after these fields were abandoned the first trees to reemerge were the loblolly pines because they grow well in direct sunlight. Loblolly forests then began to make up most of the county while the diverse mixed hardwood/ softwoods stayed in the low-lying, wet, untouched areas. While this nation was growing, it needed a timber supply to help construct cities, boats, and crates and to fulfill other construction requirements. Wicomico County and the timber industry, were extremely important because of its fast renewable timber resource and accessibility to major cities on the eastern sea board. Clear-cutting was one of the first management practices used. It was used because it allowed the industry to get as much timber as possible in a short period of time. Clear-cutting is selecting a wooded lot and cutting down all the trees and clearing away the brush. This method of forest management works well with the loblolly because its' seed cones can naturally regenerate new trees and the open sun is great for their growth. As Wicomico County and the rest of the country found out, clear-cuts could not be the only method of forest management. If everything became clear-cut then the industry would have to wait many years for the next generation to reach full maturity. Therefore, the practice of a rotating clear-cut was needed to support the timber industry. Different tracts of land had different aged stands of loblollies so that there would be a continual timber crop. Environmentally it is not a good method because it decreases the diversity of the stand because all the trees will be of the same age and a healthy forest needs a mixed forest of trees of varying ages.

In a study done by Wayne C. Zipperer of the USDA Forest Service, deforestation in Wicomico County caused by clear-cutting creates patterns in the forest cover and affects the diversity of species. According to Zipperer's study, there are five distinguishable patterns of deforestation: internal, indentation, cropping, fragmentation, removal. Each one of these patterns has a direct impact on the habitat quality of the forest patches. The interior, or the middle, of forests is important to many species for protective nesting grounds and refuge. Between the years of 1973 and 1981 the forest interior declined in Wicomico County by 3.2 square miles. This means species of birds and other animals have been forced to find other forest interiors to nest and seek refuge, thus lowering the diversity in this county. One bird that has been affected is the Red-Eyed Vireo, a Neotropical bird that breeds and nests in the mid-latitudes (Wicomico) and spends its winters in the Amazon basin of South America. In order to breed it needs deep groves of shade-trees. Wildlife also needs connected forests and habitat fragmentation is the most serious threat to having a fully biological diverse forest. The pattern of indentation is the most prevalent pattern in Wicomico County. Indentation occurs when clear-cuts are intruding into the forest interiors creating peninsulas of the forest cover. This practice promotes the loss of interior habitat and promotes forest edge habitat. Forest edge habitat is mostly used as a resting point for migrating creatures and a hunting ground for predators giving the prey less cover in which to hide.

Forests Were First Managed

For thousands of years Native Americans have been managing forests in the Northeastern United States including Wicomico County and the rest of the Eastern Shore. According to Stephen Pyne, a fire historian, Native Americans in the northeast would set fires off in forests in order to make hunting easier. According to the Native American tribes of the eastern United States fire was known as Our Grandfather Fire. Native American "economies were dependent on fire" and without fire their "economies would have collapsed". Besides burning forests to clear the way for hunting, Native Americans also used fire in the forests to obtain firewood. Although forests provided the first Europeans with timber fuel and game their views of the forest were "an obstacle to agriculture". This European ideology of the forest led to widespread deforestation during the 18th Century. While trees were being cut down to make room for farm fields, the timber cut was being sent down to the islands of Barbados and Antigua who no longer had enough timber because they used up all of their resources. During these times someone would go out into the forest with an ox-drawn cart and then load the timber onto that. Forests were slashed and burned, meaning people would girdle a tree (removing its bark) in order for the tree to die and dry. After many months once the large trees were dead and dry fire was used to clear large portions of the forests. There is actually a small town in Worcester County called Girdletree which took its name from this slash and burn technique of clearing the forests. According to Jack Wennersten an Environmental Historian, early "Chesapeake farmers and planters had little use for the forest as an aesthetic end in itself. Trees on the horizon irritated their eyes and they wanted to see bare ground".

The forest management techniques of the Native Americans and European settlers were different. The Natives used the forest to benefit themselves however left a minimum impact on the surrounding environment. The settlers viewed the forest as a both an obstacle and resource. Their views exploited the forest to fit their needs and left major impacts to the surrounding environment. Clear-cuts are also dangerous environmentally through nutrient run-off. When clear-cuts take place around bodies of water, excess nutrients in the soil runoff by erosion and into the water. The nutrients promote extra unwanted growth of aquatic vegetation that clouds the water and chokes out fish. The nutrients are allowed to run-off because the root systems of the trees that once held the soil together are no longer present and cannot prevent erosion. The extra nutrients in the water mean excess grow in aquatic vegetation that chokes out fish and other aquatic organisms.

Herbicides

With recent advances in herbicides and fertilizers forest management has turned to some fairly newer practices. Thinning and clear-cutting are still the most widely used management practices however; the use of herbicides has taken over some of the duties of prescribed burns. Although prescribed burns are still useful to eliminate forest fire fuel on the forest floor they are not needed to eliminate unwanted brush and trees. Herbicides have been used in Wicomico Country in order to allow purely loblolly pines to grow on certain areas and not allow other vegetation. In 1984 the herbicide Imazapyr, commonly know as Arsenal, was registered in the United States and in 1986 was first used in Wicomico County on pine stands after a clear-cut to promote the loblolly pine.

The first application method used with Arsenal in Wicomico County was when a tank of Arsenal was dragged around a clear cut by a skidder in an arc fashion; however this method did not last because it was not cost effective.

According to Metzger, since the skidder was too expensive, local foresters contracted helicopter companies to spray sites; this method was more time efficient and less expensive. Arsenal is applied to clear-cut areas because it stops the growth of unwanted vegetation without harming the commercially important loblolly pine. Herbicides were introduced into the United States in the mid 20th Century and their first uses were for agricultural fields. Prior to Arsenal, the herbicides Velpar and Tordon were used in forestry. Metzger stated that these two herbicides were not always consistent. Tordon and Velpar were not consistent because if it rained after an application it would run-off the desired tract. According to Anthony F. Maciorowski, Branch Chief of Ecological Effects Branch Environmental Fate and Effects Division, "Due to the extreme phytotoxicity, its (Tordon) persistence under typical environmental conditions, and its extreme propensity to leach into groundwater in all soil types, the EEB is strongly recommending against the reregistration of all active ingredients of Picloram. This conclusion is based on the extreme exceedance of the acute levels of concern for non-endangered and endangered terrestrial plants."

Arsenal has been used because it does not run-off if it rains soon after an application and it consistently kills unwanted vegetation. It has both an effect on the diversity of trees and vegetation in the forest and harms forest animals. Arsenal is considered as a non-selective broad-spectrum systematic herbicide that attacks plants stems and root systems. This means Arsenal does not differentiate which plants it kills, however it does not have any effect on the loblolly. This means endangered and threatened plant species are killed if they come in contact with Arsenal. Maryland currently has ten plants listed as threatened or endangered by the United States Fish & Wildlife Service. These plants include four threatened species: Seabeach Amaranth, Joint-Vetch, Swamp Pink, and the Small Whorled Pogonia. The six endangered species include: The Northeastern Bulrush, American Chaffseed, Smooth Coneflower, Canby's Dropwort, Sandplain Gerardia, and the Harperella. The Canby's Dropwort endangerment is directly due to pine plantations. Canby's Dropwort is native to Wicomico County and was once abundant in the area. It first became threatened when its needed wetland habitat was being lost to the construction of pine plantations. Bulldozers were used to fill in Wicomico's low-lying wetlands and converted them into working pine plantations. Since the Canby's Dropwort needs a wetland habitat in order to survive, the destruction of Wicomico's low-lying wetlands are the reason why this plant is now endangered. Now there is an endangered plant where the application of Arsenal on clear-cut sites can make it even more endangered.

Local foresters claim that the amount of Arsenal required to kill a mammal is rather large. This statement is true, large quantities of Arsenal exposed through animals by oral ingestion, exposure through skin or inhalation is needed for an animal to die. However, animals never take in enough of Arsenal to kill them. Thus it is deemed safe to wildlife, which is false. Just because death is not directly related to the herbicide does not mean it does not have other negative effects on the local wildlife. According to tests performed by Caroline Cox of the *Journal of Pesticide Reform*, Arsenal's acute toxicity has been observed as the cause of many problems with animals. Arsenal causes bleeding and congested lungs in rabbits (the tested animal); congestion was also found in the kidneys, liver and intestines of rabbits as well. Arsenal is also corrosive to the eyes and can cause irreversible damage. When Arsenal is exposed to the skin of animals it has been observed to cause reddening, scaling and crusting and also cause stomach ulcers and intestinal lesions on rabbits. Arsenal also stays persistent in the soil between 60 to 436 days depending on its application and run-off. The foresters are correct, Arsenal does not directly kill animals, however Arsenal does harm animals in a way that may lead to death in another manner.

Thinning

Often, before a clear-cut harvest another technique in forestry management is used to lessen the competition for the loblollies. According to some experts, thinning is the single most important management practice a pine plantation could receive.

The goal in a clear-cut it to get the maximum income from a particular stand in one cut, however, this is not the primary goal of thinning. The goal of thinning is to get the softwood loblolly pine stands ready to provide high value timber in the future.

Thinning is when certain trees are removed in order to encourage the growth of future trees by having them somewhat evenly distributed throughout the stand. In some cases the trees that are removed could be still used commercially and thus the thinning is known as a commercial thinning. In those cases where there is not a market for the removed trees the thinning is known as a precommercial thinning. Thinning is used because if a wooded lot is going to be used commercially the amount of trees on that land needs to be regulated. If there are too many trees growing on the lot it could affect the growth of the rest of the lot produce less desirable timber.

By thinning trees out, it benefits the remaining trees' growth rate, strength and market value. There are three distinct benefits to the tree farmer through thinning. The growth is concentrated on fewer trees allowing them to reach maturity faster meaning more value for the trees that are left. The low value timber does not take up unwanted room allowing only high value timber to continue to grow. Trees that would have died before the final harvest can be marketable and worth money. Thinning is a process that would take place before a final harvest which is often times a clear-cut. An unthinned forest has benefits for some timber manufacturers as well. When the loblolly is cut to be made into poles denser wood is more important and thinning would not be used to keep the trees tall and dense.

On a biological stance, thinning has great economic benefits for the tree farmer as well. The loblolly trees, along with other trees, are in a constant battle between each other for vital nutrients, sunlight, and water to survive. If the stand was just left alone without thinning, the trees would not grow as well because they have to battle for the water and nutrients. With a thinned forest, competition for the nutrients, sunlight and water is much less allowing each tree a better chance at growing into a mature and valuable tree. If natural thinning occurs most trees are too crowded and do not receive enough sunlight, and they just die. According to a North Carolina State University study, a stand that would have had 600 to 1,000 trees could be reduced only to a few hundred by the time maturity is reached at age forty. This displays that if a forest is thinned there would be less competition and more desired trees reach full maturity. One negative effect of thinning economically is that it is more expensive then clear-cutting. Since thinning is more expense the farmer would receive less money rather than if he just had a clear cut. However, with clear-cutting a farmer needs to wait 50-60 years before receiving any kind of money; with thinning some thinned trees are worth money and would give the farmer a money source between final harvests.

The management practice of thinning has been practiced for a long time in Wicomico's timber industry. During the early part of the 20th century state experts in forestry would come to Wicomico County to have lectures to teach farmers and lot owners how to manage their loblolly pine stands so that they could get the most value for their timber. In November of 1924 Fred B. Trenk, an Extension Forester of the University Maryland, gave demonstrations on thinning to local farmers. On April 22, 1926 Dr. F. B. Arenk of the University of Maryland gave a lecture on forestry and handling woodlots to local woodlot owners at the Salisbury Court House. During this lecture Dr. Arenk discussed the shortest possible time in which a loblolly's could be grown and what could be done to help the process. During this same period of time there were also demonstrations given by foresters to farmers on how to thin their forests. These demonstrations and lectures made an impact in the Wicomico area because it gave people economic benefits to their wooded lots. It was also important because it puts more land into active timber production, meaning the dominate forestry management was commercially related.

Prior to the 1980's thinning was an expensive and time consuming practice because it was done manually. Ron Metzger, Wicomico County forester for the State of Maryland, said the biggest change in thinning since he arrived here in the 1980's is the advancement in technology. Thinning is now done with specialized mechanical equipment that cuts down manual labour expenses and also speeds up the process. Now since thinning is a quicker process, more tracts can get thinned in a shorter amount of time creating more productive timber lands. This practice does damage to forest diversity because any competing trees or vegetation are thinned out to allow excess growth to the loblolly pine. Thinning promotes the growth of one species of tree at an even age. A forest with an even-aged single tree specie promotes a balanced habitat to very few organisms.

CONTRIBUTION OF FOREST PRODUCTS TO MILLENNIUM DEVELOPMENT GOALS

Forest products contribute, both directly and indirectly, to the attainment of the Millennium Development Goals (MDGs). Overall, across the studied communities, they contributed 8% in monetary terms to households' income, and 22% in subsistence form. Forest products are supporting local trade. It has also been studied that Uganda has 2,000-3,000 forest-based associations which span a number of different areas: forest production (timber and non-timber forest products), primary and secondary processing, enterprise support and environmental services such as ecotourism and carbon sequestrations projects.

It would have been extremely difficult for the former forest dwellers, the Abatwa, to meet their food security needs had they not been gathering honey, wild meat, root-tubers and fruits from Echuya Forest Reserve in Kisoro and Kabale Districts. They are lacking technologies to add value to these products. These examples illustrate the forest contribution to MDG1 that is the eradication of extreme poverty and hunger. In Nyantonzi, the poor men's category put it: *"A family that grows food crops and stores them in granary to last the dry season is not considered poor"*. The rugged and hilly land terrain in both districts has always frustrated health service delivery. As a response, communities traditionally organize themselves in self-help ambulance groups known as *"Engonzi"* to carry the sick to the nearest health centre. The stretcher they use is made of forest products, thus contributing to MDGs, 4, 5 and 6, with respect to reducing child mortality, improving maternal health and combating HIV/AIDS, malaria and other diseases respectively.

Owing to the growing scarcity of forest products in some places due to population pressure, households and communities are investing in tree planting as well as regulating access to forest reserves. In so doing, they are directly contributing to environmental sustainability or MDG7. For a district like Kisoro where access to water is only 43.9% and below the national average rate of 67%, the forest's regulation of water, on which communities greatly depend, equally contributes to MDG7. Use of water harvesting technologies is not widespread.

Communities' Understanding of User Rights,Duties and Benefits

The number of forest-product users was found to be many, 16 of them. The study team established from them what they considered their rights, duties and benefits related to forest product use. For each category, communities were asked to give a rating on a scale of 1 to 5, with 1 being the lowest and 5 the highest. With respect to user rights, a major finding is that the list of products to which communities consider themselves entitled is long. This is important so as to map out strategies to reconcile the divergent and sometimes conflicting interests of resource users. The potential conflicts and management challenges that may emerge are also real. To note, communities listed alternative strategies including free access (*e.g.* for firewood, herbs, water, community based tourism) while for others appreciated the need for regulated access with a permit.

A key revelation from the listing of duties is the desire for sustainability. This was expressed in many ways, including cutting only dry (dead) trees for timber, selective harvesting of crafts materials, avoidance of use of fires within forests and the use of selective traps for hunting wildlife. The fact that resource users did not attach high scores is a reflection of their failed duties. Nonetheless, it is a good indicator that communities would be willing to trade duties for benefits, a key ingredient for sustainable use resource agreements. The generally high scores attached to benefits is an indicator of the appreciation of the values that communities derive from forest products. For example, across all the four sites, benefits from the collection of firewood, herbs, water, building materials, timber, seeds, all scored high values.

Communities identified several forest-based problems, their solutions and agencies to take the lead in addressing them. They cut across all wealth and gender groups and therefore represent an important community consensus. They have been grouped under five categories namely: (i) policy and legal problems, (ii) forest management and sustainability problems, (iii) market and trade related problems, (iv) conflict problems and (v) socio-economic and political problems. It emerged that communities consider the existing forest legislation and regulations as too bureaucratic. They would prefer legislation that is accessible to all and more affordable charge system for forest products.

Unclear forest reserve boundaries, corruption of government officials and local leaders, low government support to forest activities and overexploitation undermine long term sustainability of the benefit flows from forest products to communities. Leaders should be held accountable for their misdeeds. Concern was also raised that conflict between the NFA and communities was rife, mainly because communities incur losses from raids by wild animals for which there are no direct compensation mechanisms to the affected people. The National Forestry and Tree Planting Act, 2004 is silent about this problem. Communities feel their benefits from use of forests would be enhanced if for example, their entitlements in form of shared revenue would be timely and regular. Fiscal reforms that offer incentives to tree planting and introduction of value-addition and technologies encouraging resource users to form marketing associations were advanced as some of the solutions. The "non-diplomatic" approach used by NFA staff was condemned by Communities as much as denying them affirmative action to allow them access to the more valued resources like timber. Even though some of the transactions like pit sawing require high capacity and expensive equipment, there could be a lee-way for communities to benefit through private sector community partnerships.

THE TRANSITION TO PLANTATION FORESTRY

More than 4 000 years ago humans began to change the way in which they met their food needs by moving from foraging and hunting to crude cropping and herding and, finally, to modern agricultural cropping and livestock raising. Today this transition in agriculture is largely completed, except in a few parts of the world.

It was only in the latter half of the twentieth century that humans began to make similar changes in forestry: a transition from a primitive mode of gathering forest bounty, which is created solely by nature (old-growth harvesting) to the development of the science of wood production (silviculture) (see Figure). By the middle of the twenty-first century, the transition to tree cropping will be largely completed, and the greatest part of human wood consumption will come from planted forests, most of them intensively managed.

Through tree growing, commercial wood can become a crop, as in agriculture, to be planted, tended and harvested. Tree growing allows for a choice of location and species, as well as providing the opportunity to provide intensive management. It is only with tree planting that investments in tree improvement are justified. Improvements that result in higher yields or desired traits only make sense if those gains can be captured physically and in the market. Under an intensively managed regime, trees can be grown much faster with the desired traits. Climatic conditions for this are particularly favourable in parts of the subtropical and tropical world.

The trend towards tree planting will undoubtedly continue. It is driven by two powerful forces: economics and environmental concerns. Promising economic returns on tree planting have been realized in some locations for several decades, especially in advantageous areas of subtropical and tropical South America, Asia and Africa, where biological growth rates are high. In recent decades, large areas of previously unforested land have been converted to planted forests across the globe from the Nordic countries to New Zealand and from South Africa to South America. Although much of the activity has been in the industrial world, many of the most promising opportunities are in developing regions. Many planted forests are on land that was previously in low-productivity agricultural uses. The economic returns on planted forests, especially high-yielding intensively managed forests, are sufficient to continue to induce substantial investments in plantation forestry (Sedjo, 1999a). The general trend to high-yielding planted forests is receiving additional momentum from environmental concerns which have resulted in prohibitions on harvesting from some old-growth and secondary forests and regulations that make such harvesting more expensive. As the environmental movement continues to exert pressure for the protection and setting aside of more native and natural forest areas, less of this type of forest is available for logging and the costs of obtaining wood from these sources are rising.

New forest practice acts and codes are increasing harvesting costs in natural forests in many countries. For example, Kajanus and Karjalainen (1996) calculate that new forest policies in the Nordic countries, which emphasize the preservation of biodiversity, have increased harvesting costs. Similarly, it is estimated that the recently revised forest practices code in British Columbia, Canada, which deals with roading standards, riparian zones and harvesting practices, has increased the costs of timber harvests in the province by at least 15 percent (Haley, 1996). As harvesting costs have increased, harvests have declined in the coastal mountains of British Columbia.

A dramatic example of the reduction in harvests from national forests is the drop in the United States, from roughly 60 million cubic metres annually in the late 1980s to about 15 million cubic metres today. Pressures resulting from environmental concerns are likely to be amplified through the impacts of various efforts, such as those of the Forest Stewardship Council (FSC), to audit forest management practices or certify forest products. Although auditing applies to plantation forestry as well as natural forest management, it tends to raise the cost of natural forest management relative to plantation forestry. This is especially true where the planted forests are established on lands that were previously in other uses, e.g. agriculture. The forces discouraging logging in natural forests are unlikely to go away in the foreseeable future. The pressures to reduce harvesting in old-growth and some second-growth natural forests will add to the attractiveness of the movement to invest in planted forests.

INDUSTRIAL WOOD DEMAND

Over the next several decades, the demand for industrial wood will undoubtedly increase, but not dramatically. Since the mid-1980s industrial wood demand has stagnated, with consumption of industrial wood remaining at roughly 1 500 million to 1 600 million cubic metres annually (FAO, 1984-2000). This levelling out can probably be attributed in part to the decline in production within the countries of the former Soviet Union as markets have replaced the earlier command and control economy. Increased recycling is probably another contributing factor. Since the decades immediately following the Second World War, the increase in wood consumption has been following a downwards trend. This trend towards stabilization has occurred in the face of substantial growth in the world's population and at a time when the global economy as a whole has been experiencing rapid economic growth, particularly in highly populated Asia.

Furthermore, if the latest United Nations (UN) population projections are to be believed, there is a good chance that by the middle of the twenty-first century the population of the planet will stabilize or even begin to decline (UN, 1998). In this context, it is difficult to see the basis for any dramatic acceleration in the rate of growth of industrial wood demand. By 2050, total world industrial wood demand will be higher than it is today, but not very much - perhaps 50 to 75 percent over 50 years (Sohngen, Mendelsohn and Sedjo, 1999).

SUPPLY

The predicted transition to planted forests would make the sources of wood supply dramatically different from what they are today. The author estimates that, at present, 22 percent of the world's timber harvest comes from what are essentially old-growth forests, while 34 percent originates in planted forests, and only 10 percent in fast-growing forests, which in 2000 consist of fast-growing exotics (Sedjo, 1999b).

Table. Current and Forecast Global Harvests, by Forest Management Situation.

Forest management situation	% of global industrial wood harvest	
	2000	2050
Old-growth	22	5
Second-growth, minimal management	14	10
Indigenous second-growth, managed	30	10
Industrial plantations, indigenous	24	25
Industrial plantations, fast-growing	10	50

By contrast, Table forecasts that by 2050 up to 75 percent of industrial wood will come from planted forests, and about 50 percent from fast-growing forests. The fast-growing forests may no longer consist largely of exotics; rapid improvement is being made with certain temperate species, e.g. poplar, so that a large fraction of future fast-growing forests may be made up of improved domestic species. However, in much of the subtropical developing world, exotics are likely to dominate.

Much of the planted industrial wood would come from high-yielding intensively managed forest operations. In some parts of the world, as is already the case in certain areas of the United States and Canada today, these operations may be called "fibre farms" and may operate on rotations as short as five to six years. However, most of the increased plantation output is likely to come from the newly created planted forests of the subtropics. Globally, fast-growing planted forests will encompass an area of perhaps 200 million hectares, or only about 6 to 7 percent of the world's currently forested area. Extensively managed natural forests that meet the auditing standards are likely to continue to provide a portion of the world's industrial wood, but largely for speciality products. Such changes have profound environmental implications.

The far more rapid growth associated with intensive management implies that huge volumes of wood will be produced from relatively small areas of land. Thus a trend towards planted forests and tree breeding does not imply, as some erroneously maintain, that vast areas of natural forests are to be replaced with planted forests. Accordingly most of the world's natural forests would remain for other purposes.

Indeed, it has been seen (for example, in the National Forest System of the United States) that increased demands are placed on natural forests for other purposes as the production of industrial wood from these lands declines. Often this change is driven by policy and regulations, but such policies would not be feasible in a world where demand for industrial wood is far outstripping supply.

Technique	Increase in yield (%)
Orchard mix, open pollination, first-generation	8
Family block, best mothers	11
Mass pollination (control for both male and female)	21

ALTERNATIVES TO WOOD

One of the challenges to industrial forestry is the contention that the world's fibre needs could best be met by substituting wood with annual fibrous plants, such as hemp and bagasse. However, the following factors suggest that the prospects for other plant fibres as alternatives to wood are likely to be small over the next 50 years.

First, using these types of fibre sources involves a number of economic and ecological problems. A major economic problem is that they have a specific peak harvesting period: the crop must then be stored and preserved until it is to be processed, and both of these functions incur cost. By contrast, timber can generally be harvested throughout the year, or at least throughout a larger portion of the year than an annual crop can, so the labour and capital equipment can be used year-round. In addition, wood tends to resist deterioration better than non-woody plants do.

From an environmental perspective, it is difficult to see how an annual crop, planted and harvested every year, can be more benign to the environment than a tree crop harvested only 20 years after planting. An annual crop generates 40 disturbances of the land over a 20-year period, compared with two disturbances for a tree crop. Furthermore, the energy required for annual planting and harvesting, and the greater amounts of other inputs such as fertilizer almost surely make such an approach environmentally inferior to tree plantation (Sedjo and Botkin, 1997).

NEW TECHNOLOGY: AN EFFECT ON COSTS

While the costs of harvesting natural forests are rising because of the effects of both availability and restrictions on current practices, the costs of intensively managed plantations tend to be falling because of new techniques, improved plants and the like. Costs can be lowered in forestry, as they have been in agriculture, through the introduction of appropriate technology. New technology can reduce unit costs by increasing the yield or the output per unit cost. Similarly, innovation can make possible the same production at reduced costs, again reducing the unit cost.

Lower costs are likely to have two effects. In the short term, return or profitability per unit will rise. Over a longer period, rising profitability is likely to increase production, thereby lowering prices to consumers and eroding some of the increased profitability. The short-term effect increases profitability, while the longer-term effect leads to lower prices for the consumer. In the context of forestry, where there is a high degree of substitutability between plantation and natural forest wood for most purposes, lowering the costs of plantation forestry also has the effect of providing a financial incentive for the industry to continue its shift away from higher-cost natural forestry.

TREE IMPROVEMENT AND BIOTECHNOLOGY

As already noted, tree improvement has few applications in the absence of planted forests. As planted forests become more common, the actual and potential applications of tree improvement using traditional techniques or biotechnology increase. Table shows the extent to which various breeding approaches have been able to generate increased productivity. Biotechnology is expected to serve as a useful tool in breeding. Genetic modification will probably usually be applied to superior trees obtained through traditional techniques.

Traits that are of interest to forestry and that are likely to be enhanceable through genetic engineering include herbicide tolerance, flowering control, fibre and lignin content, insect tolerance, disease tolerance, wood density, growth, stem straightness, nutrient uptake, and cold, wet and drought tolerance. Genetic modification of some of these traits, e.g. herbicide tolerance, has already been well developed in agriculture. The introduction of genetically modified organisms (GMOs) into agriculture, and now forestry, is very controversial because of a host of real or perceived risks to health, safety and the environment. It is difficult to predict the extent to which this issue will ultimately be resolved. However, it should be recognized that, globally, biotechnology need not be an "all or nothing" proposition. As with nuclear power, it may well be that some nations will utilize the technology while others will not. Thus, industrial wood from genetically altered trees could become common in parts of the world while remaining largely absent elsewhere.

The financial incentives for the utilization of biotechnology in forestry appear to be strong. Short rotation periods, which are emphasized in high-yielding plantation forestry (where rotations are typically from six to 30 years), provide an inherent financial advantage. Furthermore, the use of biotechnology, which is currently under development, to improve fibre characteristics and/or reduce lignin extraction costs would reduce processing costs, thereby improving financial returns. To the extent that costs of establishment and/or processing can be reduced or yields can be increased without increasing costs, net benefits can be achieved.

A CONFOUNDING ISSUE

Perhaps the greatest unknown for the future of forestry is the phenomenon of global warming and its possible effects (IPCC, forthcoming). Warming could not only change the global distribution of forests (although probably not greatly by 2050), but could also have an influence on the extent of forest plantations. Should forests be used as major global carbon sinks to mitigate the build-up of carbon in the atmosphere, afforestation of large areas could be undertaken as part of the mitigation process. This activity could receive further impetus should bioenergy be used on a large scale to replace fossil fuels. Finally, wood materials sequester carbon even as they are utilized, and they have a lower energy (fossil fuel) input in their production than most other materials, e.g. steel, brick or concrete. These advantages could provide additional incentives for the expansion of planted industrial forests.

Bibliography

Ashok K. Chauhan and Ajit Varma: *A Textbook of Molecular Biotechnology*, I.K. International Publication, Delhi, 2009.

B.D. Sharma and J.C. Rana: *Plant Genetic Resources of Western Himalayas : Status and Prospects*, Bishen Singh Mahendra Pal Singh, Delhi, 2005.

B.P. Singh and Umesh Srivastava: *Plant Genetic Resources in Indian Perspective Theory and Practices*, Indian Council of Agricultural Research, 2004.

D D Tewari: *Management of Nontimber Forest Product Resources of India : An Analysis of Forest Development*, International Book Distributors, Delhi, 2008.

D. Thangadurai, T. Pullaiah and Pedro A. Balatti: *Genetic Resources and Biotechnology, Vol. 2*, Regency Publication, Delhi, 2005.

Dharmendra Singh and S. Manivannan: *Genetic Resources of Horticultural Crops*, International Book Distributing Co, Delhi, 2009.

Giridhar A Kinhal and R Jagannatha Rao: *Adaptive Management of Medicinal Plants and Non Timber Forest Products : Strategies Implications*, BSMPS Publication, Delhi, 2008.

Hari Har Ram and Rakesh Yadava: *Genetic Resources and Seed Enterprises : Management and Policies (2 Vols-Set)*, New India Publishing Agency, Delhi, 2007.

John Ruane and Andrea Sonnino: *The Role of Biotechnology in Exploring and Protecting Agricultural Genetic Resources*, Discovery Publication, Delhi, 2007.

K.R. Dhiman, R. Bawa, K.K. Raina and N.P. Dohroo: *Non Timber Forest Products : Conservation Utilization and Management*, Bishen Singh Mahendra Pal Singh, Delhi, 2013.

Lingaraj Patro: *Forest Genetic Resources Conservation and Management*, Mangalam Publication, Delhi, 2012.

M.B. Shrivastava: *Timber Industries and Non-Timber Forest Products*, CBS Publication, Delhi, 2005.

M.P. Shiva and S.K. Verma: *Approaches to Sustainable Forest Management and Biodiversity Conservation : With Pivotal Role of Non Timber Forest Products*, International Book Distributors, Delhi, 2002.

M.P. Shiva: *Inventory of Forest Resources for Sustainable Management and Biodiversity Conservation : With Lists of Multipurpose Tree Species Yielding both Timber and Non-Timber Forest Products and Shrub and Herb Species of NTFP Importance*, Indus Publication, Delhi, 1998.

N.P. Todaria, B.P. Chamola and D.S. Chauhan: *Concepts in Forestry Research : Proceedings of the International Seminar on "Forests, Forest Products and Services: Research Development and Challenges Ahead"*, International Book Distributors, Delhi, 2007.

P.V.G.K. Reddy: *Genetic Resources of Indian Major Carps*, Daya Publication, Delhi, 2005.

Praveen Taank: *Forest Product and their Utilization*, Cyber Tech Publication, Delhi, 2010.

Ranjeet K. Singh: *Genetic Resources and Biotechnology*, Surendra Publication, Delhi, 2010.

Ranvir Singh; Hari Har Ram and J P Tiwari: *Role of Indigenous Germplasm in Improvement of Horticultural Crops : Proceedings of National Seminar (Organized by the Pantnagar Centre for Plant Genetic Resources, June 24-25, 2003)*, G B Pant University of Agriculture and Technology, 2003.

S P Rawat: *Non-Timber Forest Products of India*, Gene-Tech Books, Delhi, 2008.

S. Nautiyal and A.K. Kaul: *Non-Timber Forest Products of India*, Jyoti Publication, Delhi, 2003.

S.D. Sharma: *Genetic Resources of Rice in India : Past and Present*, Today and Tomorrows Publication, Delhi, 2010.

S.S. Negi: *Forest Products and their Utilization*, International Book Distributors, Delhi, 2002.

T. Krishna Murthy: *Minor Forest Products of India : Non-Timber Forest Products of India*, BS Publication, Delhi, 2010.

T.K. Mishra, S.K. Banerjee and D.C. Pal: *An Omnibus of Non-Timber Forest Products of India*, International Book Distributors, Delhi, 2004.

W.S. Lakra, A.K. Singh and P.C. Mahanta: *Fish Genetic Resources*, Narendra Publication, Delhi, 2009.

Index